PATHOLOGY AND BEHAVIOUR

PATHOLOGY AND BEHAVIOUR

Editor

Radoslav Jandaček

scitus academics

Pathology and Behaviour

Edited by **Radoslav Jandaček**

This Book is Printed on Acid-Free Paper. ♾

ISBN: 9781681179285

Library of Congress Control Number: 2017948958

SCITUS Academics LLC,

616, Corporate Way, Suite 2, 4766,

Valley Cottage, NY 10989

www.scitusacademics.com

Notice

Printed in United States of America.

Preface

Veterinary medicine deals with the prevention, diagnosis and treatment of disease, disorder and injury in non-human animals. The scope of veterinary medicine is wide, covering all animal species, both domesticated and wild, with a wide range of conditions which can affect different species. Like medical pathology, veterinary pathology is divided into two branches, anatomical pathology and clinical pathology. Other than the diagnosis of disease in food-producing animals, companion animals, zoo animals and wildlife, veterinary pathologists also have an important role in drug discovery and safety as well as scientific research. Clinical pathology is concerned with the diagnosis of disease based on the laboratory analysis of bodily fluids such as blood, urine or cavitary effusions, or tissue aspirates using the tools of chemistry, microbiology, hematology and molecular pathology. Veterinary pathologists study diseases of multiple species and are thus uniquely qualified to perform studies to advance our understanding of the cause of disease in animals and humans as well as studies to develop new methods to prevent and treat diseases.

'Pathology and Behaviour' aims to provide veterinary pathology students and pathologists with information needed to understand basic ultrastructural techniques and to identify tissues, interpret changes in tissue, and identify organisms at the ultrstructural level. This contains topics on sublethal and lethal cellular injury, inflammation and repair, thrombosis and hemodynamic disorders, immunologic and genetic disorders, and neoplasia. In addition, coverage on infectious, nutritional, and metabolic diseases, and ecosystem, environmental and forensic pathology is presented. This book is also appropriate for graduate students and researchers involved in biomedical and animal science programs that require basic understanding of pathogenesis of major disease processes.

Table of Contents

CHAPTER 1

Update on Epidemiology, Diagnosis and Control Technique of Newcastle Disease

Nesradin Yune* **and Nejash Abdela**

School of Veterinary Medicine, College of Agriculture and Veterinary Medicine, Jimma University, Jimma, Ethiopia

ABSTRACT

Newcastle disease is an acute, contagious viral disease of birds. It is one of the most important diseases which cause great economic loss in poultry industry. The virus that cause Newcastle disease is grouped under family Paramyxoviridae in the genus *Avulavirus* and species avian paramyxovirus type 1 (APMV-1) or Newcastle disease virus. Based on their virulence Newcastle disease virus can be divided in to viscerotropic, neurotropic, mesogenic and lentogenic strain. Although avian paramyxovirus type 1(APMV-1) can affect many species of birds including wild birds, Chickens are highly susceptible to the disease. The objectives of this paper were to highlight the epidemiology, diagnostic technique and control measures involved in Newcastle disease. Newcastle disease is currently distributed throughout the world including Central and South America, Asia, Middle East and Africa. APMV-1 can be transmitted by inhalation or ingestion, and birds shed these viruses in both feces and respiratory secretions. The virus can also be transmitted through direct contact with infected flock and indirect contact with contaminated materials. This disease can be diagnosed based history of disease outbreak, some pathognomic sign and laboratory test such as virus isolation, serological test and molecular technique. The latter has more important being its sensitive and rapid for diagnosis of the disease. Currently both live and killed vaccines are used in many countries to control and prevent the disease in chickens. Furthermore, strict biosecurity and separation infected once from health flock are also important to control and prevent spread of disease. Generally to make poultry free of this disease, good biosecurity and continual vaccination should be maintained.

Keywords: *Newcastle disease; Avian paramyxovirus type 1; Diagnosis; Vaccination*

1. INTRODUCTION

Newcastle disease is a highly contagious economically devastating viral disease of poultry [1]. The virus that cause Newcastle disease is grouped under family paramyxoviridae, genus *Avulavirus* and species Newcastle disease virus or avian paramyxovirus type 1 (APMV-1) [2,3]. The genus of this virus has RNA nucleotide, enveloped, single-stranded (SS) and has negative sense [4]. The avian paramyxovirus contain six structural protein matrix(M), RNA polymerase(L), phosphoprotin(P), nucleoprotein(NP), hemagglutinin neuraminidase(NH) and fusion(F) [5,6]. Protein V and W are additionally encoded by RNA editing of P protein [5]. Paramyxovirus type 1 can cause disease in birds of all types, sex and age [1].

Based on their virulence avia paramyxovirus (APMV-1) has been divided in to three or more pathotypes. Velogenic neurotropic strain typically associated with neurological and respiratory sign. Velogenic viscerotropic strain typically associated with gastrointestinal lesion. These two strains are more virulent. Mesogenic strain is moderate virulence while, lantogenic strains is the least virulence and used for vaccine preparation [3].

The disease was first observed in 1926 on the Indonesia island of Java then, later, it was found in various parts of the world [7]. Newcastle disease is endemic much of Asia, Africa and the Middle East, and some countries in Central and South America [3]. Newcastle disease can infect many species of birds, but the effects of the disease vary with different species. For example, ducks and gees are least sensitive and chickens are more sensitive [8]. Newcastle disease virus can be transmitted through direct contact with infected birds and secretion from mouth, noise and eyes of infected bird. Contaminated materials, feed and water can spread disease [9]. NCD can be diagnosed based on history of disease, sign and laboratory examination [10].

Only diagnosis based on clinical sign have not accurate because it resemble highly pathogenic avian flue [11]. Laboratory diagnosis for Newcastle disease includes virus isolation, serological and molecular test. Enzyme linked immune-sorbent assay(ELISA), virus neutralization test, and hemagglutination inhibition test, reverse-transcriptase polymerase chain reaction (RTPCR) and plaque neutralization test can be used for confirmation of the ND virus [10]. The clinical sign of Newcastle disease depends on strain of virus. Some virus strain attack respiratory system other affect nerve system or digestive system. The major clinical sign observed are loss of appetite, depression, weakness, greenish diarrhea, gasping, coughing, paralysis of wings and legs, corticoids and cyanosis of comb and wattle [12]. The effective way of controlling and preventing Newcastle disease is continual vaccination program using currently available vaccine. Implementing the effective biosecurity procedures is also very important to prevent the disease [13]. Despite, the huge economic impact of this disease there is scarcity of information. Therefore, the main objective of this review is to highlight the epidemiology, diagnostic technique and control measures involved in Newcastle disease.

2. GENERAL INFORMATION ON NEWCASTLE DISEASE

Newcastle disease is acute viral disease of many species of birds. NCD is economically the most important and cause loss in poultry industries [14]. The virus that cause Newcastle disease is grouped under family paramyxoviridae, genus *Avulavirus* and species Newcastle disease virus or avian paramyxovirus type 1 (APMV-1) [2]. Although APMV-1 to APMV-12 serotype are there; APMV-1 is the most pathogenic serotype and is also referred to as Newcastle disease virus (NDV) [3]. Newcastle virus affect both wild and domestic avian species but their severity vary between different strains [15]. Newcastle disease found all over the world. Depending on its strain and clinical sign it causes NCD virus grouped in to four phenotypes [16,17]. Those are velogenic neurotropic NDV affects nerves and velogenic viscerotropic affect visceral organ, mesogenic cause moderate sever and lantogenic wich is the least sever of new castle virus [18].

Newcastle disease was first observed in 1926 in island of Java by Kranveld [19,20]. The disease has got its name from town Tyne, England and was named Ranikhet disease in Asia and India [21,22]. As studied by Spradbrow [23] and Alexander [24] the disease was observed in the same and subsequent year in other parts of Asia (India, Philippines, Korea). In 1930 it was also identified as the same virus cause different severity or cause different clinical sign [16].

3. EPIDEMIOLOGY

3.1 Distribution and transmission

Newcastle disease was distributed throughout the world. Newcastle disease is epizootics in Central and South America, Asia, and Africa while sporadic epizootics occur in Europe [25]. In in Europe Newcastle disease was reached in 1981 then spread rapidly throughout the world [26]. It is endemic in south East Asia and cause high economic loss in commercial poultry farm [27]. Newcastle disease can infect over 240 species of birds. When infected birds are introduced into susceptible flocks all birds will be infected within two to six days. The disease is transmitted through direct contact with infected or carriers birds. Transmission can also occur through contact with secretion and excretion of infected birds and contact with contaminated materials (**FIGURE 1**). Another important route of transmission is through air [9, 8]. Fleas, rodent, insect and dog can also transmit NCD virus mechanical from infected faeces [28, 29].

3.2 Risk factor

Host risk factor: Newcastle disease can affect many species of birds. In addition to poultry, more than 230 species from more than one-half of the 50 orders of birds have been found to be susceptible to natural or experimental infections with avian paramyxoviruses. Chickens are mainly affected by Newcastle disease. Turkeys and pigeons, ducks [30,31], geese, as well as parrots and wild cormorants may also develop generalized disease, but clinical signs are rarely reported in geese and ducks [32,33]. In humans and rodents natural infection has been reported. NCD has less zoonotic importance and can cause conjunctivitis [33]. In all types of domestic poultry virulent strains of NCD virus have been found. High virulent serotype of NCD virus is not common for migratory wild birds [8].

In chicken mortality rate can reach 100%. The clinical sign observed during outbreak of NCD includes, depression, diarrhea, respiratory distress, cessation of egg production, nervous sign and death [16].

Pathogen risk factor: One of the major factors that determine pathogenicity of NCD virus is that it has ability to attach and penetrate host cells. To initiate viral infection two glycoproteins haemagglutinin neuraminidase (HN) and fusion protein (F) are required to expose as protrusions on the surface of the virion envelope [34]. Virus can survive for some weeks in all carcass of acutely infected bird or in egg and relatively stable and can be transmitted mechanically from infected material through movement of equipment and personnel. Virus is shade in all excretion and secretion. NCD virus is readily transmitted on fomites.

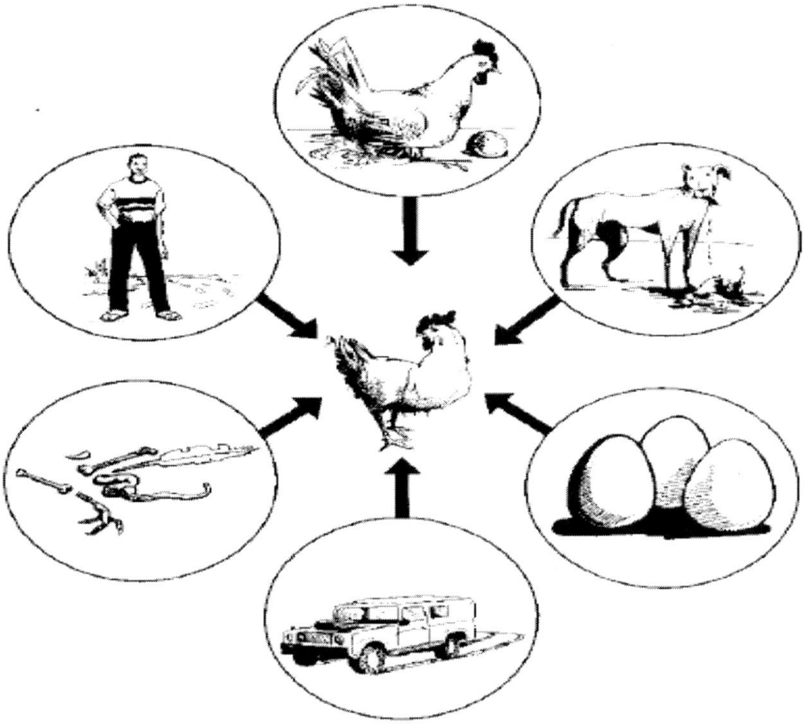

Figure 1: Newcastle disease can be transmitted from one village to another via people, vehicles, animals, baskets, hoes, cages and infected produce egg shells, feathers, bones, intestines, etc. [29].

This virus can survive for long period of time on eggshells and especially in feces, compared to an inorganic surface (filter paper) [3]. Isolate of NCD virus differs in there virulence strain and their tissue tropism in chickens [35] (**TABLE 1**).

Environmental risk factor: The environmental persistence of these viruses is highly variable, because it can be affected by many factors such as temperature, humidity, the suspending agent and exposure to light, as well as the technique used to detect the viruses. One study reported that APMV-

1 survived for up to 7 days in summer in contaminated, uncleaned poultry houses, as long as 14 days in the spring, and 30 days during the winter. Another group reported virus isolation up to 16 days after depopulation of an unvaccinated flock. However, one study found that APMV-1 remained viable for up to 255 days in a henhouse, at ambient temperatures of -11°C (12°F) to 36°C (97°F). At 23-29°C (73-84°F), APMV-1 is reported to survive in contaminated litter for 10 to 14 days, and at 20°C (68°F) in soil for 22 days. Virus has also been recovered from earthworms for 4 to 18 days and from experimentally contaminated lake water for 11 to 19 days [3].

Table 1: The four isolates group or pathotypes on the basis of virulence and tissue tropism in chickens [36, 37].

Types of Virulence Strain	Characteristics
Viscerotropic velogenic isolates	Cause severe fatal diseases characterized by hemorrhagic intestinal lesions
Neurotropic velogenic isolates	Cause acute disease characterized by nervous and Respiratory signs with high mortality
Mesogenic isolates	Cause mild disease with mortality confined to Young birds.
Lentogenic isolate	Cause mild or inapparent infection, coughing, gasping, sneezing and rales. Mortality is Negligible

3.3 Diagnosis

Newcastle disease can be diagnosed based on history, clinical sign and laboratory test. Newcastle disease clinically resembles highly pathogenic avian influenza so during outbreak rapid and accurate diagnosis is important to control and prevent dissemination of disease [11]. Laboratory diagnosis for NCD includes virus isolation, serological (enzyme-linked immune sorbent assays (ELISA), immunodiffusion test, agar gel precipitation and molecular test (Reverse transcriptionpolymerase chain reaction (RT-PCR). Isolation of the NCD virus is definitive diagnosis of NCD [38].

3.3.1 Diagnosis based on clinical sign and lesion

The clinical sign of NCD is depends on age, immune status of the host, tissue tropism and virulence of virus strain. Sudden high mortality in a flock in the absence of premonitory clinical signs occurs when susceptible species are exposed to highly virulent strain. In susceptible flocks the mortality rate in fully can reach 100% [35]. The incubation period of NCD is usually about five days. In chicken's nerve, respiratory and digestive sign may occur. The major clinical sign observed in Newcastle disease are: greenish white diarrhea, with ruffled feathers; depression in the birds and a state of prostration, a condition known as torticollis (the head turned to one side (**FIGURE 2**) and other neurological sign like paralysis of leg and wing. NCD is acute disease can cause death within 2 to 3 days [39,40].

Figure 2: NDV infected layer chicken of 29 weeks of age showing nervous signs (twisted neck and paralysis) [40].

Necropsy lesions caused by velogenic APMV-1 viruses have mainly been characterized in poultry, especially chickens [3]. Viscerotropic velogenic and neurotropic velogenic strain cause hemorrhagic lesion particularly in mucosa of the proventriculus, small intestine and ceca. In respiratory tract gross lesions are not observed. Although less likely in older birds, haemorrhages of the thymus and bursa of fabriceus may also occur [41].

3.3.2 Laboratory diagnosis

Sample for laboratory diagnosis: The appropriate sample for diagnosis in Newcastle disease includes: tissue sample (trachea, lung, spleen, soft palate, colon, bursa and brain) witch are important for hisopathology and cloacal swabs, oro-nasal swabs and Serum sample [29]. Blood sample is usually collected from wing vein. When fresh sample is collected lung and spleen it should be wrapped in plastic and placed into cold box with ice. To do serological test haemolysed or contaminated samples should not be used because it will give unreal result [29].

Virus isolation: Virus is obligate intracellular parasite that requires living cell in order to replicate. Cultured cells, eggs and laboratory animals may be used for virus isolation. To diagnosis APM-1 infection virus isolation in embryonated eggs or cell cultures serves as important for viral isolation [42]. Cloacal and Tracheal swabs are a good sample for viral isolation of NCDV. Sample from fresh faeces may also used as an alternative. Collected sample should be transported at pH 7.0-7.4 in isotonic phosphate buffered saline (PBS), containing antimicrobial drugs and media protein. Higher concentrations of antibiotic should be used when collected sample is from feaces of suspected chickens [43].

Virus isolation in culture: centrifugation of sample from feces or tissue for 10 minute at temperature 25°C to obtain supernatanted sample and measure 0.2 ml of sample and inoculate in to allantoic cavity of embryonated SPF fowl egg of 9 -11 days' incubation then incubate for 4-7 days at 35-37°C. If the tests give positive result embroynated eggs will die. Finally all eggs remained of incubation should child to 4°C and do for haemagglutination test (HA) [43].

Serological test: In the absence of vaccination, the presence of specific antibodies against the ND virus is not necessarily that it was suffering from the disease at the time of sampling, but indicates that the bird has been infected by the virus at some time. In practice, a high antibody titre is indicative of a recent infection. NDV may be employed as an antigen in a wide range of serological tests. Although numerous serological tests may be used to detect antibodies in serum they give little information on the infecting NDV strain. Two methods are used to measure antibody titres: the haemagglutination inhibition (HI) test, and the enzyme-linked immunosorbent assay (ELISA). The most commenaly used and show accurate result is haemagglutination inhibition (HI) test [41]. For both tests, it is necessary to collect blood samples from the chickens and should be taken from the wing veins [29].

Haemagglutination inhibition test: The haemagglutination inhibition (HI) test is used most widely in ND serology; its usefulness in diagnosis depends on the vaccinal immune status of the birds to be tested and on prevailing disease conditions. HI is done based on principle that the haemagglutinin on the viral envelope can bring about the clumping of red blood cells chicken and that this can be inhibited by specific antibodies [44]. Sera from species other than chicken red blood cells can also cause clumping (agglutination), so it should be removed by adsorption of the serum with chicken RBCs determining these properties. This is done by adding 0.5 ml of antisera to 0.025 ml of packed chicken RBCs, shaking gently and leaving for at least 30 minutes; the RBCs are then pelleted by centrifugation at 800 g for 2-5 minutes and the adsorbed sera are decanted. Any pretreatment of the sera is unnecessary as Chicken sera can rarely give nonspecific positive reactions in the HI test [37].

Enzyme linked immune sorbent assay: The ELISA works on the principle of recognition of anti-NDV antibodies, attached to a plate coated with viral antigen, by antibodies produced in another species against chicken antibodies [45]. Based on different strategies for detection of NCD antibodies, including, sandwich, indirect and blocking or competitive ELISAs using MAbs there are different commercial available ELISA kites. At least one kit uses a subunit antigen. Both ELISA and HI may measure antibodies to different antigens. ELISA test; depending on system used can detect antibodies to more than one antigen but, HI test is probably restricted to those directed against the HN protein. ELISAs are reproducible and have high specificity and sensitivity; they have been found to correlate well with the HI test [45].

Molecular technique: Molecular techniques such as polymerase chain reaction (PCR) and reverse transcription-polymerase chain reaction (RT-PCR) have been used for rapid and sensitive detection for Newcastle disease [46].

Polymerase chain reaction: The duplex PCR is done based on principle that it has the ability to amplify and differentiate multiple specific nucleic acids using polymerase enzymes [47,48]. However, those techniques can detect only one specific pathogen at a time [47]. PCR can detect virus following following the growth of virus in embryos in the laboratory and clinical specimens [49]. It has the potential to have high sensitivity and is now it is considered as the gold standard for nucleic acid detection [50]. However, PCR requires DNA as a template and the target viruses in this study have RNA as their nucleic acid. Therefore, RNA viruses require a reverse transcription step to produce single stranded complementary DNA (cDNA) through reverse transcriptase using a specific oligonucleotide primer and viral RNA as a template [50-52].

Reverse transcription-polymerase chain reaction: Molecular techniques like reverse transcription polymerase chain reaction (RTPCR) have been frequently used all over the world to detect viruses from the field samples. Reverse transcription polymerase chain reaction (RT-PCR) is used to detect RNA virus which is negative and single stranded RNA virus. There are two different configurations of the RT-PCR assay. In the two step RT-PCR configuration, the cDNA is synthesized in a different tube before performing PCR assay. In contrast a one step RT-PCR firstly synthesises the cDNA. The reverse transcriptase is inactivated and the polymerase is activated simultaneously and the PCR reaction is carried out in a single tube [53].

This is rapidly becoming the assay of choice. Several important steps need to be considered in developing an RT-PCR protocol. The first aspect is the RNA extraction. This needs to be an efficient process that can extract RNA from the samples even when it's in low concentrations and eliminate contaminants that will degrade the RNA [53]. Another important aspect is the gene being targeted and the choice of primers. This can have a profound effect on the efficacy of the assay [54]. Poorly design primers can result in mispriming and the amplification of nonspecific products or the formation of primer dimer [55].

Multiplex polymerase chain reaction: Multiplex PCR tests have been developed to allow simultaneous detection and differentiation of several avian viruses including NDV. These techniques have also been used experimentally to differentiate between velogenic, monogenic and lentogenic strains from chickens. It is applied simultaneously that required for avian infection including Newcastle disease for amplification and quantification of the virus. The primers that are specific for each virus are newly designated from the nucleoprotein gene of Newcastle disease virus. This technique helps mass amplification of the virus using common primers in the presence of fusion protein gene which increased the markedly sensitivity of the tests. At present, it should be noted that multiplexing RT-PCR assays aiming at broadening the range of virus detection frequently result in reduced sensitivity of the test compared with single target assays [56].

4. CONTROL TECHNIQUES OF NEW CASTLE DISEASE

4.1 Management strategies

The principal management procedures should include strict biosecurity measures which help in preventing the spread of infective material from house to house and from farm to farm [57]. Good biosecurity can protect poultry flocks from Newcastle disease. Avoid flocks not be to contact with domesticated poultry of unknown health status, any pet birds (particularly psittacines), and wild or feral birds (particularly cormorants, gulls and pigeons). Biosecurity measures include well ventilated houses, clean water supplies, minimizing travel on and off the facility, and disinfecting vehicles and equipment that enter the farm. Separation of infected from health flocks and proper disposal of died birds. Control of Pests such as insects and mice is also important for control measures of NCD. All in/all out breeding (one age group per farm), with disinfection between groups, is also advisable [3].

4.2 Vaccination

Vaccination is the most important method of controls and prevention of new castle disease. Currently, both inactivated and live vaccines for NCD are available around the world [58,59]. There is also thermo stable vaccine which was specifically developed to be used in village chicken [60]. We can use varieties of route for administration of live vaccine and schedules from hatching till grow-out [13]. Killed virus oil emulsion vaccines are administered parentally prior to the onset of egg production. Lentogenic virus vaccines are generally recommended in drinking water, by eye drop, by aerosol or intranasally. A vaccine using a heat-tolerant V4 strain has been developed for feeding to village chickens in countries where these constitute a significant proportion of poultry production [33]. Although proper vaccination protects the birds from clinical disease but it does not prevent virus replication and shedding, which results in a source of infection [61].

Live vaccine: Live vaccines are relatively cheap, sold as freeze dried, easy to administer and can be used for mass vaccination. These have been divided in to mesogenic and lentogenic groups with their preferred mode of administration being eye drop, beak intranasal installation, or dipping for lentogenic vaccines while mesogenic vaccine requires intramuscular injection. Drinking water and aerosol administrations can also be used [41].

In most countries, Hitchner B1 and La Sota vaccines are used and are derived from the mesogenic strain of NDV [41]. Some mesogenic vaccines may cause disease; particularly in young birds, especially if there is a dual infection with exacerbating organisms. Because of heat liable the live vaccines are also have disadvantage under village management system where transport and cold storage facilities are often inadequate [62].

Inactivated vaccine: Inactivated vaccine can be used situations unsuited for live vaccine and induce high level of protective antibody over long period of time. These vaccines are produced from infective allantoic fluid of virulent NDV treated with B-propiolactone or formalin to kill the virus and then mixed with adjuvant. The vaccine can administered ether subcutaneous or intramuscular injection [41].

Vaccination program: Vaccination program affect the duration of immunity. One of the most important considerations affecting vaccination programs is the level of maternal immunity in young chickens, which may vary considerably from batch to batch, farm to farm, and among individual chickens. For this reason, one of several strategies is employed. Either the birds are not vaccinated until 2-4 weeks of age when most of them will be susceptible, or 1-day-old birds are vaccinated by conjunctival instillation or by the application of a coarse spray. This will establish active infection in some birds that will persist until maternal immunity has waned. Revaccination is then carried out 2-4 weeks later [37].

5. CONCLUSION

Newcastle disease (NCD) is one of the most important viral diseases in poultry industry which can affect several species of birds. NCD is characterized by acute mortality marked by hemorrhagic lesions, respiratory and apparent or unapparent. Today Newcastle disease virus is found in most countries of the world and its transmission is through air, direct contact with infected or carrier birds, contact with secretion and excretion of infected birds and infected materials. History, observable clinical sign and laboratory examinations are the important tools in diagnosis of NCD. Good biosecurity, separation of infected ones from health and appropriate vaccination should be practiced as control and prevention method of this disease.

ACKNOWLEDGMENTS

We would like to acknowledge Jimma University for facility provision.

REFERENCES

1. Iram N, Shah MS, Ismat F, Habib M, Iqbal M, et al. (2014) Heterologous expression, characterization and evaluation of the matrix protein from Newcastle disease virus as a target for antiviral therapies. Appl Microbiol Biotechnol. 98: 1691-1701.
2. Mayo MA (2002) A summary of taxonomic changes recently approved by ICTV. Arch Virol. 147: 1655-1663.
3. Center for food security and public health (CFSPH) (2016). IOWA State University College of Veterinary Medicine. Newcastle Disease Avian Paramyxovirus-1 Infection, Goos. Paramyxovirus infection, Ranikhet disease.
4. Alexander DJ, Senne DA (2008) Newcastle disease, other avian paramyxoviruses and pneumovirus infections. In: Saif YM, Fadly AM, Glisson JR, et al. eds. Iowa State University Press, USA, Dis. Poult.,12: 75-100.
5. Cattoli G, Susta L, Terregino C, Brown C (2011) Newcastle disease review of field recognition and current methods of laboratory detection. J Vet Diagn Invest. 23: 637-656.
6. Al-Habeeb MA, Mohamed MHA, Sharawi S (2013) Detection and characterization of Newcastle disease virus in clinical samples using real time RT-PCR and melting curve analysis based on matrix and fusion genes amplification. Vet World. 6: 239-243.

7. Ashraf A, Shah MS (2014) Newcastle Disease: Present status and future challenges for developing countries. African J Microbiol Res. 8: 411-416.
8. Caupa I, Alexander DJ (2009) Avian Influenza and Newcastle Disease. A Field and Laboratory Manual. Milan: Springer-Verlag.
9. Li X, Qiu Y, Yu A, Chai T, Zhang X, et al. (2009) Degenerate primers based RT-PCR for rapid detection and differentiation of airborne chicken Newcastle disease virus in chicken houses. J Virol Methods. 158: 1-5.
10. Chaka H, Goutard F, Gil P, Abolnik C, Almeida R, et al. (2013) Serological and molecular investigation of Newcastle disease in household chicken flocks and associated markets in Eastern Shewa zone, Ethiopia. Trop Anim Health Prod 45: 705-714.
11. Khan TA, Rue CA, Rehmani SF, Ahmad A, Wasilenko JL, et al. (2010) Phylogenetic and Biological Characterization of Newcastle Disease Virus Isolates from Pakistan. J Clin Microbiol. 48: 1892-1894.
12. Pazhanivel N, Balsubramaniam GA, George VT, Mohan B (2002) Study of natural outbreak of Newcastle disease in and around Namakkal. Indian Vet J. 79: 293-294.
13. Cho S, Kwon H, Kim T, Kim JH, Yoo H, et al. (2008) Characterization of a Recombinant Newcastle Disease Virus Vaccine Strain. Clin Vaccine Immunol. 15: 1572-1579.
14. Bulbule NR, Madale DS, Meshram CD, Pardeshi RB, Chawak MM (2015) Virulence of Newcastle disease virus and diagnostic challenges. Adv Anim Vet Sci. 3: 14-21.
15. Alexander DJ (1998) Newcastle disease and other avian paramyxoviruses. In: A Laboratory Manual for the Isolation and Identification of Avian Pathogens 4th edn. Edited by Swayne DE, Glisson JR, Jackwood MW, Pearson JE, Reed WM. American Association of Avian Pathologists: Kennet Square. pp: 156-163.
16. Alexander DJ, Bell JG, Alders RG (2004) A Technology Review: Newcastle Disease. With Special Emphasis on its Effect on Village Chickens. FAO Animal Production and health Paper, FAO. p: 161.
17. Engström B, Eriksson H, Fossum O, Jansson DS (2004) Compendium poultry diseases. 2nd edn. Uppsala: National Veterinary Institute.
18. Alexander DJ (2001) Gordon Memorial Lecture. Newcastle disease. Br Poult Sci. 42: 5-22.
19. Loretu K, Mkaria J (1981) Preliminary report on Newcastle disease pathotypes in Tanzania. Tanzania Vet Bull. 3: 63-66.
20. Arifin MA, Salim SH, Mel M, Abdul Karim MI, Hassan SS (2011) Optimization of Newcastle Disease Virus Production in T-flask. Proceedings of the 2nd International Conference on Biotechnology Engineering, ICBioE'11 May 17-19, Kuala Lumpur, Malaysia.
21. Doyle TM (1927) In: Rweyemamu MM, Palaya V, Win T, Sylla D (edn). Newcastle disease vaccines for Rural Africa, Pan African Veterinary Vaccine Center, Debre Zeit, Ethiopia, pp. 7-45.
22. Narayanan MS, Parthiban M, Sathiya P, Kumanan K (2010) Molecular detection of Newcastle disease virus using Flinders Molecular detection of Newcastle disease virus using Flinders Tehnology Associates-PCR Tehnology Associates-PCR. Vet Arhiv. 80: 51-60.
23. Spradbrow PB (1987) Newcastle Disease an overview. Newcastle disease in poultry, a new feed pellet vaccine, Canberra. Aust. Centre Int. Agric Res. 5: 12-18.
24. Alexander DJ (1988) Newcastle disease: Methods of spread. Newcastle disease. Boston, Kluwer Acad. Pub. pp: 1-10.
25. Naveen KA, Singh SD, Kataria JM, Barathidasan R, Dhama K (2013) Detection and differentiation of pigeon paramyxovirus serotype-1 (PPMV-1) isolates by RT-PCR and restriction enzyme analysis. Trop Anim Health Prod. 10: 01-06.
26. Mase M, Imai K, Sanada Y, Sanada N, Yuasa N, et al. (2002). Phylogenetic analysis of newcastle disease virus genotypes isolated in Japan. J Clin Microbiol. 40: 3826-3830.
27. Munir M, Zohari S, Abbas M, Berg M (2012) Sequencing and analysis of the complete genome of Newcastle disease virus isolated from a commercial poultry farm in 2010. Archive Virol. 157: 765-768.
28. Ullah S, Ashfaque M, Rahman SU, Akhtar M, Rehman A (2004) Newcastle disease virus in the intestinal contents of broilers and layers. Pak Vet J. 24: 28.

29. Alders R, Spradbrow P (2001) Controlling Newcastle disease in village chickens: a field manual. Australian Centre for International Agricultural Research (ACIAR).
30. Zhang S, Wang X, Zhao C, Liu D, Hu Y, et al. (2011) Phylogenetic and pathotypical analysis of two virulent newcastle disease viruses isolated from domestic ducks in China. J PLoS ONE. 6: 1-9.
31. Madadgar O, Karimi V, Nazaktabar A, Kazemimanesh M, Ghafari MM, et al. (2013) A study of Newcastle disease virus obtained from exotic caged birds in Tehran between 2009 and 2010. Avian Pathol. 42: 27-31.
32. Rima B, Alexander DJ, Billeter MA, Collins PL, Kingsbury DW, et al. (1995) Paramyxoviridae. In: Murphy FA, Fauquet CM, Bishop DH, Ghabrial AW, Jarvis GP, et al. (Eds.). Virus Taxonomy. Sixth report of the international committee in taxonomy of viruses. Springer-Verlag, Wien, pp: 268-274.
33. PLAN AVE (2010) Disease Strategy Newcastle disease.
34. De Leeuw OS, Koch G, Herzog L, Ravenshorst N, Peeters BP (2005) Virulence of Differentiation of pathogenicity of avian paramyxovirus serotypes 1. Offi J Eur Uni Avian Pathol. 31: 493-499.
35. Quinn PJ, Markey BK, Carter ME, Donnelly WJ, Leonard FC (2002) Veterinary Microbiology and Microbial Disease. 4th ed., Black well publishing company: Blackwell Science, London, UK, pp: 137-143.
36. Docherty, Douglas E (1999) Viral Diseases (Field Manual of Wildlife Diseases).
37. OIE (2012) Newcastle disease. In: Manual of Diagnostic Tests and Vaccines for Terrestrial Animals. Chapter 2.3.14.
38. Saif YM, Barnes HJ, Glisson JR, Fadly AM, McDougald LR, et al. (2005) Diseases of poultry, 11th edn, pp: 66-78.
39. Beard CW, Hanson RP (1984) Newcastle disease. In: Diseases of poultry.
40. Hasan AR, Ali MH, Siddique MP, Rahman MM, Islam MA (2012) Clinical and laboratory diagnoses of newcastle and infectious bursal diseases of chickens. Bangladesh J Vet Med. 8: 131-140.
41. Alexander DJ (2003) Newcastle disease, other Paramyxoviridae and Pneumovirus Infections. In; Diseases of Poultry. 11th edn. Iowa State press, Ames. pp: 63-100.
42. Aldous EW, Alexander DJ (2001) Detection and differentiation of Newcastle disease virus (avian paramyxovirus type 1). Avian Pathol 30: 117-128.
43. Office International des Epizootics (2009) Manual of Diagnostic Testsand Vaccines for Terrestrial Animals. Newcastle disease OIE Terrestrial Manual. OIE, Paris, France, pp. 576-589.
44. Alexander DJ, Bell JG, Alders RG (1995) Newcastle disease. State Vet J. 5: 21-21.
45. Brown J, Resurreccion RS, Dickson TG (1990) The relationship between the hemagglutination-inhibition test and the enzyme-linked immunosorbent assay for the detection of antibody to Newcastle disease. Avian Dis. 34: 585-587.
46. Hewson K, Noormohammadi AH, Devlin JM, Mardani K, Ignjatovic J (2009) Rapid detection and non-subjective characterisation of infectious bronchitis virus isolates using high-resolution melt curve analysis and a mathematical model. Archives Virol. 154: 649.
47. Bellau-Pujol S, Vabret A, Legrand L, Dina J, Gouarin S, et al. (2005) Development of three multiplex RT-PCR assays for the detection of 12 respiratory RNA viruses. J Virol Methods. 126: 53-63.
48. Albert B, Bray D, Hopkins K, Johnson A, Lewis J, et al. (2004) Essential biology, edited by cell 2nd. Garland Science.
49. Creelan JL, Graham DA, McCullough SJ (2002) Detection and Differentiation of pathogenicity of avian paramyxovirus Serotype 1 from field cases using one-step reverse transcriptase- Polymerase chain reaction. Avian Pathol. 31: 493-499.
50. Mackay IM, Arden KE, Nitsche A (2002) Real-time PCR in virology. Nucl Aci Res. 30: 1292-1305.
51. Mackay IM (2004) Real-time PCR in the microbiology laboratory. Clin Micro Infect. 10: 190-1212.
52. Turner P, McLennan A, Bates A, White M (2005) Molecular Biology, 3rd edn. Bios Instant Notes, edited by Hames BD, Taylor and Francis, Leeds, UK.

53. Pfaffl MW (2004) Quantification strategies in real-time PCR. A-Z of quantitative PCR. International University Line (IUL) La Jolla, CA, USA, Freising, Germany.
54. He Q, Marjamaki M, Soini H, Mertsola J, Viljanen M (1994) Primers are decisive for sensitivity of PCR. Bio Tech. 17: 82-87.
55. Abd-Elsalam KA (2003) Bioinformatic tools and guideline for PCR primer design. African J Biotechnol. 2: 91-95.
56. Tang Q, Wanga J, Bao J, Sun H, Sun Y, et al. (2012) A multiplex RT-PCR assay for detection and differentiation of avian H3, H5, and H9 subtype influenza viruses and Newcastle disease viruses. J Virol Methods. 181: 164-169.
57. Markos T, Abdela N (2016) Epidemiology and Economic Importance of Pullorum Disease in Poultry: A Review. Global Veterinaria 17: 228-237.
58. Shim JB, So HH, Won HH, Mo I (2011) Characterization of avian paramyxovirus type-1 from migratory wild birds in chickens. Avian Pathol. 40: 565-572.
59. Xiao S, Paldurai A, Nayak B, Mirande A, Collins PL, et al. (2013) Complete genome sequence of a highly virulent Newcastle disease virus currently circulating in Mexico. Genome Announc. 1: 01-02.
60. Spadbrow (1992) Newcastle disease in village chicken. Control with thermostable oral vaccine. Proceeding of an international workshop, Kuala lumpar, Malaysia. ACIAR, Canberra.
61. Chukwudi OE, Chukwuemeka ED, Mary U (2012) Newcastle disease virus shedding among healthy commercial chickens and its epidemiological importance. Pakistan Vet J. 32: 354-356.
62. Otim MO, Mukiibi GM, Christensen H, Bisgaard M (2005) Aflatoxicosis, infectious bursal disease and immune response to Newcastle disease vaccination in rural chicken. Avian Pathol. 34: 319-323.

CHAPTER 2

Genetics of animal health and disease in cattle

Donagh P Berry[1*], Mairead L Bermingham[2], Margaret Good[3] and Simon J More[4]

[1]*Animal and Grassland Research and Innovation Centre, Teagasc, Moorepark, Co. Cork, Ireland.*

[2]*The Roslin Institute and Royal (Dick) School of Veterinary Studies, University of Edinburgh, Roslin, Midlothian EH25 9PS, UK.*

[3]*Department of Agriculture, Fisheries and Food, Kildare St., Dublin 2, Ireland.*

[4]*Centre for Veterinary Epidemiology and Risk Analysis, UCD School of Agriculture, Food Science and Veterinary Medicine, University College Dublin, Belfield, Dublin 4, Ireland.*

ABSTRACT

There have been considerable recent advancements in animal breeding and genetics relevant to disease control in cattle, which can now be utilised as part of an overall programme for improved cattle health. This review summarises the contribution of genetic makeup to differences in resistance to many diseases affecting cattle. Significant genetic variation in susceptibility to disease does exist among cattle suggesting that genetic selection for improved resistance to disease will be fruitful. Deficiencies in accurately recorded data on individual animal susceptibility to disease are, however, currently hindering the inclusion of health and disease resistance traits in national breeding goals. Developments in 'omics' technologies, such as genomic selection, may help overcome some of the limitations of traditional breeding programmes and will be especially beneficial in breeding for lowly heritable disease traits that only manifest themselves following exposure to pathogens or environmental stressors in adulthood. However, access to large databases of phenotypes on health and disease will still be necessary. This review clearly shows that genetics make a significant contribution to the overall health and resistance to disease in cattle. Therefore, breeding programmes for improved animal health and disease resistance should be seen as an integral part of any overall national disease control strategy.

1. BACKGROUND

There continues to be very significant advances in efforts to control disease in cattle, with the potential for significant improvements to both performance and welfare. These advances have included improved understanding of disease pathophysiology and epidemiology, as well as the development of products such as antibiotics and anthelmintics for improved disease control.

Concurrently, there have been considerable advances in animal breeding and genetics, relevant to animal disease control. These advances are of considerable veterinary interest, noting that observed animal performance is the outcome of the interaction between the animal's genetic makeup and the specific environment it was exposed to. Logically, therefore, improved genetics has the potential to complement current approaches to animal disease control. Improvement in animal health through genetic selection is advantageous, because genetic gain is cumulative and permanent, as the genes introduced into a population can persist for many generations. Unravelling the genetic architecture of health and disease resistance not only facilitates knowledge development on potential for breeding for improved health status but also generates knowledge for biomedical research in animals and humans including applications such as vaccine development.

The objective of this review is to summarise research studies on the genetics of animal health and disease resistance in cattle, with particular reference to studies undertaken in Irish cattle. The implications of these results in breeding for improved animal health and disease resistance are discussed.

2. GENETIC TERMINOLOGY

Prior to discussing the genetics of animal health and disease resistance, the terms commonly used by animal breeders to describe the characteristics of a population need to be explained:

2.1 Phenotype

The phenotype is simply the observed performance of an animal "in the field" (e.g., dystocia in cows or the presence or absence of infection as measured by a positive or negative diagnostic test result). A phenotype, or trait, may be continuous (also called quantitative; e.g., milk yield and growth rate) or discrete (qualitative; e.g., did or did not succumb to disease).

2.2 Genotype

The definition of genotype varies. Animal breeders commonly use genotype to describe a particular strain of animal (e.g., animals of a given breed from a particular origin). Molecular geneticists, however, commonly use genotype to describe the genetic variants (i.e., alleles) an individual possesses at a particular position in its DNA, also known as a locus.

2.3 Genetic markers

A genetic marker is a measurable variation in the DNA sequence of a population. Common types of genetic markers, generally termed polymorphisms, include microsatellites, indels (i.e., insertions or deletions of pieces of DNA), single nucleotide polymorphisms (SNPs - pronounced *"snips"*) and copy number variants (CNVs).

2.4 Quantitative trait locus (QTL)

A segment of a chromosome that has been experimentally demonstrated to be statistically associated with variation in a quantitative or complex phenotypic trait.

2.5 Heritability

Heritability summarises the proportion of phenotypic variation, or differences among a cohort of animals, attributable to genetic variation between individuals. Animal breeders commonly cite the narrow-sense heritability, denoted as h^2, which is the proportion of phenotypic variation attribute to additive genetic variation (i.e., genes passed on from one generation to the next). In the calculation of broad-sense heritability, the numerator also includes non-additive genetic variation. In this review we will only consider narrow-sense heritability estimates. Heritability varies from 0 (not heritable) to 1 (fully heritable). If the heritability is high, we can expect a large proportion of the phenotypic differences of the parents to be passed on to the progeny. Also, the larger the heritability, the greater the expressed phenotype resembles the genetic merit of the animal.

2.6 Maternal heritability

Maternal heritability is the proportion of phenotypic variation among offspring that is due to the genes expressed by the dam. A maternal heritability of calving difficulty is often quoted which encompasses for example, the size of the pelvis of the dam. The direct heritability for calving difficulty is the effect of the genes of the dam (and sire) on, for example, the size of the calf. A maternal heritability has also been reported for weaning weight [1] which includes genetic characteristics of the dam such as her milk yield.

2.7 Genetic variation

A measure of the variation or differences within a population that is due to the differences in genetic merit of the animals. More commonly, genetic variation is expressed as the genetic standard deviation (i.e., the square root of the genetic variance) within a population.

2.8 Genetic correlation

A genetic correlation describes the strength of the linear relationship between two traits due to the genetic influences on each trait. It varies from -1 (strong negative relationship) to 0 (no relationship) to +1 (strong positive relationship between two variables). Genetic correlations can be due to the same mutation affecting both traits (termed pleiotropic effect) or different mutations affecting both traits but tending to, on average, be inherited together (i.e., linked).

2.9 Estimated breeding value (EBV)

Estimated breeding value is an estimate of the genetic merit for an animal for a given trait or series of traits based on an evaluation of all available data on the performance of an animal, and close relatives, for a trait. Using traditional methods of genetic evaluation, the true breeding value (or true genetic merit) is not known. The estimates of genetic merit are generally presented as the predicted transmitting ability (PTA) in dairy cattle or expected progeny difference (EPD) in beef cattle which are the EBV divided by two (i.e., an animal only passes half its genes to its progeny).

3. BASIS OF GENETIC GAIN

Annual genetic gain for a given trait may be described by the following equation [2]:

$$\Delta G = \frac{i \cdot r \cdot \sigma_g}{L}$$

where ΔG is annual genetic gain; i is the intensity of selection; r is the accuracy with which the genetic merit of each animal is known, σ is the genetic standard deviation (i.e., the square root of the genetic variance or simply just a measure of the genetic differences among animals); and L is the generation interval.

The greater the selection intensity the greater will be the genetic gain for that trait. The accuracy of selection is affected by both the heritability of the trait and the information available on the animal itself and its relatives. Figure 1 illustrates how the accuracy of selection, ignoring pedigree contributions, increases as the number of half-sib progeny with records increases across different heritability estimates. For a given number of progeny, the accuracy will be greater for higher heritability traits. However, accuracies of near unity are still achievable even for low heritability traits if sufficient information is available. Therefore, with the appropriate breeding programme (i.e., large paternal half-sib groups) and infrastructure for the collection and storage of data, genetic gain in low heritability traits is certainly achievable if genetic variation is present. The accuracy of selection for a given trait may also be augmented by indirect selection for a correlated trait (Figure 1). The genetic standard deviation is a measure of the variation present in the population and the generation interval is the average age of the parents when its progeny are born. This is approximately 6 years in Irish dairy and beef cattle [3], which is consistent with international estimates.

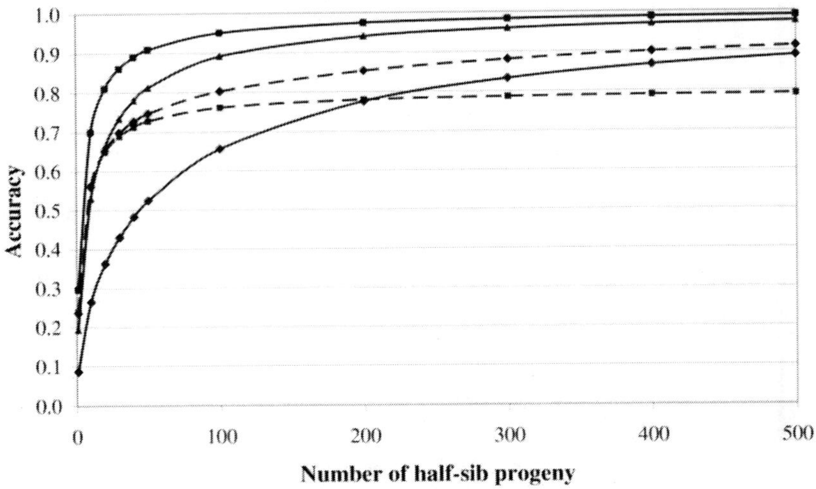

Figure 1 Accuracy of selection across different number of half-sib progeny based on direct selection where the heritability of the trait is 0.03 (diamond with continuous line), 0.15 (triangle and continuous line), 0.35 (square with continuous line) and indirect selection where the goal trait is the 0.03 heritability trait and data is available on the 0.35 heritability trait alone (square with broken line) or also available on the 0.03 heritability trait (diamond with broken line), assuming a genetic correlation of 0.80 between both traits.

4. HEALTH AND DISEASE IN CATTLE

4.1 Viral diseases

There are many bovine viral diseases in Ireland, including bovine viral diarrhoea (BVD) and several viral infections associated with the respiratory system (infectious bovine rhinotracheitis [IBR], bovine parainfluenza-3 [PI-3] and bovine viral syncytial virus [BRSV]). As yet, there has been no Irish research to quantify the genetic variation present in susceptibility to bovine viral diseases.

Breed differences in susceptibility to respiratory disease, however, have been demonstrated elsewhere [4]. Furthermore, heritability estimates different from zero have also been reported, highlighting the presence of within-breed genetic differences in susceptibility to respiratory disease. Heringstad et al. [5] estimated a heritability of 0.05 (SE = 0.018) for respiratory disease in Norwegian calves, with a genetic standard deviation of 2 percentage units. Norway is free from IBR and BVD so the most prevalent agents in respiratory disease in Norway are BRSV and PI-3.

As part of a larger scale study with 12 different breed types in the USA, Snowder et al. [4] reported a heritability of susceptibility to respiratory disease in pre-weaned calves of between 0.00 and 0.26 (SE = 0.04); the standard deviation varied from zero to 1.47 percentage units. Furthermore, a significant maternal heritability of between 0.00 and 0.13 (SE = 0.07) was observed. This significant maternal heritability indicates that, independent of the genes transmitted to the offspring from the dam, the genetic makeup of the dam herself, either through her uterine environment or post-natal influence (e.g., immunoglobulins present in colostrum), also influences the genetic susceptibility of the calf to respiratory disease. In a follow-up study, Snowder et al. [6] estimated the heritability for respiratory disease in weaned feedlot calves (average age of 176 days) of between 0.04 (SE = 0.01) and 0.08 (SE = 0.01), while the genetic standard deviation varied from 0.57 to 1.17 percentage units. In an independent US study, Schneider et al. [7] reported a heritability of resistance to respiratory disease in pre-weaning and post-weaning calves of 0.11 (SE = 0.06) and 0.07 (SE = 0.04), respectively. These estimates all suggest that there is indeed genetic variation in susceptibility to viral respiratory disease.

4.2 Udder health

Mastitis is one of the most costly diseases in dairy cattle production systems and is also likely to impact significantly on profitability in suckler beef production systems. A plethora of international scientific studies have been undertaken on the genetics of udder health, including genetic analysis of Irish Holstein-Friesian dairy cows [8]. Heritability estimates for mastitis have been variable across studies, reflecting variation in a multitude of factors such as the person undertaking the recording (i.e., farmer or veterinarian) and how they interpret the clinical signs, the completeness of data recording (i.e., some observations not recorded), as well as the pathogen and the environment, including exposure, which may influence the expression of an animal's genotype. Based on a detailed review of literature present at the time, Mrode and Swanson [9] reported a weighted average heritability of lactation average somatic cell count (SCC) and mastitis in primiparae of 0.11 (SE = 0.04) and 0.04, respectively. A recent heritability estimate for mastitis of 0.05 (SE = 0.005) in Irish Holstein-Friesian dairy cows [8] is consistent with this review. The genetic standard deviation for mastitis varies from 1.2 to 7.0 percentage units, suggesting that genetic gain can be achieved by selecting for clinical mastitis. Using data from long-term controlled experiments in Norway, Heringstad et al. [10] showed that a reduction in genetic merit to susceptibility to clinical mastitis is achievable with active selection of sires of superior genetic merit for clinical mastitis. After five generations of selection, cows actively selected for low clinical mastitis had a mean incidence of less than 5% while the incidence of clinical mastitis in cows on the selection lines for increased production was over 25% [10]. A concurrent decline in incidence of ketosis, but not retained foetal membranes (RFM), was also observed [10].

In their review, Mrode and Swanson [9] reported the mean genetic correlation between SCC and mastitis to be ~0.70; this indicates that SCC is a good genetic predictor of mastitis and can, therefore, be used in breeding programmes to augment the accuracy of selection, even where data on clinical mastitis may be available (Figure 1). Recent estimates from Ireland are consistent with these estimates [8]. Nonetheless, despite the relatively strong genetic correlation between SCC and mastitis, indicating that they are both measuring similar genetic characteristics of udder health, the correlation is not unity and, therefore, additional genetic gain can be achieved by recording clinical mastitis, even if data on SCC are available, and used in genetic evaluations.

4.3 Lameness

After mastitis, lameness is probably the next most costly disease in dairy and beef production systems. Heritability of lameness varies from 0.03 to 0.096 when scored by farmers [8, 11, 12] or by trained assessors [13]. Using data on lameness in Canadian Holstein cows scored by either farmers or veterinarians, Van Dorp et al. [14] reported a heritability for lameness of 0.16; the genetic standard deviation calculated from the provided estimates of the incidence and heritability was 9.5 percentage units. The heritability of lameness in Irish Holstein-Friesian dairy cattle based on predominantly farmer-recorded data was 0.04 (SE = 0.005) [8]. Heritability estimates for different claw disorders (interdigital dermatitis, heel erosion, sole haemorrhage, digital dermatitis, white line disease, interdigital hyperplasia, sole ulcer and chronic laminitis) varied from 0.01 (SE = 0.01) to 0.10 (SE = 0.02) in Dutch multiparous Holsteins [15]. The lowest heritability estimates were for sole ulcers and chronic laminitis, while the highest estimates were for digital dermatitis and interdigital hyperplasia.

4.4 Other bacterial diseases

A broad range of other diseases in Irish cattle can be attributed to bacterial infection. Of these, bovine tuberculosis (TB; caused by infection with *Mycobacterium bovis*) is perhaps the most problematic. There have been intensive national efforts in Ireland towards *M. bovis* eradication for many years [16].

Several studies have attempted to estimate genetic parameters for bacterial diseases. The heritability of susceptibility to *M. bovis* infection in cattle is described in two recent studies [17, 18], including one study in Irish Holstein-Friesian dairy cows [17]. Heritability estimates on the observed and liability scales varied from 0.06 to 0.18; standard errors varied from 0.012 to 0.044. The genetic standard deviation for the traits varied from 17 to 23 percentage units. Heritability estimates were based on response to the tuberculin test, and on the presence of TB lesions, as confirmed during abattoir inspection. These results clearly show that significant improvement can be made through genetic section towards TB resistance.

Few genetic parameters are available on the susceptibility to *Mycobacterium avium* subspecies *paratuberculosis* (MAP), the causative organism for Johne's disease, and these have generally been confined to analysis of antibody response to MAP. Heritability estimates vary from 0.05 to 0.15 [19, 20, 21, 22], with estimates of between 0.07 to 0.15 (SE varying from 0.024 to 0.062) for Irish Holstein-Friesian dairy cows [23]. The genetic standard deviation for serological response to MAP was calculated from the results of Berry et al. [23] as 5.4 percentage units.

As yet, no attempt has been made to estimate the heritability for susceptibility to salmonellosis in cattle. However, Wray and Sojka [24] reported that Friesian calves are more resistance to *Salmonella typhimurium* infection than Jersey calves, which suggests that genetic differences do exist. Furthermore, heritability estimates for mortality, survival time and quantitative caecal salmonella carrier state at the end of the rearing period for experimentally infected chickens was 0.12, 0.06 and 0.09, respectively [25], which suggests that some genetic variation may also exist in other species.

Like salmonellosis, no estimate of the heritability of brucellosis is available, although Templeton et al. [26] increased natural resistance to brucellosis in calves from 20% to 59% following the breeding of dams to a naturally resistant bull; this suggests that genetic variation is indeed present.

4.5 Metabolic diseases

Some research has been undertaken to investigate the heritability of metabolic diseases of cattle, including ketosis, hypocalcaemia (i.e., milk fever), hypomagnesaemia (i.e., grass tetany) and displaced abomasum. In general, heritability estimates for these metabolic diseases vary from 0.009 to 0.31. Heritability estimates for hypocalcaemia vary from 0.01 to 0.13 [11, 27, 28, 29, 30]; the standard deviation varied from 0.3 to 4.0 percentage units. The heritability for ketosis varies from 0.01 to 0.16 [28, 29, 30, 31]; the standard deviations vary from 0.9 to 1.2 percentage units. There appears to be only one heritability estimate for hypomagnesaemia in dairy cattle of 0.004 (SE = 0.004) [29]. However, heritability estimates for other health traits in that study [29], based on farmer-recorded data, were, in general, lower than estimates from other studies. Heritability estimates for displaced abomasum varies from 0.15 to 0.31 [28, 31]. There appears to be no estimate of heritability for acidosis in cattle.

The larger heritability for displaced abomasum relative to the other diseases may be surprising given the known influences of nutrition and management on each of these metabolic disorders. However, the larger heritability may not in fact be due to a truly greater influence of additive genetics on the trait but due to a more accurate recording of the trait since most will be diagnosed by a veterinarian. Furthermore, displaced abomasum is a binary trait, with no ambiguity with respect to the severity of disease.

4.6 Fertility-related conditions

Most diseases impact in some way on reproductive performance. For the purpose of this review, fertility-related diseases are confined to RFM, metritis and cystic ovaries. In a study with over half a million Norwegian Red first lactation cows [31], the heritability of cystic ovaries, metritis and RFM were 0.07 (SE = 0.01), 0.03 (SE = 0.01) and 0.06 (SE = 0.01), respectively. The heritability estimates reported by Heringstad [31] were similar to those reported in other studies for cystic ovaries of between 0.02 and 0.22 [12, 14, 28, 32], for metritis of between 0.04 and 0.06 [12, 14, 33], and for RFM of between 0.004 to 0.08 [14, 34, 35]. Furthermore, the genetic correlation between RFM expressed in first, second and third lactation were all greater than 0.54 (SEs were less than 0.062) suggesting that the same genes may be influencing this disorder irrespective of parity [30]. No correlations for either metritis or cystic ovaries across lactation are available.

Heringstad [31] documented a genetic correlation between metritis and RFM of 0.64 (SE = 0.10), indicating that selection for one will also reduce the incidence of the other. In contrast, Van Drop et al. [14] reported a near zero genetic correlation (0.06) between the two traits; no standard error was provided but the estimate was based on 12,471 lactations compared to over half a million by Heringstad [31]. Heringstad [31] failed to identify a correlation between metritis and cystic ovaries (0.14; SE = 0.15) agreeing with Van Dorp et al. [14] who reported a correlation of 0.03. Although Pösö and Mäntysaari [33] reported a moderate to strong positive correlation between metritis and cystic ovaries (0.59), the associated standard error was large (0.33) suggesting it was not different from zero. The genetic correlation between cystic ovaries and RFM was close to zero [14], although Heringstad [31] reported a negative correlation of -0.26 (SE = 0.13) between the traits. Large standard errors were associated with the correlations, due primarily to the low heritability of the respective traits, thereby increasing the uncertainty of the results. However, the lack of any strong genetic correlation between the disease conditions suggests that they are under different genetic control and need to be selected for individually.

4.7 Endo- and ecto-parasites

Although much of the research on the genetics of susceptibility to both endo- and ecto-parasites has been undertaken in sheep [36], there is evidence suggesting genetic variation exists among cattle in susceptibility to parasites. Pan et al. [37] reported a heritability of 0.084 to 0.124 (SE = 0.025 to 0.042) for susceptibility to *Neospora caninum* in Canadian Holstein cattle. Heritability estimates for gastrointestinal nematode burden, measured as eggs per gram of faeces or larvae per gram of faeces, ranged from 0.00 to 0.25 (SE = 0.02 to 0.05) in Dutch Holstein-Friesian cattle [38]; the average was 0.09. Davis [39] in a review of cattle studies reported an average heritability of resistance to ticks of 0.31, a heritability for resistance to buffalo fly of 0.21, and a heritability, excluding the reported estimate greater than one (heritability cannot be greater than one), of 0.32 for faecal egg count.

5. IMPACT OF PREVIOUS AND CURRENT BREEDING GOALS ON ANIMAL HEALTH AND SUSCEPTIBILITY TO DISEASE

It is not possible to accurately quantify genetic trends for most health traits in most countries because accurate records and data on animal health are not available, certainly not routinely. Nonetheless, past genetic trends may be predicted based on estimated genetic correlations with production traits, and the impact of these correlations quantified using knowledge on past breeding goals.

With several notable exceptions, particularly from the Nordic countries, breeding programmes of the past selected aggressively for increased milk production [40]. Current international breeding goals now generally include functional traits such as fertility, udder health, type traits and functional survival [40]. Nonetheless, a large proportion of the relative emphasis in international breeding goals is still on milk production and, therefore, genetic trends can be approximated based on genetic correlations with milk production.

Few estimates are currently available on the genetic correlation between milk production and several infectious diseases. Only one study has reported a genetic correlation between 305-day milk yield and serological response to MAP [23]. Dependent on the parity of animal and dataset used, the genetic correlation varied from -0.44 to 0.07 (standard errors varied from 0.083 to 0.228) in that population of Irish Holstein-Friesian dairy cows. A genetic correlation of -0.48 (SE = 0.13) was observed between 305-day milk yield and bovine tuberculosis in multiparous UK Holstein-Friesian dairy cattle [18], while the genetic correlations with milk, fat and protein yield in Irish Holstein-Friesian cattle varied from -0.31 to 0.39 (standard errors varied from 0.13 to 0.20) [41]. This suggests that selection for increased milk production alone may increase genetic susceptibility to MAP and possibly also bovine TB.

Table 1 summarises the genetic correlation, across a range of international studies from different breeds of dairy cattle, between milk production and several diseases. Although there were a few exceptions, selection for increased milk production alone without any cognisance of other traits is expected to increase the incidence of mastitis, lameness, cystic ovaries, ketosis and metritis. The impact on RFM is less clear.

Table 1 Genetic correlations between various health traits and 305-day milk yield in dairy cattle

SCC[1]	Mastitis	Lameness	Milk Fever	Cystic ovaries	Metritis	Ketosis	RFM[2]	DA[3]	Country	Reference
0.06 to 0.82									Multiple	[9][4]
-0.97 to 0.48									Multiple	[9][5]
	0.21 (0.06)	0.29 (0.11)	0.19 (0.06)						UK	[11]
	0.09	0.02		0.10	-0.05	0.12			US	[12][6]
	0.15	0.24		0.23	0.02		-0.28		Canada	[14]
			0.27 (0.17)				0.26 (0.23)		US	[27]
	0.37			-0.14					Canada	[28]
		0.27[7]	-0.67	-0.06		0.77		-0.04	Canada	[28]
	0.35 (0.10)		0.27 (0.12)	0.70 (0.15)					UK	[29]
				0.34 (0.11)					Netherlands	[32]
				0.42 (0.08)	0.68 (0.23)				Finland	[33]
0.10 (0.08)	0.46 (0.09)								Finland	[56]
-0.11 to 0.00 (0.10 to 0.11)	0.35 to 0.61 (0.10 to 0.11)								Finland	[56]
0.15 (0.06)	0.45 (0.09)								France	[57]
0.198 (0.110)		0.056 to 0.34 (0.12 to 0.15)[7]							Germany	[58]
	0.51					0.65			Norway	[59]
	0.29 (0.08)								Germany	[60]
			-0.49						Norway	[61]

[1]Somatic cell count; [2]Retained foetal membranes; [2]Displaced abomasum; [3]Based on nadir calcium levels around parturition; [4]Based on a review of 7 scientific papers; [5]Based on a review of 9 scientific papers; [6]Estimates based on product moment correlations between estimated breeding values· [7]Lameness based on claw and hoof disorders or culling for leg problems.

6. LIMITATIONS OF THE CURRENT METHODOLOGY IN ANIMAL BREEDING

Traditional genetic evaluations use sophisticated statistical methodology to estimate the genetic merit of an animal, and require large datasets containing information on animals, their pedigree, sibs and progeny. These methods of genetic evaluation are not without their limitations, particularly when seeking to breed for improved resistance to disease. Examples of the limitations include:

- the phenotype measured contains error (i.e., variable diagnostic accuracy depending on the farmer or the veterinarian),
- the phenotype may not be measurable in both genders (e.g., mastitis in cows),
- performance in adult cattle cannot be measured in calves, thereby delaying the time lag to identification of genetically different animals for adult performance,
- antagonistic or unfavourable genetic correlations between traits of interest cannot be easily resolved (e.g., health and milk production), and
- genotype by environment interactions may exist (i.e., environment influences the expression of an animal's genes), which may complicate the statistical analyses.

Furthermore, accurate estimates of genetic merit require large and expensive progeny testing schemes. This is particularly true for low heritability traits where large progeny group sizes with recorded health phenotypes are necessary to obtain accurate estimates of genetic merit. New technologies provide an opportunity to reduce or alleviate some of these limitations.

7. USEFULNESS OF NEW 'OMICS' TECHNOLOGIES IN BREEDING FOR IMPROVED ANIMAL HEALTH

Advancements in 'omics' technologies and how the emerging data are interpreted and combined have the potential to overcome, or at least minimise, some of the aforementioned disadvantages of the traditional methods of genetic evaluation. The recent advancements in 'omics' technologies and their applications to animal breeding have been discussed in detail by Berry et al. [42]. Marker-assisted selection, combining the traditional genetic evaluation of an animal with its genotype at several loci, was originally proposed as a method of incorporating genomic information into genetic evaluations. However, few markers have been commercialised based on associations with quantitative traits related to animal health. The use of genetic markers for pre-screening animals for single gene genetic disorders or traits has, however, been successfully used in animal breeding [43].

Genomic selection [44] is now replacing the traditional methodology of genetic evaluation in dairy cattle, and is thought to be superior to marker-assisted selection. Genomic selection will also soon be implemented in beef cattle. Unlike marker-assisted selection which uses data on only tens of genetic markers, genomic selection is based on associating thousands of genetic markers, spread densely across the genomes of several thousand animals, with the range of phenotypes available on those animals. The end result is the estimate of the association between each marker and each phenotype. Miniaturised genotyping platforms with individual detection elements capable of simultaneously assaying up to 777,000 distinct SNP genetic markers, which are evenly distributed across the bovine genome are now commercially available (Illumina Inc., San Diego, CA). However, whole genome sequencing on large numbers of animals will soon become feasible [45], increasing the number of useful markers dramatically. Once each marker effect is known, then an animal with no available phenotype (e.g., a disease free newborn calf) can be genotyped and its direct genomic value (DGV) estimated based solely on its genotype. This is usually integrated with the traditional genetic evaluation of the animal to generate a genomic-estimated breeding value (GEBV). The advantage of such a method is that accurate estimates of genetic merit can be achieved exploiting knowledge of the genotype of the animal [46], even if the animal is very young. Reliability estimates of up to 50% have been reported in the Irish genomic selection-breeding programme for Holstein-Friesian dairy cattle in which no phenotypic data were available, which are higher than the approximate 30% reliability achieved previously using traditional methodology [47]. Nonetheless, large quantities of data are still required to accurately estimate the effect of each genetic marker (Figure 2, based on equations provided by Daetwyler et al. [48]).

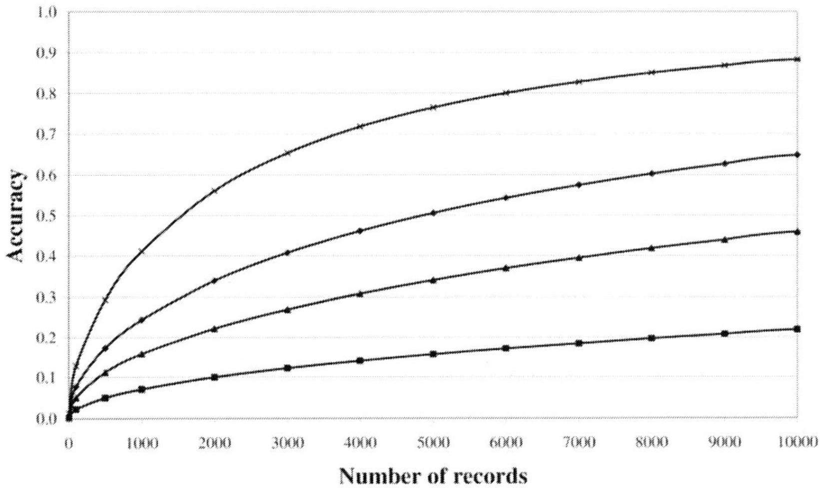

Figure 2 Accuracy of genomic selection for a trait within a heritability of 0.03 (square), 0.15 (triangle), 0.35 (diamond) and 0.90 (x) across different numbers of animals genotyped and phenotyped (Daetwyler et al. [47]).

8. CONCERNS ABOUT BREEDING FOR IMPROVED RESISTANCE

Several concerns may be raised about the possible repercussions or limitations of breeding for improved animal health and resistance to disease, some of which have already been discussed in depth by Stear et al. [49]. Concerns include:

- sustainability of breeding for resistance in the face of continuous evolution of pathogens,
- the feasibility of selection for traits where little genetic variation may exist,
- impact of selection for resistance to a given disease on the genetic resistance to other diseases,
- the impact of selecting on a phenotype thought to reflect resistance to a disease, on the actual genetic predisposition to the disease (e.g., tuberculin test as a reflection of susceptibility to TB or somatic cell count as a reflection of resistance to mastitis),

- impact of selection for disease resistance on genetic gain in other traits affecting profitability, and
- the cost-benefit of developing a breeding programme to select for improved resistance to disease.

There is some concern that the rates of evolution of pathogen genomes are significantly more rapid that those of their vertebrate hosts [50]. In other words, pathogens will evolve faster than genetic gain in the host for resistance, thereby making any breeding programme unsustainable. Stear et al. [49] suggests that one potential mechanism to overcome such a concern would be to breed for improved resilience or tolerance to infection which would impose no selective pressure on the pathogen to evolve unlike breeding for resistance to the pathogen. However, there are certain conditions where improved tolerance of infection would not be beneficial in all circumstances. For example, improving tolerance to bovine tuberculosis would have little, if any, benefit in countries implementing eradication programmes for bovine tuberculosis where infected animals are culled.

An argument is sometimes used that progression from infection to disease is purely environmental or that the heritability of disease resistance is too low for breeding to be fruitful. However, this review clearly shows genetic variation in resistance to the majority of diseases in cattle. Therefore, if sufficient phenotypic data is available, then the impact of low heritability on genetic response is minimised. Furthermore, new technologies such as genomic selection, if exploited, will further reduce the impact of low heritability on genetic gain.

Interactions between diseases are expected [10, 15] and, therefore, there is concern that selection for resistance to a given disease may increase susceptibility to other diseases. However, such a dilemma only materialises if antagonistic genetic correlations exist between traits. It is well known that sick animals are, on average, phenotypically more susceptible to infection with other diseases. This is substantiated by positive genetic correlations between many disease traits (discussed previously in this review). Nonetheless, accurate routine recording of disease incidence should be undertaken and monitored to ensure no unfavourable trends exist, as well as providing data to estimate genetic correlations with other health and disease traits, or more importantly expected response to selection based on the breeding goal. One option in an attempt to minimise any potential antagonistic effects could be to select for general immunity or tolerance to infectious diseases of major economic importance.

Due primarily to a lack of routinely available data for a given phenotype of interest, biological predictors are sometimes used as proxies. Such proxies include reaction to the tuberculin test as a predictor of susceptibility to tuberculosis, serological response to MAP as a predictor of susceptibility to

MAP, and more commonly somatic cell count as a predictor of clinical mastitis. Although the correlations amongst most predictors and the traits being predicted are strong [9, 17], they are not unity. This suggests that firstly additional genetic gain could be achievable by measuring the goal trait itself, but secondly selection on the proxy could have unfavourable repercussions for genetic susceptibility to the goal trait itself. For example, concern has been expressed that cows with low somatic cell count may be less able to respond to pathogenic challenges. Studies which experimentally induced intramammary infections reported that low somatic cell count was associated with increased risk or severity of infection [51, 52]. However, studies investigating the relationship between somatic cell count and naturally occurring clinical mastitis failed to reveal any association between low somatic cell count and increased risk of clinical mastitis [53]. Nonetheless, the implications are not clear and should not be ignored when implementing breeding programmes for resistance using an indicator trait.

Response to selection for a given trait within a breeding goal is a function of, amongst others, the relative weight placed on that trait, the quantity of information available on that trait (which affects the accuracy of selection) and the phenotypic and genetic relationships with other traits within the breeding goal. The inclusion of a trait antagonistically correlated with performance within a breeding goal will reduce the genetic gain in the performance trait. Most studies, already discussed, have documented an antagonistic correlation between both animal health and resistance to disease and milk production (Table 1), implying that placing selection pressure on health and disease resistance will reduce genetic gain in milk production. However, selection on a balanced breeding goal will increase genetic gain in overall profitability if each of the traits is weighted appropriately by their respective economic values.

Routine access to accurate phenotypes is vital to achieving sustainable genetic gain in animal health and disease resistance. However, for rare diseases in particular, a cost-benefit analysis of generating sufficient data to estimate breeding values and then incorporate these values into a breeding programme should be evaluated relative to embarking on a project to control the disease through other means (e.g., vaccination). This is particularly true since expected responses to genetic selection will take a long time to develop although they are cumulative and permanent. Nonetheless, the development of 'omics' technology and associated analysis of the data may ameliorate the necessity for continuous routine access to such data. Furthermore, when evaluating the economic benefit of breeding for disease resistance, some account must be taken of the impact of selection on its effects on the transmission of disease through the population and, therefore, the challenge faced by all animals including those not selected for resistance [54].

9. FUTURE RESEARCH

One of the biggest constraints of breeding for improved animal health or resistance to disease is routine access to accurate phenotypes (i.e., measurements) of health traits. Therefore, research must be undertaken on the development of inexpensive and humane methods of either challenging animals for disease resistance to generate phenotypes (e.g., the tuberculin test) or the development of accurate bio-markers that can be readily measured in large numbers of animals at a relatively low cost. Furthermore, increased collaboration between veterinarians and animal breeders on the definition and collection of the relevant phenotypes, as well as the most appropriate statistical model, based on biological soundness, is vital to achieving genetic gain.

Not alone do developments in animal breeding and the relevant 'omics' disciplines aid in breeding for increased resistance to infection, but elucidation of the genome of the host, coupled with the availability of the whole genome of the pathogen and how that interacts with the genome of the host, will help in determining the mode of infection and thus the development of effective vaccines or prophylactic treatments.

10. CONCLUSIONS

There is overwhelming evidence that genetics make a significant contribution to the health and resistance to disease in cattle, and that the tools for simultaneous selection on these traits and other performance traits are available. Furthermore, because of differences in exposure rates as well as the lack of complete sensitivity and specificity of tests, heritability estimates for health and disease resistance traits discussed in this review are likely to be underestimates of the true heritability [55]. Additionally, new developments in 'omics' technologies provide a considerable resource that can be exploited to further increase genetic gain, especially in health and disease traits. Nonetheless, resistance to most diseases will be governed by a large number of genes, and mutations within genes. Therefore, absolute resistance is unlikely, and genetics alone is not the solution to improved animal health. Rather, it should be seen as an integral part of an overall programme for improved cattle health, both at farm level and national level.

REFERENCES

1. McHugh N, Evans RD, Amer PR, Fahey AG, Berry DP: Genetic parameters for cattle price and live-weight from routinely collected data at livestock auctions and commercial farms. J Anim Sci. 2011, 89: 29-39. 10.2527/jas.2010-3044.
2. Rendel J, Robertson A: Estimation of genetic gain in milk yield by selection in a closed herd of dairy cattle. J Genetics. 1950, 50: 1-8. 10.1007/BF02986789.
3. McParland S, Kearney JF, Rath M, Berry DP: Inbreeding trends and pedigree analysis of Irish dairy and beef cattle populations. J Anim Sci. 2007, 85: 322-331. 10.2527/jas.2006-367.
4. Snowder GD, Van Vleck LD, Cundiff LV, Bennett GL: Influence of breed, heterozygosity, and disease incidence on estimates of variance components of respiratory disease in preweaned beef calves. J Anim Sci. 2005, 83: 1247-1261.
5. Heringstad B, Chang YM, Gianola D, Østerås O: Genetic analysis of respiratory disease in Norwegian Red calves. J Dairy Sci. 2008, 91: 367-370. 10.3168/jds.2007-0365.
6. Snowder GD, Van Vleck LD, Cundiff LV, Bennett GL: Bovine respiratory disease in feedlot cattle: Environmental, genetic, and economic factors. J Anim Sci. 2006, 84: 1999-2008. 10.2527/jas.2006-046.
7. Schneider MJ, Tait RG, Ruble MV, Busby WD, Reecy JM: Evaluation of fixed sources of variation and estimation of genetic parameters for incidence of bovine respiratory disease in preweaned calves and feedlot cattle. J Anim Sci. 2010, 88: 1220-1228. 10.2527/jas.2008-1755.
8. Berry DP, Kearney JF, Twomey K, Cromie AR, Evans RD: Revision of the fertility and health genetic evaluations for Irish Holstein-Friesian dairy cattle. Proceedings of the 9th World Congress on Genetics Applied to Livestock production. Leipzig, Germany. 2010, August 1-6 2010
9. Mrode RA, Swanson GJT: Genetic and statistical properties of somatic cell count and its suitability as an indirect means of reducing the incidence of mastitis in dairy cattle. Anim Breed Abstr. 1996, 64: 847-857.
10. Heringstad B, Klemetsdal G, Steine T: Selection responses for disease resistance in two selection experiments with Norwegian Red cows. J Dairy Sci. 2007, 90: 2419-2426. 10.3168/jds.2006-805.
11. Pryce JE, Veerkamp RF, Thompson R, Hill WG, Simm G: Genetic aspects of common health disorders and measures of fertility in Holstein Friesian dairy cattle. Anim Sci. 1997, 65: 353-360. 10.1017/S1357729800008559.
12. Zwald NR, Weigel KA, Chang YM, Welper RD, Clay JS: Genetic selection for health traits using producer recorded data. II Genetic correlations, disease probabilities, and relationships with existing traits. J Dairy Sci. 2004, 87: 4295-4302. 10.3168/jds.S0022-0302(04)73574-2.
13. Boettcher PJ, Dekkers JCM, Warnick LD, Wells SJ: Genetic analysis of clinical lameness in dairy cattle. J Dairy Sci. 1998, 81: 1148-1156. 10.3168/jds.S0022-0302(98)75677-2.
14. Van Dorp TE, Dekkers JCM, Martin SW, Noodhuizen JPTM: Genetic parameters of health disorders, and relationships with 305-day milk yield and conformation traits of registered Holstein cows. J Dairy Sci. 1998, 81: 2264-2270. 10.3168/jds.S0022-0302(98)75806-0.
15. Van der Waaij EH, Holzhauer M, Ellen E, Kamphuis C, de Jong G: Genetic parameters for claw disorders in Dutch dairy cattle and correlations with conformation traits. J Dairy Sci. 2005, 88: 3672-3678. 10.3168/jds.S0022-0302(05)73053-8.
16. More SJ, Good M: The tuberculosis eradication programme in Ireland: a review of scientific and policy advances since 1988. Vet Microbiol. 2006, 112: 239-51. 10.1016/j.vetmic.2005.11.022.
17. Bermingham ML, More SJ, Good M, Cromie AR, Mullowney P, Higgins IM, Berry DP: Genetics of tuberculosis in Irish Holstein-Friesian dairy herds. J Dairy Sci. 2009, 92: 3447-3456. 10.3168/jds.2008-1848.

18. Brotherstone S, White IMS, Coffey M, Downs SH, Mitchell AP, Clifton-Hadley RS, More SJ, Good M, Woolliams JA: Evidence of genetic resistance of cattle to infection with Mycobacterium bovis. J Dairy Sci. 2010, 93: 1234-1242. 10.3168/jds.2009-2609.
19. Gonda MG, Chang YM, Shook GE, Collins MT, Kirkpatrick BW: Genetic variation of Mycobacterium avium ssp. paratuberculosis infection in US Holsteins. J Dairy Sci. 2006, 89: 1804-1812. 10.3168/jds.S0022-0302(06)72249-4.
20. Hinger M, Brandt H, Erhardt G: Heritability estimates for antibody response to Mycobacterium avium subspecies paratuberculosis in German Holstein cattle. J Dairy Sci. 2008, 91: 3237-3244. 10.3168/jds.2008-1021.
21. Koets AP, Adugna G, Janss LLG, van Weering HJ, Kalis CHJ, Wentink GH, Rutten VPMG, Schukken YH: Genetic variation of susceptibility to Mycobacterium avium subsp. paratuberculosis infection in dairy cattle. J Dairy Sci. 2009, 83: 2702-2708. 10.3168/jds.S0022-0302(00)75164-2.
22. Mortensen M, Nielsen SS, Berg P: Genetic variation and heritability of the antibody response to Mycobacterium avium subspecies paratuberculosis in Danish Holstein cows. J Dairy Sci. 2004, 87: 2108-2113. 10.3168/jds.S0022-0302(04)70029-6.
23. Berry DP, Good M, Mullowney P, Cromie AR, More SJ: Genetic variation in serological response to Mycobacterium avium subspecies paratuberculosis and its association with performance in Irish Holstein-Friesian dairy cows. Livest Sci. 2010, 131: 102-107. 10.1016/j.livsci.2010.03.007.
24. Wray C, Sojka WJ: Experimental Salmonella typhimurium infection in calves. Res Vet Sci. 1978, 25: 139-143.
25. Janss LLG, Bolder NM: Heritabilities of and genetic relationships between salmonella resistance traits in broilers. J Anim Sci. 2000, 78: 2287-2291.
26. Templeton JW, Estes DM, Price RE, Smith R, Adams LG: Immunogenetics of natural resistance to bovine brucellosis. Proceedings of the 4th World Congress on Genetics Applied to Livestock Production. 1990, 396-399.
27. Thompson JR: Genetic interrelationships of parturition problems and production. J Dairy Sci. 1984, 67: 628-635. 10.3168/jds.S0022-0302(84)81348-X.
28. Uribe HA, Kennedy BW, Martin SW, Kelton DF: Genetic parameters for common health disorders of Holstein cows. J Dairy Sci. 1995, 78: 421-430. 10.3168/jds.S0022-0302(95)76651-6.
29. Kadarmideen HN, Thompson R, Simm G: Linear and threshold model genetic parameters for disease, fertility and milk production in dairy cattle. Anim Sci. 2000, 71: 411-419.
30. Heringstad B, Chang YM, Gianola D, Klemetsdal G: Genetic analysis of clinical mastitis, milk fever, ketosis, and retained placenta in three lactations of Norwegian Red Cows. J Dairy Sci. 2005, 88: 3273-3281. 10.3168/jds.S0022-0302(05)73010-1.
31. Heringstad B: Genetic analysis of fertility-related diseases and disorders in Norwegian red cows. J Dairy Sci. 2010, 93: 2751-2756. 10.3168/jds.2009-2879.
32. Hooijer GA, Lubbers RBF, Ducro BJ, van Arendonk JAM, Kaal-Lansbergen LMTE, van der Lende T: Genetic parameters for cystic ovarian disease in Dutch Black and White dairy cattle. J Dairy Sci. 2001, 84: 286-291. 10.3168/jds.S0022-0302(01)74478-5.
33. Pösö J, Mäntysaari EA: Genetic relationships between reproductive disorders, operational days open and milk yield. Livest Prod Sci. 1996, 48: 41-48.
34. Koeck A, Egger-Danner C, Fuerst C, Obritzhauser W, Fuerst-Waltl B: Genetic analysis of reproductive disorders and their relationship to fertility and milk yield in Austrian Fleckveih dual purpose cows. J Dairy Sci. 2010, 93: 2185-2194. 10.3168/jds.2009-2570.
35. Lyons DT, Freeman AE, Kuck AL: Genetics of health traits in Holstein cattle. J Dairy Sci. 1991, 74: 1092-1100. 10.3168/jds.S0022-0302(91)78260-X.
36. Morris CA, Watson TG, Bisset SA, Vlassoff A, Douch PGC: Breeding sheep in New Zealand for resistance or resilience to nematode parasites. Breeding for Resistance to Infectious Diseases in Small Ruminants. Edited by: Gray GD, Woolaston RR, Eaton BT. 1995, ACIAR Monograph series, Canberra, 77-98.
37. Pan Y, Jansen GB, Duffield TF, Hietala S, Kleton D, Lin CY, Peregrine AS: Genetic susceptibility to Neospora canium infection in Holstein cattle in Ontario. J Dairy Sci. 2004, 87: 3967-3975. 10.3168/jds.S0022-0302(04)73537-7.

38. Cooppieters W, Mes THM, Druet T, Farnir F, Tamma N, Schrooten C, Cornelissen AWCA, Georges M, Ploeger HW: Mapping QTL influencing gastrointestinal nematode burden in Dutch Holstein-Friesian dairy cattle. BMC Genomics. 2009, 10: 96-10.1186/1471-2164-10-96.
39. Davis GP: Genetic parameters for tropical beef cattle in Northern Australia: a review. Aust J Agric Res. 1993, 44: 179-198. 10.1071/AR9930179.
40. Miglior F, Muir BL, Van Doormaal BJ: Selection indices in Holstein cattle of various countries. J Dairy Sci. 2005, 88: 1255-1263. 10.3168/jds.S0022-0302(05)72792-2.
41. Bermingham ML, More SJ, Good M, Cromie AR, Higgins IM, Berry DP: Genetic correlations between measures of Mycobacterium bovis infection and economically important traits in Irish Holstein-Friesian dairy cows. J Dairy Sci. 2010, 93: 5413-5422. 10.3168/jds.2009-2925.
42. Berry DP, Meade KG, Mullen MP, Butler S, Diskin MG, Morris D, Creevey C: The integration of 'omic' disciplines and systems biology in animal breeding in cattle. Animal. 2011, 5: 493-505. 10.1017/S1751731110002120.
43. Dekkers JCM: Commercial application of marker- and gene-assisted selection in livestock: Strategies and lessons. J Anim Sci. 2004, 82: E313-328.
44. Meuwissen THE, Hayes BJ, Goddard ME: Prediction of total genetic value using genome-wide dense marker maps. Genetics. 2001, 157: 1819-1829.
45. Mardis ER: Next-generation DNA sequencing methods. Annu Rev Genomics Hum Genet. 2008, 9: 387-402. 10.1146/annurev.genom.9.081307.164359.
46. VanRaden PM, Van Tassell CP, Wiggans GR, Sonstegard TS, Schnabel RD, Taylor JF, Schenkel FS: Invited Review: Reliability of genomic predictions for North American Holstein bulls. J Dairy Sci. 2009, 92: 16-24. 10.3168/jds.2008-1514.
47. Berry DP, Kearney JF, Harris BL: Genomic selection in Ireland. Interbull Bulletin 2009. Uppsala Sweden. 2009, 29-34.
48. Daetwyler H, Villanueva B, Woolliams JA: Accuracy of predicting the genetic risk of disease using a genome-wide approach. PLOS One. 2008, 3: e3395-10.1371/ journal. pone.0003395.
49. Stear MJ, Bishop SC, Mallard BA, Raadsma H: The sustainability, feasibility and desirability of breeding livestock for disease resistance. Res Vet Sci. 2001, 71: 1-7. 10.1053/rvsc.2001.0496.
50. Jackson RW, Johnson LJ, Clarke SR, Arnold DL: Bacterial pathogen evolution: breaking news. Trends Genet. 2011, 27: 32-40. 10.1016/j.tig.2010.10.001.
51. Schukken YH, Mallard BA, Dekkers JCM, Leslie KE, Stear MJ: Genetic impact on the risk of intramammary infection following Staphylococcus aureus challenge. J Dairy Sci. 1994, 82: 2393-2401. 10.3168/jds.S0022-0302(99)75490-1.
52. Shuster DE, Lee EK, Kehlri ME: Bacterial growth, inflammatory cytokine production, and neutrophil recruitment during coliform clinical mastitis in cows within ten days after calving, compared with at mid lactation. Amer J Vet Res. 1996, 57: 1569-1575.
53. Philipsson J, Ral G, Berglund B: Somatic-cell count as a selection criterion for clinical mastitis resistance in dairy cattle. Livest Prod Sci. 1995, 41: 195-200. 10.1016/0301-6226(94)00067-H.
54. MacKenzie K, Bishop SC: A discrete-time epidemiological model to quantify selection for disease resistance. Anim Sci. 1999, 69: 543-551.
55. Bishop SC, Woolliams JA: On the genetic interpretation of disease data. PLoS One. 2010, 5: e8940-10.1371/journal.pone.0008940.
56. Pösö J, Mäntysaari EA: Relationship between clinical mastitis, somatic cell score, and production for the first three lactations of Finnish Ayrshire. J Dairy Sci. 1996, 79: 1284-1291.
57. Rupp R, Boichard D: Genetic parameters for clinical mastitis, somatic cell score, production, udder type traits, and milking ease in first lactation Holsteins. J Dairy Sci. 1999, 82: 2198-2204. 10.3168/jds.S0022-0302(99)75465-2.
58. Koenig S, Sharifi AR, Wentrot H, Landmann D, Eise M, Simianer H: Genetic parameters of claw and foot disorders estimated with logistic models. J Dairy Sci. 2005, 88: 316-3325. 10.3168/jds.S0022-0302(05)73015-0.

59. Simianer H, Solbu H, Schaeffer LR: Estimated genetic correlations between disease and yield traits in dairy cattle. J Dairy Sci. 1991, 74: 4358-4365. 10.3168/jds.S0022-0302(91)78632-3.
60. Hinrichs D, Stamer E, Junge W, Kalm E: Genetic analyses of mastitis data using animal threshold models and genetic correlation with production traits. J Dairy Sci. 2005, 88: 2260-2268. 10.3168/jds.S0022-0302(05)72902-7.
61. Tveit B, Svendsen M, Hove K: Heritability of hypocalcemia at first parturition in Norwegian cattle: genetic correlations with yield and weight. J Dairy Sci. 1991, 74: 3561-3567. 10.3168/jds.S0022-0302(91)78548-2.

CHAPTER 3

Foot-and-mouth disease: past, present and future

Syed M Jamal and Graham J Belsham

National Veterinary Institute, Technical University of Denmark, Lindholm, 4771 Kalvehave, Denmark

ABSTRACT

Foot-and-mouth disease (FMD) is a highly contagious disease of cloven-hoofed animals including cattle, pigs, sheep and many wildlife species. It can cause enormous economic losses when incursions occur into countries which are normally disease free. In addition, it has long-term effects within countries where the disease is endemic due to reduced animal productivity and the restrictions on international trade in animal products. The disease is caused by infection with foot-and-mouth disease virus (FMDV), a picornavirus. Seven different serotypes (and numerous variants) of FMDV have been identified. Some serotypes have a restricted geographical distribution, e.g. Asia-1, whereas others, notably serotype O, occur in many different regions. There is no cross-protection between serotypes and sometimes protection conferred by vaccines even of the same serotype can be limited. Thus it is important to characterize the viruses that are circulating if vaccination is being used for disease control. This review describes current methods for the detection and characterization of FMDVs. Sequence information is increasingly being used for identifying the source of outbreaks. In addition such information can be used to understand antigenic change within virus strains. The challenges and opportunities for improving the control of the disease within endemic settings, with a focus on Eurasia, are discussed, including the role of the FAO/EuFMD/OIE Progressive Control Pathway. Better control of the disease in endemic areas reduces the risk of incursions into disease-free regions.

1. INTRODUCTION

The earliest description of probable foot-and-mouth disease (FMD) in cattle was made by an Italian monk, Hieronymus Fracastorius, in Venice in 1514. The affected animals refused their feed, the oral mucosa showed redness and the animals had vesicles in the oral cavity and on their feet. Most of the affected animals eventually recovered. This description, made 500 years ago, shows a strong resemblance to that of FMD when seen currently. FMD is considered one of the most important diseases of cloven-hoofed animals; it affects cattle, buffaloes, pigs, sheep, goats and about 70 wildlife species, e.g. African buffaloes. The disease has been present in almost every part of the world where livestock are kept. More than 100 countries are still affected by FMD worldwide and distribution of the disease roughly reflects economic development. The more developed countries have eradicated the disease. However, an incursion of the disease into the normally disease-free countries can cause enormous economic losses. The disease is caused by a single stranded positive sense RNA virus, foot-and-mouth disease virus (FMDV), belonging to the genus *Aphthovirus* within the family *Picornaviridae*. This review includes a description of the properties of the virus and the systems for detecting and characterizing FMD outbreaks. This information is then used to describe the current distribution of the disease/virus and how the FAO/EuFMD/OIE Progressive Control Pathway can assist in disease control within endemic countries and hence reduce the risk of incursions into disease free regions. The major focus here is on Eurasia but other regions are also considered when appropriate.

2. STRUCTURE OF FMDV

The FMDV particle is roughly spherical in shape and about 25–30 nm in diameter. It consists of the RNA genome surrounded by a protein shell or capsid. The capsid is composed of 60 copies of the capsomers. Each capsomer consists of four structural polypeptides, VP1, VP2, VP3 and VP4. The VP1, VP2 and VP3 are exposed on the surface of the virus while VP4 is located internally. The protein coat surrounds a single stranded, positive sense RNA genome about 8400 nucleotides (nt) in length. The RNA includes three separate parts i.e. the 5′ untranslated region (5′ UTR), a long coding region and the 3′ untranslated region (3′ UTR) (Figure 1). A small protein (24 or 25 residues long), termed VPg, which is encoded by the 3B portion of the viral genome region, is covalently linked to the 5′ end of the genome.

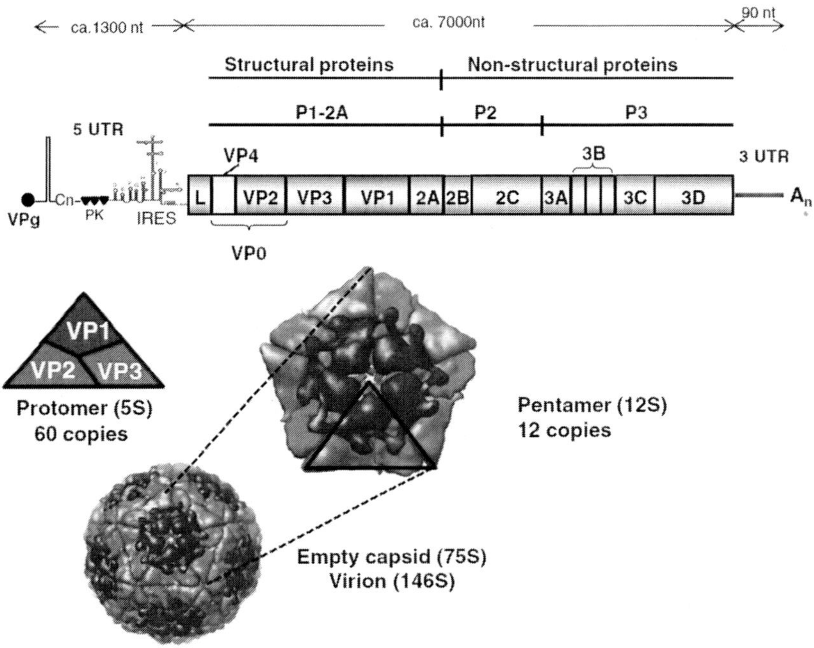

Figure 1 Genome organization of fmdv and the structure of virus. The FMDV genome includes a single large ORF, indicated by the shaded rectangle. The regions within the rectangle indicate the individual proteins. The 5′ UTR includes several distinct structural elements including: a poly(C) tract (Cn), 3 or 4 pseudoknots (PK) and the internal ribosome entry site (IRES). The VPg peptide is made in 3 different forms (encoded by the $3B_{1\text{-}3}$) and each acts as the primer for RNA synthesis so each RNA genome, when synthesized, is covalently linked to a VPg. The assembly of virus particles from protomeric and pentameric subunits is indicated. Assembled virus particles contain a single copy of the viral RNA and 60 copies of the 4 different capsid proteins (VP1-VP4). Self-assembly of empty capsid particles, lacking the RNA genome, can also occur. The VP4 protein is internal.

The 5′ UTR is about 1300 nt in length [1] and consists of an S fragment at its 5′ end, a poly C tract, a series of RNA pseudoknot structures, a *cis*-acting replication element (*cre*) (also known as the 3B-uridylylation site (*bus*)), and the internal ribosome entry site (IRES). The S fragment is 360 nt in length and is predicted to fold to form a large hairpin structure. The poly C tract is of variable length (150–250 nt) but is comprised of over 90% C residues. The function of the pseudoknots is unknown. The *cre*/*bus* is a stable stem-loop element of about 55 nucleotides and contains a conserved motif (AAACA) which acts as a template for uridylylation of VPg ($3B_{1-3}$) by the viral RNA polymerase. Thus the *cre/bus* is involved in the initiation of RNA replication. The IRES is about 450 nt in length and is responsible for cap-independent initiation of viral protein synthesis [1].

The coding region follows the 5′ UTR. It is the major portion of the viral genome and is about 7000 nt in length. It encodes a large polyprotein which is then cleaved by viral proteases to form four different structural and eleven different non-structural proteins plus a variety of precursors, some of which have distinct functions. After translation, initially four primary products are formed, namely, L^{pro}, P1-2A, P2 and P3. The Leader protease (L^{pro}) is the N-terminal component of the polyprotein. The L coding region contains two separate AUG initiation codons (usually 84 nucleotides apart) that result in the generation of two different L proteins, termed Lab and Lb. The L^{pro} is responsible for the inhibition of host cell protein synthesis by inducing the cleavage of the host protein, eIF4G, which is a translation initiation factor [1] that is required for the translation of the capped cellular mRNAs. As a result, FMDV RNA can freely use the host cell's protein synthesis machinery for its own protein synthesis since the FMDV IRES can function with the residual C-terminal fragment of eIF4G [1]. The P1-2A capsid precursor is cleaved by the 3C protease ($3C^{pro}$) to produce 1AB (VP0), 1C (VP3) and 1D (VP1) (plus 2A) and during encapsidation of the genome the VP0 is cleaved to make VP4 and VP2 (note 2A is a very short peptide, <20 residues). The VP4 is entirely internal within the virus particle whereas VP1, VP2 and VP3 are surface exposed and contribute to the antigenic properties of the virus [2, 3]. The VP1 contains at least two important immunogenic sites, the G-H loop (at amino acid positions 141–160) and the C-terminus (residues 200–213). The G-H loop includes an arginine-glycine-aspartic acid (RGD) motif, which is required for attachment of the virus to the host cell via an integrin receptor [4, 5]. Integrins are a group of α-β heterodimeric glycoproteins which are located on the cell surface; some 15 α and 8 β sub-units combine to form 20 different α-β heterodimers. The αvβ6 heterodimer is a receptor for the extracellular matrix proteins whose expression is restricted to epithelial cells and it also binds to FMDV through interaction with the RGD motif [6]. The virus, however, can also infect cells in an RGD-independent manner using alternative molecules e.g. heparan sulphate proteoglycans receptors [7].

The nucleotide sequences of the VP1 coding region have been used for genetic characterization of FMDV strains because of their significance for virus attachment and entry, protective immunity and serotype specificity. VP1 sequence based phylogenetic analyses have been used widely to deduce evolutionary dynamics, epidemiological relationships among the genetic lineages and in the tracing of the origin and movement of outbreak strains [8, 9, 10, 11, 12].

The P2 and P3 regions of the polyprotein are processed to the non-structural proteins (NSPs) [1]. The P2 region generates the proteins 2B and 2C while the P3 region is cleaved to form the proteins 3A, three distinct copies of VPg ($3B_{1-3}$), $3C^{pro}$ and $3D^{pol}$. The P2 and P3 encoded proteins are involved in protein processing ($3C^{pro}$) and genome replication (2B, 2C, 3A, $3B_{1-3}$ (VPg) and $3D^{pol}$). The $3C^{pro}$ is responsible for cleavage of P1-2A into VP0, VP1, VP3 as well as the formation of the different non-structural proteins.

The 3′ UTR is much shorter than the 5′ UTR. It is about 90 nucleotides long and folds to form a specific stem-loop structure, followed by a polyA tract of variable length [13]. The 3′ UTR must play an important role in viral genome replication.

3. SEROTYPES OF FMDV

FMD virus exists as seven different serologically distinct types. Serotypes O and A were initially discovered by Vallee and Carre [14]. They showed that cattle that had recovered from clinical disease due to an FMD virus which originated in France became re-infected almost immediately when mixed with animals infected with FMD virus that originated in Germany. They named these serotypes after their place of origin; O for the department of Oise in France and A for Allemagne (the French word for Germany). Their work was extended by Waldmann and Trautwein [15] with the discovery of a third serotype which was named serotype C. Later three additional serotypes were identified in samples originating from South Africa and they were named as Southern African Territories 1, 2 and 3 (SAT1, SAT2, SAT3) [16]. The seventh serotype, Asia-1, was initially detected in a sample collected from a water buffalo at Okara, Punjab, Pakistan in 1954 [17]. Examination of an extensive number of samples from across the world have failed to reveal the existence of another serotype although there are many different "sub-types", some of which are quite distinct from other strains of the same serotype.

4. DIAGNOSIS OF FMD

Due to the rapidity of spread of FMD and the serious economic consequences that can arise from an outbreak, prompt, sensitive and specific laboratory diagnosis and identification of the serotype of the viruses involved in disease outbreaks is essential. The disease is diagnosed based on clinical signs, including high temperature, excessive salivation, formation of vesicles on the oral mucosa, on the nose plus the inter-digital spaces and coronary bands on the feet. However, the clinical signs can be confused with other diseases (e.g. vesicular stomatitis and swine vesicular disease) and thus laboratory based diagnosis is also necessary. Furthermore, there is no cross protection between the serotypes and the serotype of a virus involved in an outbreak cannot be ascertained on the basis of clinical signs. Thus determination of the serotype involved in field outbreaks has to be established within laboratories to permit proper control/vaccination programs to be followed. Various techniques have been used to diagnose the disease and to ascertain the serotype of the virus. The current methods are described below:

4.1. Neutralization test

The virus neutralization test (VNT) is currently considered as the "gold standard" for detection of antibodies to structural proteins of FMDV and is a prescribed test for import/export certification of animals/animal products [18]. However, as various primary cells and cell lines with variable degrees of sensitivities are used in the VNTs, they are more prone to variability than other serological tests. Furthermore, VNT is slower, subject to contamination and requires restrictive biocontainment facilities in contrast to other serological tests which can use inactivated viruses as antigens.

4.2. Enzyme linked immunosorbent assay (ELISA)

The complement fixation test (CFT) was the test of choice for diagnosis of FMD and virus typing until the 1970s and is still used in some endemic areas. However, in order to overcome the problems of its low sensitivity and difficulty in interpretation of its results due to pro- and anti-complement activities, enzyme linked immunosorbent assays (ELISAs) for antigen detection and virus typing were desired. Roeder and Le Blanc Smith [19] established suitable assays by using high titre antisera raised in rabbits and guinea pigs against purified 146S FMDV particles for antigen capture and detection, respectively. The assays were found to be 125 times more sensitive than the CFT and are still routinely used for the diagnosis of FMD and for virus typing. The ELISA, however, gives positive results with only about 70-80% of epithelial suspensions that contain virus due to a lack of sensitivity. Thus the virus may have to be propagated in tissue culture and subsequently tested in ELISA to detect the virus and ascertain the serotype.

Monoclonal antibody (MAb)-based ELISAs have also been developed for diagnosis of FMD and virus typing [20, 21]. Recently, a sandwich ELISA using recombinant integrin $\alpha v\beta 6$ (a receptor for FMDV) for virus capture and serotype-specific monoclonal antibodies as detecting reagents was compared with the conventional polyclonal antibody-based sandwich ELISAs for the identification and serotyping of all the seven types of FMDV. The integrin/MAb ELISA recognized FMDVs of wide antigenic and molecular diversity from all seven serotypes. Although some FMDVs could not be detected, the assay showed greater specificity than the conventional polyclonal ELISA while retaining test sensitivity [22].

4.3. Virus isolation

As indicated above, the presence of relatively high levels of FMDV antigen in vesicular material can be detected by ELISA. However, when the virus concentration is too low to be detected by ELISA, then it has to be propagated in susceptible cell cultures. Primary cell cultures (such as bovine thyroid cells and porcine or ovine kidney cells) or cell lines (such as BHK or IBRS2) are considered to be generally suitable for isolation of FMDV [18]. However, the production of consistent quality, ready-for-use primary cells is laborious, time-consuming and expensive. Furthermore, virus isolation requires the presence of infectious virus, which depends on sample quality. Up to 4 days may be required to demonstrate the presence of virus, especially when the levels of virus are low (thus it also takes 4 days to be confident, using this methodology, that no virus is present). Moreover, some FMDVs fail to grow in a specific cell type. Thus the absence of apparent growth does not guarantee absence of the virus and therefore samples collected from a suspected case of FMD should be subjected to further investigations, e.g. using another testing system. Additional disadvantages include the problems associated with obtaining and maintaining a regular supply of cells; possible contamination of cell cultures and the necessity to confirm any apparent virus growth by ELISA. These issues may delay the initiation of control measures to contain outbreaks.

4.4. Reverse transcription-polymerase chain reaction (RT-PCR)

The reverse transcription-polymerase chain reaction (RT-PCR) has been shown to be a useful tool for the diagnosis of FMD as it offers the advantages of fast, sensitive and reliable diagnosis. A variety of RT-PCR methods have been reported in recent years for the early detection of FMDV RNA in epithelium, cell culture isolates and other tissues using universal primers for all seven serotypes [23]. Typing of FMDV by RT-PCR was first demonstrated by Rodriguez et al. [24] for the differentiation of the serotypes O, A and C. Serotype specific primers have since been designed for the detection of all seven FMDV serotypes by RT-PCR [25, 26]. Primers designed for these assays target various regions of the virus genome, including the 5′ UTR, the open reading frame and the 3′ UTR. However, evaluation of available sets of primers, designed for universal and serotype-specific diagnosis of FMDV, on a variety of field samples, representing all the seven serotypes of FMDV, has shown that no single primer set is capable of diagnosing the disease or typing of the virus. In order to improve the diagnostic sensitivity of RT-PCR, multiplex assays, incorporating more than one set of primers have been developed [27, 28]. However, differentiation/serotyping could only be made for certain groups of serotypes or individual isolates. Thus the conventional RT-PCR is not sufficiently sensitive and specific to replace methods using virus propagation in cell culture and ELISA.

Recently, real time/quantitative RT-PCR (rRT-PCR) methods have been developed which do not require post-PCR processing (e.g. gel analysis) and the signals can be monitored directly as the target cDNA is being amplified. Other advantages of the rRT-PCR include high throughput capability and the ability to quantify the genetic material in the starting sample. A TaqMan assay has been shown to be very robust and as effective for primary detection of FMDV as virus isolation in conjunction with antigen ELISA [29]. Currently, two different rRT-PCR TaqMan assays are in common use, one targeting the internal ribosomal entry site (IRES) within the 5′ UTR [30] and the second targeting the 3D (RNA polymerase) coding sequence [31].

The speed and accuracy of detection of the rRT-PCR assay was further improved by coupling the assays with robotic methods for extraction of nucleic acid from the samples and for set up of the assays. This has made the assay highly suitable for the diagnosis of the primary index case and for use in an ongoing outbreak. The rRT-PCR assays are currently used as a routine test for FMD diagnosis and quantification of the virus in many developed countries. However, these assays are not designed to discriminate between serotypes of FMDV. Although these assays exploit highly conserved regions across all the seven serotypes of FMDV, serotype biases still exist within both of these assays. The 5′ UTR assay has been shown to be more sensitive in detecting serotype A viruses, whereas, the 3D assay has greater sensitivity for detecting the SAT viruses [32]. Additionally each of the assays failed to detect a small number of FMDV isolates due to the presence of nucleotide mismatches within the region targeted by the probes. Thus, no single stand-alone assay is capable of detecting FMDV with 100% sensitivity. Recently, Tam and colleagues [33] reported fluorescence-based multiplex rRT-PCR assays for the detection of FMDV and virus typing. The assay was found to have greater sensitivity for detection but some cross-reactivity between some serotypes was also noted. Further work is in progress to develop rRT-PCR assays for serotyping of the virus.

4.5. Reverse transcription loop-mediated isothermal amplification (RT-LAMP)

The development of portable equipment for rRT-PCR has enabled molecular diagnosis of FMD possible in the field. This approach, however, requires expensive and fragile instruments and relies on precision thermocycling. Thus other approaches, like loop-mediated amplification (LAMP), were developed, which enable the tests to be conducted in the field using inexpensive tools. LAMP amplifies specific nucleotide sequences at a constant temperature and thus does not require a thermocycler. The assay is based on the principle of DNA amplification by an autocycling strand displacement reaction. The assay is performed using a set of two specially designed inner primers and two outer primers and a DNA polymerase with high strand displacement activity [34]. The primers recognize 6 independent target sequences in the initial stage and 4 independent sequences during the later stages of the LAMP reaction. The reaction is carried out in less than an hour using a standard water bath or heating block and the results can be visualized with the naked eye. The advantages of its simple operation, rapid reaction and potential for visual interpretation without instrumentation make the technique attractive for field use in endemic countries. The RT-LAMP assay that has been developed for FMDV detection can be used in a high throughput system [35]. This assay has, however, not yet been extensively evaluated for its ability to replace or supplement the techniques currently in use.

4.6. Chromatographic strip test

Virus isolation combined with ELISA and RT-PCR assays are reliable and accurate for the diagnosis of FMD but the shipment of samples from the field to the laboratory and the poor quality/amount of submitted samples can result in hindrance of early diagnosis of the disease. A rapid and specific test for disease diagnosis at the site of a suspected outbreak may allow timely implementation of control measures. A MAb-based chromatographic strip test for FMD diagnosis was developed by Reid et al. [36]. The test was found to be at least as sensitive as the conventional antigen ELISA for the detection of FMDV in epithelial suspensions tested and had an equivalent 100% sensitivity on the cell culture supernatants of FMDV serotypes O, A, C and Asia-1. Further research is underway to develop chromatographic strip tests capable of ascertaining the serotype of the virus.

4.7. Differentiation between infected and vaccinated animals (DIVA)

Detection of animals that have been infected with FMDV is of considerable importance for the control of FMD especially in a previously FMD free country or in a country with sporadic outbreaks. Both previously infected and vaccinated animals can have neutralizing antibodies in their sera, but it is important for trade purposes to be able to distinguish previously infected animals from those that have just been vaccinated against the disease. This is because a high proportion (up to 50%) of animals infected with FMDV can become "carriers", these are defined as animals which continue to have infectious virus present within the oropharynx more than 28 days post-infection. The animals are clinically normal and can maintain this state for a long period (ca. 2–3 years in cattle). It is possible that such animals can act as a source of infection for other animals, indeed evidence for this process exists for African buffalo, however it has not been demonstrated in experimental studies with cattle alone (see [37] and [38] for more detailed descriptions). Viral replication during infection results in the production of both structural (SP) and nonstructural (NSP) proteins. Like the SPs, some NSPs are immunogenic [39]. Vaccines consisting of purified preparations of inactivated 146S virions induce antibodies almost exclusively against the SP of the virus (at least after a small number of vaccinations). Thus it can be possible to discriminate between infected and vaccinated animals based on the detection of antibodies to NSPs.

Differentiation of infection from vaccination based on the antibodies to NSP has been reported using either panels of proteins or the individual proteins 2C or 3ABC. Early assays for the detection of anti-NSP antibodies relied on radioimmunoprecipitation [40] or enzyme linked immunoelectrotransfer blot assays [41]. However, these assays are not suited for rapid examination of large numbers of sera and thus alternative techniques like ELISA were developed. Several ELISAs based on the detection of antibodies to various NSPs of FMDV have been established [42, 43]. However, these ELISAs used species-specific conjugates, making simultaneous examination of sera from different species difficult. Thus there was a need for an assay which enabled simultaneous testing of sera from different species. Sørensen et al. [44] developed a blocking ELISA which was species independent; using baculovirus expressed FMDV NSPs as antigen and polyclonal antibodies produced in guinea pigs as capture and detecting antibodies. The polyclonal antibodies were later replaced with monoclonal antibodies [45] for high throughput. As antibodies to NSP persist for long periods [46], positive animals are not necessarily still infected although they can be carriers (see above). However, tests for detection of NSP antibodies cannot be used for detection of carrier animals as some persistently infected animals do not show sero-conversion against NSPs [47], the carrier animal status may occur in previously vaccinated animals in which only limited virus replication occurs. Moreover, no serological tests are currently available that can differentiate between FMDV carriers and other animals that show a serological response to FMDV. Furthermore, it can be difficult to differentiate between vaccinated and previously-infected animals if the vaccine used has been prepared from cell culture supernatant (i.e. not purified) and/or contains varying degrees of contaminating viral NSPs especially if multiple vaccinations have occurred.

5. CHARACTERIZATION OF FMDV BELOW THE LEVEL OF SEROTYPE (STRAINS/SUBTYPES)

Within each serotype of FMDV, there is a spectrum of variants with their own antigenic, biological and epidemiological characteristics. In some cases there is poor cross-protection between variants within a serotype and thus characterization of sub-serotypes/strains becomes necessary to ensure selection of appropriate vaccines to control an outbreak. Initially strains were characterized based on their performance in cross-protection tests in animals. Subtype variants were distinguished by the fact that immunization against one subtype did not confer the same level of immunity to another variant of the same serotype as to the homologous strain. Animals vaccinated with one strain withstood homologous challenge but were only partially protected against challenge with a heterologous strain [48, 49]. Quantitative cross-protection tests based on determination of heterologous and homologous 50% protective dose (PD_{50}) values have been derived [50]. However, such tests are very costly for routine use, time-consuming and are subject to considerable variation due to differences in individual animal susceptibility. Thus other methods have been developed.

Antigenic characterization was used to compare field viruses with vaccine strains by determination of the serological relationship (r_1 value) using hyperimmune sera in ELISA [51] or in VNTs using cell culture [52]. Using VNTs, r_1 values of ≥ 0.3 have been shown to reflect a close antigenic relationship between the field isolates and vaccine strains, indicative of good protection by the vaccine, whereas values < 0.3 reflect a more distant antigenic relationship, indicating that the vaccine is unlikely to protect against the field isolates. Antigenic characterization using serological tests like VNT and ELISA, using defined sera/MAbs, are useful in showing antigenic diversity but they are unable to characterize strains individually and cannot be used to trace the origin of an outbreak.

Nucleotide sequence analysis has now become the definitive technique for characterization of FMDV strains. This technique was first used for the study of the epidemiology of FMD by Beck and Strohmaier [53], who investigated the origin of FMDV outbreaks in Europe over a 20 year period. The first phylogenetic analysis of FMDV using nucleotide coding sequences for VP1 was reported by Dopazo et al. [54]. Since then a number of studies have been published on nucleotide sequence analysis for all the seven serotypes of FMDV (e.g., [8, 9, 12, 55, 56]), some of these have used complete genome sequences for tracing of outbreaks [8, 12].

The serotypes of FMDV have on average 86% nucleotide sequence identity to each other across the whole genome [57] but the VP1 coding region is substantially more variable and shows only about 50-70% identity [58]. Serotypes O, A, C and Asia-1 have been further classified into genotypes based on up to 15% difference in VP1 coding sequences [9, 58]. Thus, FMD serotype O viruses from around the world have been classified into eight genetically and geographically distinct genotypes, called topotypes, on the basis of the VP1 coding sequence [11, 58]. The eight topotypes are Middle East-South Asia (ME-SA), South–East Asia (SEA), Cathay, Indonesia-1, Indonesia-2, East Africa, West Africa and Europe-South America (Euro-SA). Among these, the first five are circulating in Asia, of which the dominant one is ME-SA [58]. Four different lineages within the ME-SA topotype have been defined on the basis of phylogenetic relationships and a nucleotide sequence difference (within the VP1 coding sequence) of > 7.5% [58, 59]. Within the ME-SA topotype, a pandemic strain of FMDV, designated as the PanAsia lineage, spread vigorously and is discussed in detail below.

FMDV serotype A is genetically and antigenically the most diverse of the Eurasian serotypes [58]. Up to 32 sub-types had been defined by the late 1970s [60, 61] though this system of sub-typing was later discontinued. More recently, serotype A FMDVs, from the whole world, have been classified into 26 genotypes based on > 15% difference in VP1 coding sequence [62]. The serotype A viruses form three geographically distinct genotypes (topotypes) i.e. Asia, Europe-South America (Euro-SA) and Africa. The level of nucleotide differences, within the VP1 coding region, between serotype A viruses belonging to different continental topotypes is up to ~24% [62]. Topotype Asia is the most prevalent in the Middle East and South Asian region and exists in various lineages e.g. A_{15}, A_{22}, A-IRN87, A-IRN96, A-IRN99, A-Iran05, etc. The A-Iran05 lineage is currently dominant in the West Eurasian region and has evolved into different sublineages [9, 63]. Viruses belonging to the A-Iran05$^{BAR\text{-}08}$ sublineage are antigenically distinct from the vaccine strain, A22/Iraq [9].

FMDV serotype Asia-1 is considered to be genetically and antigenically the least diverse serotype [58]. Indeed, Ansell et al. [64] reported that 44 serotype Asia-1 FMDVs isolated between 1954 and 1990 throughout Asia were less variable compared to other FMDV serotypes in their VP1 coding sequences. It is noteworthy, however, that the RGDLXXL receptor binding motif of serotype Asia-1 viruses was more variable when compared to serotypes O and A of FMDV [9, 56, 65]. Previous reports have classified FMDV serotype Asia-1 in different ways. For example, Ansell et al. [64] grouped serotype Asia-1 viruses isolated throughout Asia between 1952 to 1992 into 18 groups, while Mohapatra et al. [66] classified serotype Asia-1 viruses from India, sampled over the last two decades, into seven lineages and Valarcher et al. [67] divided these FMDVs from 2003–2007 into six Groups. An additional Group of Asia-1 viruses, designated as Group-VII [56] and later named as Sindh-08 by WRL-FMD, has also been recently reported, which is currently circulating in the West Eurasian region. Viruses belonging to the Group-VII are not efficiently neutralized by antisera raised against the Asia-1/Shamir and Asia-1/Ind/8/79 vaccine strains [56]. Outbreaks due to this novel Group have also been reported in animals given vaccine based on the Asia-1/Shamir strain. A homologous vaccine prepared from an isolate (Asia-1/TUR/ 11) responsible for a field outbreak in Turkey in 2011, has been found to be effective to contain the spread of the Group-VII (Sindh-08) viruses [68].

Unfortunately, there are no uniform criteria or nomenclature for these classifications. As indicated, serotypes O and A FMDVs have been classified into lineages but serotype Asia-1 FMDVs have been classified into Groups (see [56, 67]).

6. GEOGRAPHICAL DISTRIBUTION OF FMD

The serotypes of FMDV are not distributed uniformly around the world. The serotype O, A and C viruses have had the widest distribution and have been responsible for outbreaks in Europe, America, Asia and Africa. However, the last reported outbreak due to serotype C FMDV was in Ethiopia during 2005 [69] and so serotype C viruses may no longer exist outside of laboratories. The SAT1-3 viruses are normally restricted to sub-Saharan Africa. However, there have been some limited outbreaks due to SAT1 viruses in the Middle East between 1962–1965 and 1969–1970 and then in Greece in 1962 [58]. Similarly, there have been reports of minor incursions of the serotype SAT2 in Yemen in 1990 and in Kuwait and Saudi Arabia in 2000 [70]. More recently, FMD outbreaks due to serotype SAT2 spread from sub-Saharan Africa through northern African countries (Egypt and Libya) and into Palestine [71]. This serotype was also detected in Bahrain. The serotype Asia-1 FMDVs are generally confined to Asia, except for two incursions into Greece, one in 1984 and a second in 2000.

Although considerable information is available on the virus, the disease and vaccines, FMD still affects extensive areas of the world. FMD free countries have introduced a number of measures to retain their status because of the heavy economic losses resulting from this disease. The USA has experienced FMD nine times since 1870. Each time the disease was eradicated with strict slaughter and quarantine control procedures. The last FMD outbreak in the USA occurred near Montebello, California, in 1929. Infected hogs contracted the disease after being fed swill with meat scraps from a tourist steamship coming from Argentina. Since then, the USA has had restrictions on importation of susceptible animals and livestock products from countries where FMD is present. Canada has been free of FMD since 1952 and Mexico since 1953.

Historically, FMD in Australia was documented in the early 1800s and early 1870s [72]. In 1871–1872, at least five episodes of FMD were documented in cattle that originated from the UK, either bound for Australia, in quarantine or in a bull which had landed two months earlier [72]. The last outbreak of FMD in Australia was reported in 1872.

The Pandemic serotype O virus (designated as the PanAsia strain) belongs to the ME-SA topotype which has spread rapidly and vigorously [73]. This lineage replaced the other lineages of FMDV previously circulating in the Middle East [74]. This lineage has been responsible for disease outbreaks everywhere in the world where FMD is endemic or sporadic except for South America and been responsible for incursions into previously disease-free countries. The PanAsia lineage was first detected in India in 1982 [59] but was confined to India until 1990. Its predominance in field outbreaks in India was, however, noticed from 1996 onwards [59]. It spread northwards to Nepal in 1990 and again in 1997–1999 and to Bhutan in 1998 and also towards the west, into Bahrain, Kuwait, Saudi Arabia, Syria, Yemen, Iran and Lebanon in 1998 and to UAE, Israel and Turkey in 1999 [73]. The lineage spread further to China in 1999 and into South East Asian countries causing outbreaks in Thailand in 1999, Malaysia and Laos (PDR) in 2000 and Vietnam in 2002. The virus also caused disease outbreaks in South Korea [75] and Japan in 2000 [76]. These two countries were previously free of FMD since 1934 and 1908, respectively. South Korea faced outbreaks again in 2002 and then in 2010. The 2002 outbreaks were caused by serotype O virus, belonging to the PanAsia lineage, whereas, both serotype O (topotype SEA, lineage MYA-98) and A (topotype ASIA, genotype SEA, lineage MYA-97) viruses were responsible for the 2010–2011 outbreaks [77]. South Korea appears to have had 3 independent introductions of the virus in 2010. Firstly there was an incursion of FMDV serotype A (A/SEA/MYA/97) in January, 2010. The disease was controlled using a stamping out policy [78]. There had been no reported outbreaks caused by serotype A in eastern Asia since 1973. This incursion was followed by second introduction of FMDV in April, 2010, in this case serotype O (O/SEA/MYA/98). South Korea was declared free without vaccination by the OIE on 27th September, 2010 after implementing a stamping out policy [79]. The third incursion, due to O/SEA/MYA/98, took place in November, 2010 and then spread throughout the country. Initially, a national stamping out policy was implemented for all animals on farms with FMDV infected animals. However, a nationwide vaccination-to-live policy was adopted later using emergency vaccination. Vaccination was effective in controlling the disease. A total of 3.48 million FMD susceptible animals were culled [80]. Similarly, Japan was hit by FMD ten years after the previous outbreak [81], FMDV type O, belonging to the MYA-98 lineage within the SEA topotype, was detected on 20thApril, 2010 at a beef feeding farm in southern Japan. The disease spread to the surrounding areas. Emergency vaccination was started on 22nd May, 2010 within the infected zones. All the vaccinated animals were subsequently destroyed. During the three-month FMD epidemic, a total of 292 infected farms were detected and 290,000 animals were destroyed as a control measure [82]. The VP1 sequence data indicate that mainland Southeast Asia is the source of FMDV serotypes O and A in Eastern Asia [80, 83].

In 2000, the O-PanAsia virus was detected in Uzbekistan, Mongolia, Armenia, Georgia and Russia and then in 2001 in Kyrgyzstan and during 2001–2003 in Tajikistan [84]. In 2000, the virus spread to KwaZulu-natal Province in South Africa. This was the first recorded outbreak in that country due to serotype O and the first since 1957 in this region of Southern Africa [85, 86]. The outbreak was attributed to the feeding of swill to pigs from a ship which had originated from Asia. More recently, outbreaks due to SAT1 FMDV were reported in 2011 and further FMD outbreaks due to SAT2 have occurred in South Africa in 2012 while outbreaks due to SAT1 have been reported again in 2013 [87].

In 2001, the PanAsia virus was introduced into Europe where it caused disease outbreaks in the UK and was then spread to Ireland, France and the Netherlands. The outbreaks of FMD in United Kingdom in 2001 were the first since a single case in 1981 [88]. Before the 1981 case, the UK was hit by a major outbreak of FMD in 1967 and 442 000 animals were slaughtered to achieve eradication of the disease [89]. The 2001 outbreaks were controlled using a stamping out policy in which 6.5 million infected and in-contact animals were killed. The total economic losses due to the 2001 outbreak were estimated at between USD 12.3-13.8 billion [90]. Another outbreak hit the UK in 2007 and it was later identified that the disease was due to escape of virus from either the vaccine production facility or the Institute for Animal Health, both sited in Pirbright [8].

Currently, the O-PanAsia-II^{ANT-10} strain is responsible for extensive outbreaks in the whole Middle East and South Asian region. This strain was responsible for FMD in Pakistan in 2009, which spread westwards and caused disease in Turkey, Israel, Libya and Bulgaria [12, 91]. Bulgaria experienced an outbreak of FMD at the end of 2010, the first case of the disease was detected in wild boar. FMDV serotype O virus strain O-PanAsia-II[ANT-10] was found to be responsible for the outbreaks at 11 separate sites which occurred in two separate phases [12, 92]. The country had remained free of FMD for 14 years since its previous outbreak in 1996. The disease also affected cattle, buffalo, sheep and goats. A stamping out policy was used and a total of 1372 animals were destroyed. Seropositive wildlife were detected near the outbreaks but the disease has not been maintained by them since no new cases of FMD were detected after April, 2011 and the country was declared free of FMD in July, 2011 [87].

7. FMD VIRUS POOLS

Despite the propensity and opportunities for spread of FMDVs into new regions, comparison of the VP1 coding nucleotide sequences reveals a tendency for similar viruses to recur in the same geographical area.

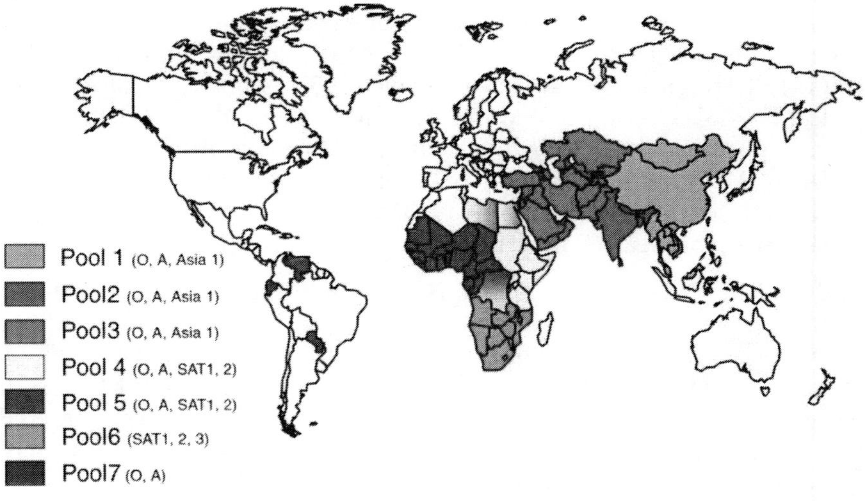

Figure 2 Geographical distribution of seven pools of foot-and mouth disease viruses. Serotype O FMDV is the most widely distributed serotype of the virus (in 6 of the 7 indicated virus pools) whereas, in contrast, SAT3 is only present in pool 6 (within southern Africa). The Asia-1, SAT1 and SAT2 serotypes also have quite limited geographical distribution. However, individual countries can have multiple serotypes in circulation at the same time and hence it is necessary to be able to determine which serotype is responsible for an outbreak if vaccination is to be used. Countries which are normally free of the disease (marked in yellow) can still suffer incursions of the virus which can have high economic costs.

This tendency apparently reflects some degree of ecological isolation, likely reflecting patterns of animal movement and trade or specific wildlife reservoirs (e.g. African buffalo) within a region. Based on genetic and antigenic analyses, FMDVs throughout the world have been sub-divided into seven regional pools [93, 94]. Certain countries share viruses belonging to two different pools, for example, Egypt and Libya (Figure 2). Virus circulation and evolution within these regional virus pools result in changing needs for appropriate vaccine selection.

8. PROGRESSIVE CONTROL PATHWAY FOR FMD

In response to the repeated epidemic events and requests for assistance by affected and at risk countries in West Eurasia, FAO/EuFMD convened a meeting of 14 directly affected countries in Shiraz, Iran in 2008 at which a long term regional approach for progressive control of FMD in the region was developed, known as the West Eurasia Regional Roadmap [95]. It was the first time that a Progressive Control Pathway for FMD (PCP-FMD) was established to determine national progress and to develop national and regional actions plans and support. Several FAO projects supported both national and regional PCP activities, such as improved FMD laboratory networking (WELNET) and epidemiological support. The activities implemented since the 2008 meeting have allowed the detection of three epidemics of regional (West Eurasian) significance in the past 4 years. These epidemics were caused by the serotype A-Iran-05$^{BAR\text{-}08}$ sublineage in 2008 [9], the type O PanAsia-II$^{ANT\text{-}10}$ strains in 2009–12 [12, 91, 92], and the serotype Asia-1 (Group-VII) viruses [56] of 2008–12, all of which involved east to west migration of the virus and to some extent involved Central Asian countries. The rapidity and frequency of incursions in the past 4 years presents a major problem for disease control, particularly when the strains involved are poorly matched to the routinely used vaccines [9, 56].

The PCP-FMD has now been adopted as a joint tool between FAO/EuFMD/OIE [94]. This pathway includes a set of criteria to assist FMD endemic countries to progressively reduce the level of FMDV circulation and to mitigate the impact of FMD [96]. The PCP-FMD consists of a set of FMD control activity stages (Figure 3). Each stage has well-defined outcomes which can be achieved through a variety of activities. The specific activities required to achieve the outcomes have, however, not been prescribed. This non-prescriptive approach has the advantage that each country can adopt an approach according to its national/regional requirements and capabilities to achieve the outcomes. Understanding the "local" epidemiology of FMD and the active monitoring of the virus circulation are the foundations of the PCP-FMD and activities to meet these requirements are required in all stages [96]. Determination of the factors responsible for maintenance and spread of the disease and knowledge about circulating subtypes of FMDV are essential for effective control of the disease.

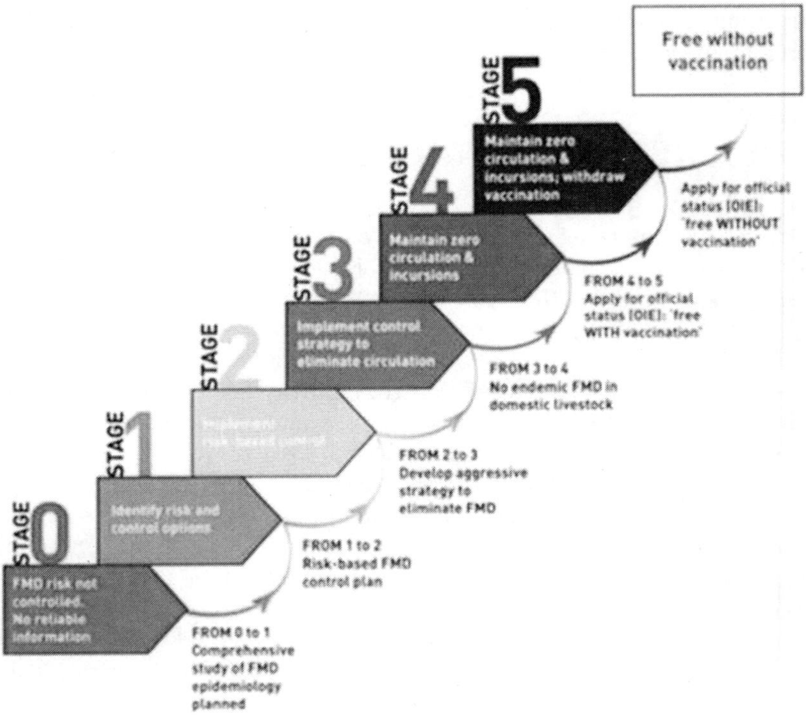

Figure 3 The fao/eufmd/oie progressive control pathway for fmd**.** The status of countries on the PCP-FMD is evaluated according to defined criteria. Countries with endemic disease are in stages 0 to 3 while countries with no endemic disease within livestock are at stage 4 or above. The image was kindly supplied by EuFMD.

ELISAs for detection of antibodies to FMDV NSPs are generally used to monitor virus circulation as this test can discriminate between vaccinated and infected animals [44, 45] if purified vaccines are used. The interpretation of NSP test results is, however, complicated in FMD endemic countries, where animals can be exposed to multiple serotypes and where non-purified vaccines are frequently used. Moreover, anti-NSP antibodies can persist for a long period [46] and do not necessarily indicate recent

FMDV infection. Inadequate laboratory diagnostic capacity and its implications for the PCP-FMD in Eastern Africa have been described recently by Namatovu et al. [97]. In order to monitor FMDV circulation effectively, each participating country should, therefore, have sufficient FMD diagnostic and surveillance capacity. Progression along the PCP is based on an annual assessment of evidence-based activities related to FMD epidemiology and control measures undertaken by the country. The organizational infrastructure of each country is required to be progressively strengthened to ensure that the activities required to monitor and control FMD are implemented [98].

Countries where the disease is endemic with no reliable information on the disease status, are classified as in stage 0. In order to move from stage 0 to 1, a comprehensive study on the epidemiology of FMD is required to be planned [96]. No country in the West Eurasian region is now in stage 0 [94], the majority of the countries are presently in stage 1 and are expected to enter stage 2 by the end of 2013, whereas the Turkish Thrace region is currently in stage 4. Stage 1 assists in identifying appropriate control options. Countries in stage 1 are in the process of developing their control strategies in at least one animal production sector based on a comprehensive assessment of the epidemiology and control options. Progression from stage 1 to 2 requires a risk based FMD control plan. Thus stage 2 involves the implementation of the chosen policy. Countries in stage 2 have implemented a risk-based FMD control strategy that aims to reduce disease in at least one animal production sector. In order to move from stage 2 to 3, an aggressive strategy to eliminate FMD needs to be developed. Countries in stage 3 have adopted a control plan to progressively reduce/eliminate virus circulation in at least one region/production system [96]. This requires very significant national capacity and ongoing investment including the ability to ensure maintenance of sufficient herd immunity in critical populations to prevent FMD virus circulation.

Moving to stage 4 requires that FMD is controlled to an extent that it is not endemic in domestic livestock. If a country decides to continue along the FMD-PCP to stage 4 and beyond, it may ask the OIE for endorsement of its national FMD eradication program. Progression to stage 4 would thus indicate attaining officially recognized FMD free status with vaccination by the OIE for the whole or part of the country. Countries in stage 4 have maintained zero circulation with no incursions of FMD [96].

Vaccination plays a vital role in controlling FMD for countries in stages 2–4. Normally vaccine quality control is determined by the producer and vaccine batches are only released when they pass the quality control parameters. However, lack of maintenance of the cold chain before, during or after transport/importation may reduce the vaccine efficacy, even if the vaccine initially contained sufficient payload of serotype(s), matching with the circulating strain(s). Furthermore, as the dose–response relationship in FMD vaccination is influenced by the serotype and type of adjuvant present in the vaccine [99], both the antigen pay-load and the quality of the adjuvant in formulated vaccine need to be established. An independent quality control of the formulated vaccine is therefore necessary for effective control of the disease [100]. Vaccination alone may well not be able to contain the disease unless it is coupled with restrictions on animal movement. Control of animal movements is, however, complicated by many factors including social customs [52], religious festivals [17], trade of animals in live animal markets [10] and both formal and informal animal movement. Progression from Stage 4 to 5, and from Stage 5 to Pathway completion, would be through the existing official OIE recognition processes of freedom from FMD with or without vaccination, respectively.

In order to target resources available for surveillance and control of the disease, each stage of the PCP requires risk assessment and risk management activities. Socio-economic approaches including impact assessment, value chain analyses and cost-benefit analyses of intervention play an important role in animal disease control. However, integration of these approaches within animal disease control strategies is still in its infancy and remains a challenge.

The PCP-FMD approach has now been adopted for different regions including the West Eurasia roadmap 2020 (since 2008), the 2020 Roadmap for FMD Control in South-East Asia and China (SEACFMD), Southern Africa Roadmap (since 2011), Eastern Africa Roadmap (since 2012) and the Plan Hemisférico de Erradicación de la Fiebre Aftosa (PHEFA) for South America [94]. Thus the PCP is expected to form the backbone of the global FAO/OIE strategy for the control of FMD.

The West Eurasia roadmap 2020 projects that this region will reach stage 3 by 2020. The annual assessment of individual countries within West Eurasia, however, shows that the majority of countries did not reach the expected earlier stages at the projected times. It therefore seems unlikely that all the countries within West Eurasia will reach stage 3 by 2020.

9. CONCLUSIONS/RECOMMENDATIONS

Despite considerable information being available about the virus, the disease and vaccines, FMD remains a major threat to the livestock industry world-wide. New sublineages of FMDV continue to evolve to produce novel strains which sometimes break through vaccine-induced immunity and can result in major epidemics. This warrants the need for continued surveillance, vaccine matching and vaccine quality control. Vaccination alone is unlikely to control the disease unless it is coupled with animal movement control. Animal identification systems and animal movement controls are therefore also needed to be in place for effective control of the disease.

The majority of countries in Asia and Africa, where the disease is endemic, are deficient in knowledge about the circulating sub-types of FMDV due to deficiencies in submitting outbreak samples to reference laboratories. The samples originating from these countries are often not collected in systematic way and are therefore not necessarily representative of the particular production systems/areas for which the PCP is intended. Each country must build capacity in diagnostics, epidemiology and economics in order to improve the information on the nature of FMDV circulation in that specific environment. Many of the countries participating in the PCP do not have the capacity to design and implement appropriate studies for determination of risk factors and for carrying out risk assessment and socio-economic studies. Furthermore, the non-descriptive approach in the PCP, in which the outcomes are defined but the means to achieve the outcomes are not specified, can itself be a hurdle in identification of risk factors for a particular environment/region. Therefore, these countries should be supported by helping them to design appropriate studies for determination of the relevant risk factors and for conducting risk assessment, risk management and socio-economic studies. The laboratory data should be combined with field epidemiological information for meaningful interpretation. Regional cooperation including timely information on the FMD outbreaks particularly in border areas, timely data/information sharing and simultaneous adoption of control measures are required for effective control of this transboundary disease. Early detection and early responses should be prepared in order to contain the disease effectively. Sufficient quality and quantity of assured FMD vaccines with matching serotypes needs to be made available for inducing sufficient herd immunity for further progression from stage 3 to stage 4 and preparations to achieve this can take some time.

ACKNOWLEDGEMENTS

The authors are thankful to Giancarlo Ferrari for commenting on the manuscript.

REFERENCES

1. Belsham GJ: Translation and replication of FMDV RNA. Curr Top Microbiol Immunol. 2005, 288: 43-70.
2. Lea S, Hernández J, Blakemore W, Brocchi E, Curry S, Domingo E, Fry E, Abu-Ghazaleh R, King A, Newman J, Stuart D, Mateu MG: The structure and antigenicity of a type C foot-and-mouth disease virus. Structure. 1994, 2: 123-139. 10.1016/S0969-2126(00)00014-9.
3. Thomas AA, Woortmeijer RJ, Puijk W, Barteling SJ: Antigenic sites on foot-and-mouth disease virus type A10. J Virol. 1988, 62: 2782-2789.
4. Fox G, Parry NR, Barnett PV, McGinn B, Rowlands DJ, Brown F: The cell attachment site on foot-and-mouth disease virus includes the amino acid sequence RGD (arginine-glycine-aspartic acid). J Gen Virol. 1989, 70: 625-637. 10.1099/0022-1317-70-3-625.
5. Jackson T, Sharma A, Ghazaleh RA, Blakemore WE, Ellard FM, Simmons DL, Newman JWI, Stuart DI, King AMQ: Arginine-glycine-aspartic acid-specific binding by foot-and-mouth disease viruses to the purified integrin αvβ3 in vitro. J Virol. 1997, 71: 8357-8361.
6. Monaghan P, Gold S, Simpson J, Zhang Z, Weinreb PH, Violette SM, Alexandersen S, Jackson T: The alpha(v)beta6 integrin receptor for Foot-and-mouth disease virus is expressed constitutively on the epithelial cells targeted in cattle. J Gen Virol. 2005, 86: 2769-2780. 10.1099/vir.0.81172-0.
7. Mason PW, Rieder E, Baxt B: RGD sequence of foot-and-mouth disease virus is essential for infecting cells via the natural receptor but can be bypassed by an antibody-dependent enhancement pathway. Proc Natl Acad Sci U S A. 1994, 91: 1932-1936. 10.1073/pnas.91.5.1932.
8. Cottam EM, Wadsworth J, Shaw AE, Rowlands RJ, Goatley L, Maan S, Maan NS, Mertens PP, Ebert K, Li Y, Ryan ED, Juleff N, Ferris NP, Wilesmith JW, Haydon DT, King DP, Paton DJ, Knowles NJ: Transmission pathways of foot-and-mouth disease virus in the United Kingdom in 2007. PLoS Pathog. 2008, 4: e1000050-10.1371/journal.ppat.1000050.
9. Jamal SM, Ferrari G, Ahmed S, Normann P, Curry S, Belsham GJ: Evolutionary analysis of serotype A foot-and-mouth disease viruses circulating in Pakistan and Afghanistan during 2002–2009. J Gen Virol. 2011, 92: 2849-2864. 10.1099/vir.0.035626-0.
10. Jamal SM, Ferrari G, Hussain M, Nawroz AH, Aslami AA, Khan E, Murvatulloev S, Ahmed S, Belsham GJ: Detection and genetic characterization of foot-and-mouth disease viruses in samples from clinically healthy animals in endemic settings. Transbound Emerg Dis. 2012, 59: 429-440. 10.1111/j.1865-1682.2011.01295.x.
11. Samuel AR, Knowles NJ: Foot-and-mouth disease type O viruses exhibit genetically and geographically distinct evolutionary lineages (topotypes). J Gen Virol. 2001, 82: 609-621.
12. Valdazo-González B, Polihronova L, Alexandrov T, Normann P, Knowles N, Hammond J, Georgiev G, Özyörük F, Sumption K, Belsham GJ, King DP: Reconstruction of the transmission history of RNA virus outbreaks using full genome sequences: foot-and-mouth disease virus in Bulgaria in 2011. PLoS One. 2012, 7: e49650-10.1371/journal.pone.0049650.
13. Dorsch-Hasler K, Yogo Y, Wimmer E: Replication of picornaviruses. I. Evidence from in vitro RNA synthesis that poly (A) of the poliovirus genome is genetically coded. J Virol. 1975, 16: 1512-1517.
14. Vallée H, Carré H: Sur la pluralité du virus aphteux. C R Hebd Acad Sci Paris. 1922, 174: 1498-1500. (in French)
15. Waldmann O, Trautwein K: Experimentalle untersuchungen ueber die pluralitet des maul-und klauenseuche virus. Berlin Tierarztl Wochenschr. 1926, 42: 569-571. (in German)
16. Brooksby JB: The virus of foot-and-mouth disease. Adv Virus Res. 1958, 5: 1-37.
17. Brooksby JB, Rogers J: Methods used in typing the virus of Foot-and-mouth disease at Pirbright, 1950–55. Methods of typing and cultivation of foot-and-mouth disease virus: Project No. 208. 1957, Paris: European Productivity Agency of the Organization of European Cooperation (OEEC)

18. OIE: Foot-and-mouth disease. Manual of diagnostic tests and vaccines for terrestrial animals (mammals, birds and bees). Volume 1. 2012, Paris, France: World Organization for Animal Health (OIE), 7
19. Roeder PL, Le Blanc Smith PM: Detection and typing of foot-and-mouth disease virus by enzyme linked immunosorbent assay: a sensitive, rapid and reliable technique for primary diagnosis. Res Vet Sci. 1987, 43: 225-232.
20. Morioka K, Fukai K, Yoshida K, Yamazoe R, Onozato H, Ohashi S, Tsuda T, Sakamoto K: Foot-and-mouth disease virus antigen detection enzyme-linked immunosorbent assay using multiserotype-reactive monoclonal antibodies. J Clin Microbiol. 2009, 47: 3663-3668. 10.1128/JCM.00695-09.
21. Chen H, Peng Y, Zhang Y, Liu X: Detection of foot-and-mouth disease serotype O by ELISA using a monoclonal antibody. Hybridoma (Larchmt). 2012, 31: 462-464. 10. 1089/hyb.2012.0081.
22. Ferris NP, Grazioli S, Hutchings GH, Brocchi E: Validation of a recombinant integrin αvβ6/monoclonal antibody based antigen ELISA for the diagnosis of foot-and-mouth disease. J Virol Methods. 2011, 175: 253-260. 10.1016/j.jviromet.2011.05.026.
23. Meyer RF, Brown CC, House C, House JA, Molitor TW: Rapid and sensitive detection of foot-and-mouth disease virus in tissues by enzymatic RNA amplification of the polymerase gene. J Virol Methods. 1991, 34: 161-172. 10.1016/0166-0934(91)90096-I.
24. Rodriguez A, Martínez-Salas E, Dopazo J, Dávila M, Sáiz JC, Sobrino F: Primer design for specific diagnosis by PCR of highly variable RNA viruses: typing of foot-and-mouth disease virus. Virology. 1992, 189: 363-367. 10.1016/0042-6822(92)90717-4.
25. Vangrysperre W, De Clercq K: Rapid and sensitive polymerase chain reaction based detection and typing of foot-and-mouth disease virus in clinical samples and cell culture isolates, combined with a simultaneous differentiation with other genomically and/or symptomatically related viruses. Arch Virol. 1996, 141: 331-344. 10.1007/BF01718403.
26. Callens M, De Clercq K: Differentiation of the seven serotypes of foot-and-mouth disease virus by reverse transcriptase polymerase chain reaction. J Virol Methods. 1997, 67: 35-44. 10.1016/S0166-0934(97)00074-8.
27. Bao HF, Li D, Guo JH, Lu ZJ, Chen YL, Liu ZX, Liu XT, Xie QG: A highly sensitive and specific multiplex RT-PCR to detect foot-and-mouth disease virus in tissue and food samples. Arch Virol. 2008, 153: 205-209. 10.1007/s00705-007-1082-y.
28. Giridharan P, Hemadri D, Tosh C, Sanyal A, Bandyopadhyay SK: Development and evaluation of a multiplex PCR for differentiation of foot-and-mouth disease virus strains native to India. J Virol Methods. 2005, 126: 1-11. 10.1016/j.jviromet.2005.01.015.
29. Reid SM, Grierson SS, Ferris NP, Hutchings GH, Alexandersen S: Evaluation of automated RT-PCR to accelerate the laboratory diagnosis of foot-and-mouth disease virus. J Virol Methods. 2003, 107: 129-139. 10.1016/S0166-0934(02)00210-0.
30. Reid SM, Ferris NP, Hutchings GH, Zhang Z, Belsham GJ, Alexandersen S: Detection of all seven serotypes of foot-and-mouth disease virus by real-time, fluorogenic reverse transcription polymerase chain reaction assay. J Virol Methods. 2002, 105: 67-80. 10.1016/S0166-0934(02)00081-2.
31. Callahan JD, Brown F, Osorio FA, Sur JH, Kramer E, Long GW, Lubroth J, Ellis SJ, Shoulars KS, Gaffney KL, Rock DL, Nelson WM: Use of a portable real-time reverse transcriptase-polymerase chain reaction assay for rapid detection of foot-and-mouth disease virus. Am Vet Med Assoc. 2002, 220: 1636-1642. 10.2460/javma.2002.220.1636.
32. King DP, Ferris NP, Shaw AE, Reid SM, Hutchings GH, Giuffre AC, Robida JM, Callahan JD, Nelson WM, Beckham TR: Detection of foot-and-mouth disease virus: comparative diagnostic sensitivity of two independent real-time reverse transcription-polymerase chain reaction assays. J Vet Diagn Invest. 2006, 18: 93-97. 10.1177/10406 387 0601800114.
33. Tam S, Clavijo A, Engelhard EK, Thurmond MC: Fluorescence-based multiplex real-time RT-PCR arrays for the detection and serotype determination of foot-and-mouth disease virus. J Virol Methods. 2009, 161: 183-191. 10.1016/j.jviromet.2009.04.033.

34. Notomi T, Okayama H, Masubuchi H, Yonekawa T, Watanabe K, Amino N, Hase T: Loop-mediated isothermal amplification of DNA. Nucleic Acids Res. 2000, 28: E63-10.1093/nar/28.12.e63.
35. Dukes JP, King DP, Alexandersen S: Novel reverse transcription loop-mediated isothermal amplification for rapid detection of foot-and-mouth disease virus. Arch Virol. 2006, 151: 1093-1106. 10.1007/s00705-005-0708-5.
36. Reid SM, Ferris NP, Brüning A, Hutchings GH, Kowalska Z, Akerblom L: Development of a rapid chromatographic strip test for the pen-side detection of foot-and-mouth disease virus antigen. J Virol Methods. 2001, 96: 189-202. 10.1016/S0166-0934(01)00334-2.
37. Alexandersen S, Zhang Z, Donaldson AI, Garland AJ: The pathogenesis and diagnosis of foot-and-mouth disease. J Comp Pathol. 2003, 129: 1-36. 10.1016/S0021-9975(03) 00041-0.
38. Tenzin , Dekker A, Vernooij H, Bouma A, Stegeman A: Rate of foot-and-mouth disease virus transmission by carriers quantified from experimental data. Risk Anal. 2008, 28: 303-309. 10.1111/j.1539-6924.2008.01020.x.
39. Tesar M, Berger HG, Marquardt O: Serological probes for some foot-and-mouth disease nonstructural proteins. Virus Genes. 1989, 3: 29-44. 10.1007/BF00301985.
40. Berger HG, Sraub OC, Ahl R, Tesar M, Marquardt O: Identification of foot-and-mouth disease virus replication in vaccinated cattle by antibodies to non-structural virus proteins. Vaccine. 1990, 8: 213-216. 10.1016/0264-410X(90)90048-Q.
41. Bergmann IE, de Mello PA, Neitzert E, Beck E, Gomez I: Diagnosis of persistent aphthovirus infection and its differentiation from vaccination response in cattle by use of enzyme linked immunoelectrotransfer blot analysis with bioengineered nonstructural viral antigens. Am J Vet Res. 1989, 54: 825-831.
42. Bergmann IE, Malirat V, Neitzert E, Beck E, Panizzutti N, Sanchez C, Falczuk A: Improvement of a serodiagnostic strategy for foot-and-mouth disease virus surveillance in cattle under systematic vaccination: a combined system of an indirect ELISA-3ABC with an enzyme-linked immunoelectrotransfer blot assay. Arch Virol. 2000, 145: 473-489. 10.1007/s007050050040.
43. Shen F, Chen PD, Walfield AM, Ye J, House J, Brown F, Wang CY: Differentiation of convalescent animals from those vaccinated against foot-and-mouth disease by a peptide ELISA. Vaccine. 1999, 17: 3039-3049. 10.1016/S0264-410X(99)00148-6.
44. Sørensen KJ, Madsen KG, Madsen ES, Salt JS, Nqindi J, Mackay DKJ: Differentiation of infection from vaccination in foot-and-mouth disease by the detection of antibodies to the non-structural proteins 3D, 3AB and 3ABC in ELISA using antigens expressed in baculovirus. Arch Virol. 1998, 143: 1461-1476. 10.1007/s007050050390.
45. Sørensen KJ, de Stricker K, Dyrting KC, Grazioli S, Haas B: Differentiation of foot-and-mouth disease virus infected animals from vaccinated animals using a blocking ELISA based on baculovirus expressed FMDV 3ABC antigen and a 3ABC monoclonal antibody. Arch Virol. 2005, 150: 805-814. 10.1007/s00705-004-0455-z.
46. Paton DJ, Ferris NP, Hutchings GH, Li Y, Swabey K, Keel P, Hamblin P, King DP, Reid SM, Ebert K, Parida S, Savva S, Georgiou K, Kakoyiannis C: Investigations into the cause of foot-and-mouth disease virus seropositive small ruminants in Cyprus during 2007. Transbound Emerg Dis. 2009, 56: 321-328. 10.1111/j.1865-1682.2009.01088.x.
47. Brocchi E, Bergmann I, Dekker A, Paton D, Sammin D, Greiner M, Grazioli S, De Simone F, Yadin H, Haas B, Bulut N, Malirat V, Neitzert E, Goris N, Parida S, Sørensen K, De Clercq K: Comparative evaluation of six ELISAs for the detection of antibodies to the non-structural proteins of foot-and-mouth disease virus. Vaccine. 2006, 24: 6966-6979. 10.1016/j.vaccine.2006.04.050.
48. Hedger RS, Herniman KA: The Middle East foot-and-mouth disease type A epizootic 1964–65. I. Epizootiology and antigenic studies. Bull Off Int Epizoot. 1966, 65: 303-
49. Hyslop NSG, Davie J, Carter SP: Antigenic differences between strains of foot-and-mouth disease virus of type SAT 1. J Hyg. 1963, 61: 217-230. 10.1017/S0022172400020933.
50. Muntiu N, Dohotaru V, Bercan A, Tomescu A: Cross-protection product in an attempt of quantitative antigenic differentiation of two foot-and-mouth disease virus strains type O in cattle. Bull Off Int Epiz. 1965, 63: 1627-1641.

51. Kitching RP, Rendle R, Ferris NP: Rapid correlation between field isolates and vaccine strains of foot-and-mouth disease virus. Vaccine. 1988, 6: 403-408. 10.1016/0264-410X(88)90139-9.
52. Rweyemamu M: Foot-and-mouth disease control strategies in Africa. Prev Vet Med. 1984, 2: 329-340. 10.1016/0167-5877(84)90076-X.
53. Beck E, Strohmaier K: Sub-typing of European foot-and-mouth disease virus strains by nucleotide sequence determination. J Virol. 1987, 61: 1621-1629.
54. Dopazo J, Sobrino F, Palma EL, Domingo E, Moya A: Gene encoding capsid protein VP1 of foot-and-mouth disease virus: a quasispecies model of molecular evolution. Proc Natl Acad Sci U S A. 1988, 85: 6811-6815. 10.1073/pnas.85.18.6811.
55. Ayelet G, Mahapatra M, Gelaye E, Egziabher BG, Rufeal T, Sahle M, Ferris NP, Wadsworth J, Hutchings GH, Knowles NJ: Genetic characterization of foot-and-mouth disease viruses, Ethiopia, 1981–2007. Emerg Infect Dis. 2009, 15: 1409-1417. 10.3201/eid1509.090091.
56. Jamal SM, Ferrari G, Ahmed S, Normann P, Belsham GJ: Molecular characterization of serotype Asia-1 foot-and-mouth disease viruses in Pakistan and Afghanistan; emergence of a new genetic Group and evidence for a novel recombinant virus. Infect Genet Evol. 2011, 11: 2049-2062. 10.1016/j.meegid.2011.09.015.
57. Fry EE, Stuart DI, Rowlands JD: The structure of foot-and-mouth disease virus. Curr Top Microbiol Immunol. 2005, 288: 71-101.
58. Knowles NJ, Samuel AR: Molecular epidemiology of foot-and-mouth disease virus. Virus Res. 2003, 91: 65-80. 10.1016/S0168-1702(02)00260-5.
59. Hemadri D, Tosh C, Sanyal A, Venkataramanan R: Emergence of a new strain of type O foot-and-mouth disease virus: its phylogenetic and evolutionary relationship with the PanAsia pandemic strain. Virus Genes. 2002, 25: 23-34. 10.1023/A:1020165923805.
60. Davie J: A complement fixation technique for the quantitative measurement of antigenic differences between strains of foot-and-mouth disease virus. J Hyg Camb. 1964, 62: 407-411.
61. Pereira HG: Subtyping of foot-and-mouth disease virus. Dev Biol Stand. 1976, 167–174: 167-174.
62. Mohapatra JK, Subramaniam S, Pandey LK, Pawar SS, De A, Das B, Sanyal A, Pattnaik B: Phylogenetic structure of serotype A foot-and-mouth disease virus: Global diversity and the Indian perspective. J Gen Virol. 2011, 92: 873-879. 10.1099/vir.0.028555-0.
63. Knowles NJ, Nazem Shirazi MH, Wadsworth J, Swabey KG, Stirling JM, Statham RJ, Li Y, Hutchings GH, Ferris NP, Parlak U, Ozyörük F, Sumption KJ, King DP: Recent spread of a new strain (A-Iran-05) of foot-and-mouth disease virus type A in the Middle East. Transbound Emerg Dis. 2009, 56: 157-169. 10.1111/j.1865-1682.2009.01074.x.
64. Ansell DM, Samuel AR, Carpenter WC, Knowles NJ: Genetic relationships between foot-and-mouth disease type Asia1 viruses. Epidemiol Infect. 1994, 112: 213-324. 10.1017/S0950268800057587.
65. Jamal SM, Ferrari G, Ahmed S, Normann P, Belsham GJ: Genetic diversity of foot-and-mouth disease serotype O viruses from Pakistan and Afghanistan, 1997–2009. Infect Genet Evol. 2011, 11: 1229-1238. 10.1016/j.meegid.2011.03.006.
66. Mohapatra JK, Sanyal A, Hemadri D, Tosh C, Biswas S, Knowles NJ, Rasool TJ, Bandyopashyay SK, Pattnaik B: Comparative genomics of serotype Asia 1 foot-and-mouth disease virus isolates from India sampled over the last two decades. Virus Res. 2008, 136: 16-29. 10.1016/j.virusres.2008.04.010.
67. Valarcher JF, Knowles NJ, Zakharov V, Scherbakov A, Zhang Z, Shang YJ, Liu ZX, Liu XT, Sanyal A, Hemadri D, Tosh C, Rasool TJ, Pattnaik B, Schumann KR, Beckham TR, Linchongsubongkoch W, Ferris NP, Roeder PL, Paton DJ: Multiple origins of foot-and-mouth disease virus serotype Asia 1 outbreaks, 2003–2007. Emerg Infect Dis. 2009, 15: 1046-1051. 10.3201/eid1507.081621.
68. Knight-Jones T, Bulut N, Fine P, Gubbins S, Sumption K, Paton D: FMD Asia-1 vaccine effectiveness in Turkey. Appliance of science in the progressive control of FMD. 2012, Jerez de la Frontera, Spain: Open session of the EuFMD, 70-

69. Rweyemamu M, Roeder P, Mackay D, Sumption K, Brownlie J, Leforban Y, Valarcher J-F, Knowles NJ, Saraiva V: Epidemiological patterns of foot-and-mouth disease worldwide. Transbound Emerg Dis. 2008, 55: 57-72. 10.1111/j.1865-1682.2007.01013.x.
70. Grubman MJ, Baxt B: Foot-and-mouth disease. Clin Microbiol Rev. 2004, 17: 465-493. 10.1128/CMR.17.2.465-493.2004.
71. Valdazo-González B, Knowles NJ, Hammond J, King DP: Genome sequences of SAT 2 foot-and-mouth disease viruses from Egypt and Palestinian Autonomous Territories (Gaza Strip). J Virol. 2012, 86: 8901-8902. 10.1128/JVI.01231-12.
72. Fisher JR: Foot-and-mouth disease in Australia. Aust Vet J. 1984, 16: 58-161.
73. Knowles NJ, Samuel AR, Davies PR, Midgley RJ, Valarcher JF: Pandemic strain of foot-and-mouth disease virus serotype O. Emerg Infect Dis. 2005, 11: 1887-1893. 10.3201/eid1112.050908.
74. Samuel AR, Knowles NJ, Kitching RP, Hafez SM: Molecular analysis of foot-and-mouth disease type O viruses isolated in Saudi Arabia between 1983 and 1995. Epidemiol Infect. 1997, 119: 381-389. 10.1017/S0950268897008285.
75. Shin JH, Sohn HJ, Choi KS, Kwon BJ, Choi CU, Kim JH, Hwang EK, Park JH, Kim JY, Choi SH, Kim OK: Identification and isolation of foot-and-mouth disease virus from primary suspect cases in Korea in 2000. J Vet Med Sci. 2003, 65: 1-7. 10.1292/jvms.65.1.
76. Sakamoto K, Kanno T, Yamakawa M, Yoshida K, Yamazoe R, Murakami Y: Isolation of foot-and-mouth disease virus from Japanese black cattle in Miyazaki Prefecture, Japan, 2000. J Vet Med Sci. 2000, 64: 91-94.
77. Yoon SH, Lee KN, Park JH, Kim H: Molecular epidemiology of foot-and-mouth disease virus serotypes A and O with emphasis on Korean isolates: temporal and spatial dynamics. Arch Virol. 2011, 156: 817-826. 10.1007/s00705-011-0921-3.
78. Park JH, Lee KN, Ko YJ, Kim SM, Lee HS, Park JY, Yeh JY, Kim MJ, Lee YH, Sohn HJ, Cho IS, Kim B: Diagnosis and control measures of the 2010 outbreak of foot-and-mouth disease A type in the Republic of Korea. Transbound Emerg Dis. 2013, 60: 188-192. 10.1111/j.1865-1682.2012.01333.x.
79. Park JH, Lee KN, Ko YJ, Kim SM, Lee HS, Park JY, Yeh JY, Kim MJ, Lee YH, Sohn HJ, Moon JS, Cho IS, Kim B: Outbreaks and diagnosis of foot-and-mouth disease serotype O in the Republic of Korea, April–June 2010. Transbound Emerg Dis. in press
80. Park JH, Lee KN, Ko YJ, Kim SM, Lee HS, Shin YK, Sohn HJ, Park JY, Yeh JY, Lee YH, Kim MJ, Joo YS, Yoon H, Yoon SS, Cho IS, Kim B: Control of foot-and-mouth disease during 2010–2011 epidemic, South Korea. Emerg Infect Dis. 2013, 19: 655-659. 10.3201/eid1904.121320.
81. Nishiura H, Omori R: An epidemiological analysis of the foot-and-mouth disease epidemic in Miyazaki, Japan, 2010. Transbound Emerg Dis. 2010, 57: 396-403. 10.1111/j.1865-1682.2010.01162.x.
82. Tsutsui T, Yamamoto T: The foot-and-mouth disease epidemic in Japan, 2010. New tools and challenges for progressive control of FMD Open Session of the Research Group of the Standing Technical Committee on FMD. 2010, Vienna, Austria: EuFMD/FAO, Rome, Italy
83. Knowles NJ, He J, Shang Y, Wadsworth J, Valdazo-González B, Onosato H, Fukai K, Morioka K, Yoshida K, Cho IS, Kim SM, Park JH, Lee KN, Luk G, Borisov V, Scherbakov A, Timina A, Bold D, Nguyen T, Paton DJ, Hammond JM, Liu X, King DP: Southeast Asian foot-and-mouth disease viruses in Eastern Asia. Emerg Infect Dis. 2012, 18: 499-501. 10.3201/eid1803.110908.
84. Scherbakov A, Timina A, Borisov V: Phylogenetic study of FMDV isolates collected in Russia, CIS-countries and Mongolia in 2000–2007. The global control of FMD – Tools, ideas and ideals Open session of the Research Group of the Standing Technical Committee on FMD. 2008, Erice, Italy: EuFMD/FAO, Rome, Italy
85. Brückner GK, Vosloo W, Du Plessis BJ, Kloeck PE, Connoway L, Ekron MD, Weaver DB, Dickason CJ, Schreuder FJ, Marais T, Mogajane ME: Foot and mouth disease: the experience of South Africa. Rev Sci Tech. 2002, 21: 751-764.
86. Sangare O, Bastos AD, Marquardt O, Venter EH, Vosloo W, Thomson GR: Molecular epidemiology of serotype O foot-and-mouth disease virus with emphasis on West and South Africa. Virus Genes. 2001, 22: 345-351. 10.1023/A:1011178626292.

87. OIE (World Organization for Animal Health) / Official Disease Status/ FMD. [http://www.oie.int/en/animal-health-in-the-world/official-disease-status/fmd/en-fmd-carte/]
88. Knowles NJ, Samuel AR, Davies PR, Kitching RP, Donaldson AI: Outbreak of foot-and-mouth disease virus serotype O in the UK caused by a pandemic strain. Vet Rec. 2001, 148: 258-259.
89. Segarra AE, Rawson JM: Foot-and-mouth Disease: A Threat to U.S. Agriculture. CRS Report for Congress. 2001, USA: Congressional Research Service. The Library of Congress, 6-
90. Thompson D, Muriel P, Russell D, Osborne P, Bromley A, Rowland M, Creigh-Tyte S, Brown C: Economic costs of the foot-and-mouth disease outbreak in the United Kingdom in 2001. Rev Sci Tech. 2002, 21: 675-687.
91. Brito BP, Perez AM, Jamal SM, Belsham GJ, Pauszek SJ, Ahmed Z, Rodriguez L: Foot-and-mouth disease virus serotype O phylodynamics: genetic variability associated to epidemiological factors in Pakistan. Transbound Emerg Dis. in press
92. Valdazo-González B, Knowles NJ, Wadsworth J, King DP, Hammond JM, Ozyörük F, Firat-Saraç M, Parlak U, Polyhronova L, Georgiev GK: Foot-and-mouth disease in Bulgaria. Vet Rec. 2011, 168: 247-
93. Di Nardo A, Knowles NJ, Paton DJ: Combining livestock trade patterns with phylogenetics to help understand the spread of foot and mouth disease in sub-Saharan Africa, the Middle East and Southeast Asia. Rev Sci Tech. 2011, 30: 63-85.
94. Sumption K, Domenech J, Ferrari G: Progressive control of FMD on a global scale. Vet Rec. 2012, 170: 637-639. 10.1136/vr.e4180.
95. FAO: Development of a roadmap for the progressive control of foot-and-mouth disease in West Eurasia. Report of a workshop held in Shiraz, Islamic Republic of Iran during November 9–13. 2008, Rome, Italy: Food and Agriculture Organization of the United Nations
96. FAO: Progressive control pathway for foot-and-mouth disease control (PCP-FMD). 2011, Rome, Italy: Food and Agriculture Organization of the United Nations
97. Namatovu A, Wekesa S, Tjørnehø K, Dhikusooka M, Muwanika V, Siegsmund H, Ayebazibwe C: Laboratory capacity for diagnosis of foot-and-mouth disease in Eastern Africa: implications for the progressive control pathway. BMC Vet Res. 2013, 9: 19- 10.1186/1746-6148-9-19.
98. Bartels C, DeHaan N, McLaws M: PCP stage 1–3. Results, technical developments and issues. 2012, Jerez de la Frontera, Spain: Appliance of science in the progressive control of FMD Open session of the EuFMD, 30-
99. Jamal SM, Bouma A, van den Broek J, Stegeman A, Chenard G, Dekker A: Foot-and-mouth disease vaccine potency testing: The influence of method of fractionation, serotype, type of adjuvant, valency and type of virus culture on the dose–response curve in cattle. Vaccine. 2008, 26: 6317-6321. 10.1016/j.vaccine.2008.09.021.
100. Jamal SM, Shah SI, Ali Q, Mehmood A, Afzal M, Afzal M, Dekker A: Proper quality control of formulated foot-and-mouth disease vaccines in countries with prophylactic vaccination is necessary. Transbound Emerg Dis. in press

CHAPTER 4

Advances in Ileitis Control, Diagnosis, Epidemiology and the Economic Impacts of Disease in Commercial Pig Herds

Alison M. Collins

Department of Primary Industries, New South Wales, EMAI, PMB 4008, Narellan, NSW 2568, Australia; Tel.: +61-2-46406-306; Fax: +61-2-46406-384

ABSTRACT

Proliferative enteropathy, commonly known as "ileitis" continues to be a significant production-limiting disease in pig herds throughout the world. The disease can be controlled with a combination of vaccination and antibiotic medication. However, pressure from consumers to reduce antibiotic use in livestock industries highlights the need to better understand the epidemiology of ileitis, the mechanisms of immunity, and to identify management factors that can reduce the load of *Lawsonia intracellularis* in both pigs and the environment. New diagnostic assays and economic modelling of ileitis will help producers target optimal treatment times and minimize the production losses associated with ileitis. This review aims to outline the current advances in disease diagnosis, epidemiology, control strategies and the economic impact of both clinical and sub-clinical disease.

Keywords: animal diseases; epidemiology; economic impact; ileitis; proliferative enteropathy; *Lawsonia intracellularis*; transmission; immunity; diagnosis

1. INTRODUCTION

Ileitis is the common name for the wasting disease in pigs more accurately named porcine proliferative enteropathy (PPE). Clinical signs include reduced weight gains and diarrhea in growing pigs. Finisher or breeding animals can be affected with the more acute proliferative hemorrhagic enteropathy (PHE), losing blood in their feces due to intestinal hemorrhage. The proliferative enteropathies encompass a group of conditions (necrotic enteritis, regional ileitis, PHE and porcine intestinal adenomatosis) with similar pathogenesis. Gross pathology presents as a thickened mucosa in the distal small intestine and proximal large intestine. Histopathology is characterized by proliferation of crypt enterocytes, loss of goblet cells and intracellular bacteria in the apical cytoplasm of proliferating enterocytes [1]. The obligate intracellular bacterium, *Lawsonia intracellularis*, is the etiologic agent of all forms of PPE [2].

The characteristic gross and histological lesions of proliferative enteropathy (PE) were first described in North American weaned pigs in 1931 [3]. Subsequent reports described lesions of PPE in Europe, Asia, Australasia and America over the next 40 years [4,5,6,7,8,9,10]. However it was not until 1973 that the presence of intracellular bacteria within proliferative lesions was first described [11]. It took another 20 years for the etiologic agent, *L. intracellularis*, to be cultured in mammalian cells and for Koch's postulates to be fulfilled with the reproduction of PE in pigs [2,12].

Pigs of all ages, and pigs housed in a range of management systems, including farrow to finish farms and multiple site production farms are affected with PE. However, it is predominantly in grower and finisher pigs where disease control strategies are focused. Antibiotics were commonly used to treat or prevent PE, and practices developed where pigs were medicated continuously from about nine weeks of age until slaughter. While no data has been published to suggest that *L. intracellularis* develops resistance to commonly used antibiotics, the overuse of antibiotics could lead to antibiotic resistance in other bacteria. The aim of this review is to detail the advances in PE epidemiology, diagnosis and control strategies that will help reduce the reliance on antibiotics to control PE.

2. THE PREVALENCE OF PROLIFERATIVE ENTEROPATHY

The estimated prevalence of disease and *L. intracellularis* infection depends on the diagnostic method used. Slaughter checks relying on gross pathology have suggested that somewhere between 5% and 20% of pigs are affected with PE [13,14,15,16,17]. However, slaughter checks significantly underestimate the incidence of PE in grower pigs because PE lesions resolve within two to four weeks of clinical signs being present [18]. Producers and veterinarians surveyed in the UK and Australia estimated that clinical disease was prevalent in somewhere between 30% and 56% of herds [19,20]. Serological surveys of finisher pigs estimated that more than 80%, and up to 100% of pigs are infected with *L. intracellularis* [21,22,23]. In contrast, detection of *L. intracellularis* DNA in feces (by polymerase chain reaction) suggests that between 4.5% and 45% of herds are infected with *L. intracellularis* and the prevalence of infected animals ranges between 3.3% and 12% [24,25,26,27,28,29]. The diagnostic tests also vary in their ability to diagnose clinical disease, sub-clinical disease or *L. intracellularis* infection. The ability to interpret the results of infection prevalence studies in herds will help producers target disease control strategies to the right age group of pigs, and reduce the unnecessary use of antibiotics. Strategies to ensure the rational use of antibiotics are outlined in section 4.1.

3. DIAGNOSIS OF PROLIFERATIVE ENTEROPATHY

Early diagnosis of PE relied on gross pathology at the abattoir (thickening of the ileal mucosa) and histopathology; visualizing the proliferative epithelium and intracellular curved bacteria in the apical cytoplasm of cells with a Warthin Starry silver stain, and later with a specific monoclonal antibody to *L. intracellularis* [30,31]. Necropsies to diagnose PE on farm were only undertaken on sick pigs, not expected to recover, so other diagnostics were needed. The development of ante-mortem diagnostic assays for *L. intracellularis* infection was aided by the culture of *L. intracellularis*.

3.1. Serological Assays

Relatively impure *L. intracellularis* proteins, extracted from PE affected tissue, were initially used in both an ELISA and an indirect fluorescent antibody test (IFAT) to detect serum IgG antibodies to *L. intracellularis* [32,33]. The IFAT was able to overcome high background problems observed in the early ELISA by visualizing specific antigen-antibody reactions under the microscope, allowing differentiation between fluorescing *L. intracellularis* and non-specific reactions. Culture of *L. intracellularis* allowed the development of whole cell immunoperoxidase antibody tests (IPMA) [34] and ELISA tests using *L. intracellularis* outer membrane proteins or lipopolysaccharides (LPS) as antigens [35,36], but most of these assays have remained as research tools. A blocking ELISA for serum IgG to *L. intracellularis* (bioScreen Ileitis antibody ELISA; Munster, Germany) is now commercially available and routinely used to determine the timing and prevalence of *L. intracellularis* infection. Serology can help producers and veterinarians identify risk factors and the efficacy and optimal timing for PE control measures including vaccination, medication and improved hygiene [37,38,39]. The sensitivity of this blocking ELISA was between 72% and 90.5% and the specificity was between 83% and 93%, using sera from both field and experimental challenge trials [40,41]. Regardless of which serological assay is used, *L. intracellularis* antibodies are not routinely detected until about 21 days post challenge, after clinical signs and fecal shedding are already observed, so serology is an indicator of previous infection [42,43]. The IFAT, IPMA and LPS ELISA have been modified to measure *L. intracellularis* specific IgA and IgM antibodies in sera and in mucosal scrapings, so they continue to be used to elucidate the immune response to both vaccination and virulent challenge [44,45,46].

3.2. Molecular Assays

The development of DNA based diagnostic assays such as hybridization and the polymerase chain reaction (PCR) did not rely upon the culture of *L. intracellularis*. Hybridization probes were produced from short fragments of *L. intracellularis* DNA that were cloned [47] and used to detect *L. intracellularis* DNA in pig faeces. Later, short fragments of *L. intracellularis* DNA were amplified by PCR using oligonucleotides synthesized to detect *L. intracellularis* DNA [48,49,50,51]. The relatively short duration of *L. intracellularis* excretion (up to three weeks) and the higher cost of PCR (compared to serology) has meant that the PCR has been used primarily to detect infection in diarrheic pigs, and is less useful for monitoring PE control in herds. The PCR assay cannot discriminate between live and dead *L. intracellularis* in feces, so some reports of PCR positive pigs detected at a single time point make it difficult to interpret whether the pigs were actively infected at the time or if *L. intracellularis* was transiting through their digestive tracts.

Although fecal PCR is both more sensitive and specific than other diagnostic assays, it was not quantitative until recently, partly because the amplification of *L. intracellularis* DNA can be inhibited by the presence of large amounts of competing non-target DNA [52], bile salts and bilirubin in feces [53]. The interpretation of fecal PCR results has also been confused by the sensitivity of the assay. Detection of *L. intracellularis* DNA does not prove that *L. intracellularis* is the cause of diarrhea. The development of a multiplex PCR assay that can detect *L. intracellularis* and *Brachyspira* spp. in the same sample helped identify multiple causes of diarrhea in the same animal [54,55].

Recently a number of quantitative real time PCR (qPCR) assays have been developed for *L. intracellularis* [56,57,58,59]. Sensitivity of these new assays was aided by advances in DNA extraction protocols with magnetic particle processors capable of capturing nucleic acids and removing them from inhibitors in the feces. These qPCR assays can calculate *L. intracellularis* numbers in feces relative to standards seeded with known numbers of *L. intracellularis*, providing a reproducible linear range of 10^8 to 10^4 bacteria/g feces. The number of *L. intracellularis* excreted in pig feces (determined by qPCR) is between 10^4 and 10^8 *L. intracellularis* per gram of feces depending on whether pigs are clinically or sub-clinically affected with PE [56,57,58,59], with clinically affected pigs excreting more than 10^7 *L. intracellularis* per gram of faeces.

L. intracellularis excretion levels correlated well with antibody titres, the severity of histopathology lesions, the duration of *L. intracellularis* excretion and the severity of diarrhea [60,61]. A weaker inverse correlation exists between the qPCR and production losses associated with PE, which may be explained by other factors that impact on pig growth. However, a one $\log_{10}$ unit increase in the number of *L. intracellularis* increases the risk for low growth rate by two times [61]. The critical number or threshold of *L. intracellularis* that causes reduced growth rates in experimentally infected pigs is reported to be above 10^7 *L. intracellularis* per gram of feces [62]. Once the number of *L. intracellularis* shed by pigs increased from 10^7 to 10^8per gram of feces, average daily gain was reduced by 131 g/day. Identifying a critical threshold of infection across a wide range of commercial management systems will allow producers to use the qPCR to target ileitis treatment options.

Preliminary studies have indicated that fecal samples can be pooled prior to qPCR analysis to help reduce costs and to encourage producers to use the qPCR as a herd health monitoring test. Theoretical pooling of samples indicated that the qPCR was still accurate when pools of 10 individual samples were used [63]. However, pools of five fecal samples provided more accurate results than pools of 10 feces when the qPCR assays were performed [64].

The diagnosis of *L. intracellularis* infection in real time by qPCR enables producers to quantify the efficacy of PE control measures, as reduced numbers of *L. intracellularis* correlate with reduced severity of PE. Data from experimental challenge trials shows the significant reduction in the number of *L. intracellularis* shed in pigs vaccinated with Enterisol® Ileitis or medicated with antibiotics [46,65]. The qPCR can also be used to quantify the efficacy of disinfection and hygiene protocols [39] and the biosecurity risk associated with rodents near piggeries [58]. Avoiding the over-use of antibiotics for the control of PE relies on the correct interpretation of all of the new diagnostic assays. Quantitative assays identify the severity of disease and help producers decide on what level of intervention is needed to control PE. Qualitative assays such as serology provide producers with information on the optimal timing of disease control measures and the prevalence of exposure to *L. intracellularis* in the herd.

4. CONTROL OF ILEITIS

The development of experimental PE challenge models and diagnostics to monitor *L. intracellularis* infection enabled antibiotic efficacy studies to be performed in the absence of concurrent infections commonly found on farm. Prior to this, veterinarians relied on monitoring clinical signs of PE or pathology at slaughter to determine whether treatment strategies were effective. However, diarrhea and poor growth could be caused by other factors and suspect gross pathology at abattoir was only proven by histopathology in less than 11% of cases [14]. These new disease challenge models enabled the more rational use of antibiotics for PE control, and allowed evaluation of other potential control strategies including immunization and improved hygiene.

4.1. Antibiotic Medication

The efficacy of antibiotics such as olaquindox, to control PE was originally identified by field veterinarians trying to control outbreaks of PE [66]. Later experimental challenge trials demonstrated that high doses of tiamulin, tylosin, chlortetracycline, lincomycin and olaquindox were able to treat pigs with PE, reducing clinical signs, histological lesions of PE and the duration of fecal shedding of *L. intracellularis* [67,68,69,70,71,72,73,74]. High doses of tiamulin, tylosin, olaquindox and chlortetracycline could also prevent *L. intracellularis* infection if given continuously in feed [67,70,74]. However, once these antibiotics were removed, the pigs remained immunologically naïve and were susceptible to a later *L. intracellularis* challenge [75]. In addition, continuous use of antibiotics may increase the potential for other bacteria to develop resistance.

Public concern over the development of antibiotic resistance by bacteria led to significant pressure world-wide to find alternative strategies to control PE and to reduce the prophylactic use of antibiotics. Protective immunity to PE was first demonstrated in a minimal disease piggery with two sequential outbreaks of the hemorrhagic form of ileitis (PHE). Pigs that survived the first outbreak were protected from a subsequent disease outbreak [76]. The authors proposed that immunity could be induced by bringing young animals into contact with the source of infection for three weeks, prior to the addition of antibiotics to terminate infection.

Protective immunity was later demonstrated in experimental challenge models, where pigs previously challenged orally with *L. intracellularis* were protected from re-infection on a subsequent challenge [43,45,77]. The failure to detect *L. intracellularis* in the feces of re-challenged pigs by PCR indicated that virulent, live *L. intracellularis* in the second inoculum were not able to colonize mucosal cells in previously challenged pigs. Therapeutic strategies were designed to allow *L. intracellularis* infection and immunity to re-infection, with antibiotics used to avoid clinical disease. Antibiotics were removed from pig diets for 12 to 18 days and then pulsed in feed or in water at high doses for two to four days every two to three weeks to suppress disease. The period without antibiotics allowed pigs to become infected with *L. intracellularis* but also to develop immunity to re-infection [78].

Antibiotics, along with extensive cleaning of pens and depopulation of younger pigs were also used in attempts to eradicate PE, by eliminating *L. intracellularis* from the pig and the environment [79,80,81,82]. Eradication of clinical signs of PE appeared possible on smaller farms, but very few studies were able to demonstrate complete eradiation of *L. intracellularis* from the herd and the environment over time. Biosecurity needed to be maintained at a very high level to ensure continued eradication of *L intracellularis* [82]. Outbreaks of severe disease often followed when *L. intracellularis* was accidentally introduced into a naïve herd [80].

4.2. Improved Hygiene

L. intracellularis can survive (outside of the pig) in conventional pens for at least two weeks at temperatures between 9 °C and 18 °C [39,50].

Management practices that improve biosecurity and hygiene, including all-in-all-out production and minimal mixing of pigs, are associated with a lower risk of *L. intracellularis* infection [83,84,85]. The housing and flooring in pens may also impact on the survival of *L. intracellularis*. It may be expected that pigs reared on deep litter systems would be exposed to greater numbers of pathogens, including *L. intracellularis*, because deep litter systems are only cleaned and/or disinfected between batches of pigs, compared with conventional pens where manure is removed more regularly. However, pigs grown out on deep litter systems had a similar prevalence of *L. intracellularis* infection as cohorts raised on traditional concrete-based systems [86], and grower and finisher pigs reared on partially slatted floors had a higher risk of diarrhea than pigs housed on solid concrete floors [87]. Power hosing concrete slatted floors with cold water, followed by disinfection with potassium peroxymonosulfate (Virkon S) and drying time was able to significantly reduce the survival and transmission of *L. intracellularis* from the environment to naïve pigs [39]. The efficacies of a wide range of other disinfectants against *L. intracellularis* have been tested by *in vitro* culture methods. The iodophore povidone iodine and quaternary ammonium compounds were both effective against *L. intracellularis* in *in vitro* studies, in the absence of faecal material [50,88]. However, the efficacy of disinfectants needs to be studied in pig pens with residual manure, which can inactivate the disinfectant. Reducing the number of *L. intracellularis* in pens between batches of pigs reduces the severity of disease in subsequent batches, but the aim is not to remove all *L. intracellularis*, or naïve herds may develop, which are highly susceptible to outbreaks of PHE, the more acute form of PE.

4.3. Vaccination and Induced Immunity

Oral vaccination with a commercial avirulent live *L. intracellularis* vaccine (Enterisol® Ileitis) protects pigs from clinical disease; significantly reducing microscopic lesions of PE following a virulent heterologous *L. intracellularis* challenge [89]. Vaccination did not completely prevent fecal shedding of *L. intracellularis* following virulent challenge, but did significantly reduce the number of *L. intracellularis* shed in feces and the duration of shedding in vaccinated pigs [42,89,90]. In many commercial herds antibiotic medication was no longer required to control PE in vaccinated herds. However, producers believed that in some straw-bedded systems *L. intracellularis* infection was harder to suppress with vaccination alone [91]. In these cases, antibiotics were used in tandem with vaccination.

As expected for an intracellular bacterium, *L. intracellularis* infection induces both specific humoral (IgA, IgG) and cell mediated (IFN-γ) immune responses in serum and the intestinal mucosa. The immune response to both natural challenge and oral vaccination is dose-dependent [43,44,45,46]. The absence of an anamnestic serum IgA response in re-challenged pigs, along with an increased gamma IFN response, led to the speculation that cell-mediated immune responses alone were likely to be mediators of protective immunity to *L. intracellularis* [45]. However, accumulation of IgA in proliferating crypt cells and in the cell debris of the crypt lumen of PE-affected pigs, where *L. intracellularis* also reside [92], suggests that IgA is also involved in a protective immune response to *L. intracellularis.* Induction of mucosal immunity is believed to depend upon presentation of the antigen at the mucosal surface, but protection against other intracellular pathogens (*Salmonella enteriditis*) following systemic vaccination [93] suggest that systemic vaccination may protect pigs from *L. intracellularis* challenge. In early studies, disease severity was significantly reduced in pigs vaccinated twice intramuscularly (IM) with *L. intracellularis* bacterin or recombinant GroEL-like protein [94]. More recently, IM vaccinations, with *L. intracellularis* bacterin or recombinant outer membrane proteins of *L. intracellularis* (19/21 and 37 kDa antigens), protected pigs from PE [95]. In comparative studies with the commercial live *L. intracellularis* vaccine (Enterisol® Ileitis), vaccination (10-times dose) administered orally or intra-muscularly induced a similar protective immune response, with systemic and mucosal *L. intracellularis*-specific antibodies and mucosal cytokines [46]. These inactivated *L. intracellularis* vaccines appear to protect against infection if sufficient *L. intracellularis* antibodies are present to prevent infection at the mucosal. Alternate adjuvants or higher doses of *L. intracellularis* antigens may increase the efficacy of systemic vaccination routes. To date no detectable immune marker for protection has been identified, so veterinarians have to rely on detection of exposure to *L. intracellularis* (serum IgG antibodies) as a proxy for immunity.

The need to remove all medication for three days before and after vaccination was initially seen as an obstacle to the uptake of an oral vaccine [91]. The vaccine was registered for weaner pigs that may have concurrent infections requiring antibiotic medication. While vaccination pre-weaning may overcome this obstacle, the potential for maternal antibodies to interfere with vaccination was a concern [91]. While maternally acquired antibodies are vital in protecting piglets from pathogen invasion, they may also interfere with oral vaccination. Maternal antibodies ingested by piglets bind to antigens found in the intestinal lumen, form complexes and are then internalized by antigen-presenting cells. Both B and T cells are stimulated in the piglet, though B cell stimulation is inhibited. Modifying the vaccination route may overcome the potential interference of maternal antibodies. Earlier studies demonstrated that IM vaccination could induce a protective immune response in *L. intracellularis* challenged pigs [46,94,95]. However, the Enterisol® Ileitis vaccine is not registered for this use.

5. PRODUCTION AND ECONOMIC IMPACTS OF ILEITIS

The clinical impact of *L. intracellularis* infection depends on the dose of bacteria ingested by pigs [43,96], but also on the expression of bacterial virulence genes [97], stimulation of the host's immune system [98] and environmental factors. The pig's diet, genetics, immune response, intestinal microflora and intestinal health are all likely to impact on the severity of disease.

Clinical signs of ileitis including diarrhea are evident for one to three weeks, reducing feed conversion efficiency (FCE) by up to 50% [99]. Most pigs recover from clinical disease within four to six weeks but chronic disease can affect pig growth over 10 weeks. Significant variation in weight gains has also been observed in pigs affected by PE, with reductions in average daily gain (ADG) ranging from 17% to 84% of the gain of unaffected pigs [99]. This variation in growth within a batch of pigs can lead to increased days to slaughter or increased back fat on larger animals and ultimately economic losses for the producer [100]. Simulation of two different clinical outbreaks of PE, (reduced feed intake, ADG, feed conversion efficiency, increased P2 back fat) indicated that clinical ileitis costs between $7 and $13 AUD for every pig [100,101].

Many pigs are sub-clinically affected, as demonstrated by the extremely high prevalence of seropositive pigs relative to the small proportion of pigs with PE lesions [17,21,23]. Diarrhea is rarely observed in pigs sub-clinically affected with PE, but reductions in ADG between 9% and 42%, and reduced FCE between 6% and 37% have been reported [99,102]. The economic impact of sub-clinical PE is difficult to estimate because many producers are unaware that sub-clinical PE is present in their herd. Experimental reproduction of sub-clinical ileitis in a group of grower pigs caused significantly decreased feed intake (4 kg) over a three week period, but did not reduce pig weights or weight gains in the same period, though variation in pig weights was significantly increased in infected pigs relative to non-infected pigs [103]. In the three weeks after infection peaked, *L. intracellularis* infected pigs increased weight more quickly than in the early infection period, but muscle deposition was significantly reduced [104]. This may in part explain the increased P2 back fat depth reported in previous studies, although no significant differences in body composition, measured with a CT scanner, were observed between sub-clinically and uninfected animals [104]. Economic modeling of these production losses on a herd basis in the absence of control strategies resulted in an $8.33 AUD reduction in revenue per pig in a herd with 80% of pigs sub-clinically affected [100].

6. EPIDEMIOLOGY OF ILEITIS

6.1. Transmission of Infection between Pigs and the Environment

The occurrence of PE in segregated early weaning systems indicated that pigs are either infected from the sow pre-weaning, from the environment, or from contact with pigs or other species of animals, birds or insects. Transmission of infection between in-contact pigs was demonstrated in natural outbreaks of PE when animals were recently mixed, or when new animals were introduced [6,10] and also in experimental challenge trials [50,105]. The transmission of *L. intracellularis* to naïve pigs is primarily via the ingestion of contaminated feces. Pigs affected with PE shed large numbers of *L. intracellularis*, enough to infect naïve pigs in contact with them [43,50,105,106]. Epidemiology studies have identified frequent mixing of pigs, the introduction of new pigs and continuous flow production systems as risk factors for PE [22,87].

The detection of *L. intracellularis* in pig feces coincides with the presence of clinical signs of PE [43,107]. However, in some experimental challenge trials, pigs continued to shed *L. intracellularis* after clinical signs of PE ceased, [42,108], suggesting a carrier state similar to Salmonella infections, where stressed carrier pigs re-commence excretion of Salmonella [109]. *L. intracellularis* bacteria were found in the tonsils of pigs clinically affected with PE [110], however, a carrier state of *L. intracellularis* infection has not been demonstrated. Pigs allowed to recover naturally from PE do not continue to shed detectable numbers of *L. intracellularis*, and are not able to transmit infection to naïve, in-contact pigs [111]. Treatment of recovered pigs with the corticosteroid dexamethasone likewise did not lead to detectable *L. intracellularis* excretion, or transmission of infection to naïve in-contact pigs. The development of immunity to *L. intracellularis* infection, even when pigs were challenged with a much higher secondary dose (increased by four logs), also indicates that *L. intracellularis* does not readily reappear from a carrier state [43,90].

Elimination of *L. intracellularis* from pigs and the environment is important in reducing the transmission of infection to naïve pigs. Chlortetracycline in feed (400 ppm) was able to rapidly terminate *L. intracellularis* excretion relative to tiamulin (200 ppm), oxytetracycline (400 ppm) and non-medicated pigs [65].

6.2. Transmission of L. intracellularis from Sows to Their Progeny

The detection of *L. intracellularis* in the feces of a small number of suckling pigs in endemically-infected herds has led researchers to speculate that sows may be a source of infection to their piglets [108,112,113]. However, the detection of antibodies to *L. intracellularis* in three to six week old pigs, and the occurrence of PE predominantly in pigs greater than six weeks of age [32], suggests that piglets are more likely protected from *L. intracellularis* by either antibodies or lymphocytes from the sow in colostrum or later by IgA in milk. In late lactation, circulating antibodies in the sow are concentrated into pre-colostrum, along with circulating IgA-secreting lymphocytes (plasma calls) which "home" to the mammary gland [114]. Piglets can passively absorb maternal antibodies during the first 24 to 36 h of life [115]. Lymphocytes from the birth sow can be absorbed by her piglets in the first week of life [116].

The ability of sows to protect piglets against *L. intracellularis* infection will therefore depend on the level of circulating antibodies in the sow and the transfer of protective immune secretions and cells to her progeny. Three week old piglets suckling on a recently infected sow were immune to challenge with *L. intracellularis*, in contrast to their weaned litter mates that were infected with the same *L. intracellularis* challenge [117]. Four weeks after the protected piglets were weaned they were susceptible to a virulent challenge; whereas their previously infected litter mates were now immune. This passive protection did not persist after pigs were weaned. Sows with no detectable *L. intracellularis* antibodies at farrowing, but from an endemically infected herd, were only able to partially protect their suckling progeny. Clinical signs were reduced, but piglets excreted *L. intracellularis* and raised significantly reduced serum IgG titres to *L. intracellularis* [117]. The ability of sows with higher *L. intracellularis* antibody titres to protect their suckling piglets was also demonstrated in field studies. The progeny of 20 seropositive and 20 seronegative gilts were challenged with *L. intracellularis* at two weeks of age and remained suckling for an additional three weeks. Only 21% of piglets on seropositive gilts had *L.*

intracellularis antibodies three weeks post challenge, relative to 86% of piglets from seronegative gilts [112].

6.3. External Vectors in Transmission of L. intracellularis

The main source of *L. intracellularis* in the piggery is infected pigs, and transmission is via ingestion of contaminated feces [50,105]. However, outbreaks of PE occur in naïve herds, suggesting that other sources of *L. intracellularis* may exist. A wide range of domestic and wildlife species can become infected with *L. intracellularis* and some also develop PE [118,119,120,121,122,123,124,125,126]. Birds, insects, rodents and feral pigs are considered the most likely external vectors for PE in pigs due to their proximity to commercial pig units. *L. intracellularis* infection of rodents has been investigated in both field and experimental challenge studies. Laboratory rats and mice challenged with porcine strains of *L. intracellularis* have been infected, raised persistent immune responses and developed caecal and colonic lesions of PE [58,127,128,129]. In addition, *L. intracellularis* DNA was detected in the intestinal tissue of wild mice and rats trapped on pig farms where PE was endemic [58,130]. A high proportion of wild rats trapped on pig farms with endemic PE were excreting *L. intracellularis* (≥70.6%), but more significantly, a small proportion of these rats excreted extremely high numbers of *L. intracellularis* (10^{10}/g of feces) [58]. Rodents may therefore be an important reservoir of *L. intracellularis* on pig farms, as less than 1 g of rodent feces would need to be ingested by a naïve pig for transmission of infection to occur.

To date the only report of *L. intracellularis* infection in birds relates to ratites including the ostrich and emu [123,124]. As these species are not commonly associated with pig herds, they are unlikely to pose a significant infection risk. Experimental challenge of sparrows with porcine isolates of *L. intracellularis* failed to demonstrate colonization or disease associated with *L. intracellularis* [131]. Conversely, invertebrates collected from a geographical spread of UK pig herds indicated that *L. intracellularis* DNA was frequently detected in house and hover flies, but not in cockroaches [126]. Small differences in DNA repeats (four microsatellite markers) indicated that the *L. intracellularis* isolate found in pigs couldn't be differentiated from the fly isolate. However, it's not possible to say whether the flies are just a measure of *L. intracellularis* infection in the herd or a potential source of infection to naïve pigs.

7. CONCLUSIONS

Veterinarians and pig producers have benefitted from the recent knowledge gained in the diagnosis of *L. intracellularis* infection, transmission and survival of *L. intracellularis* and PE control strategies. This also translates to benefits for society, as these recent advances will enable producers to control PE with a reduced reliance on antibiotics.

The development of quantitative diagnostic assays enable producers to measure the severity of *L. intracellularis* infection in pig herds in real time, providing them with opportunities to fine tune treatments and ultimately reduce costs associated with disease control. In the absence of an immune correlate for protection, producers should continue to regularly monitor their herds by serology to ensure that *L. intracellularis* infection is widespread, as a proxy for immunity.

Knowledge of the potential sources of *L. intracellularis* in pigs, external vectors and the environment enables producers and veterinarians to develop hygiene and biosecurity procedures, including rodent eradication programs, to prevent outbreaks of PE. Management strategies that reduce the spread and number of *L. intracellularis* in the pig herd include all-in-all-out production, quarantine areas for newly introduced pigs, minimal mixing of pigs and effective disinfection of pens or sheds between batches.

Vaccination, either with the current live oral vaccine or with a new recombinant protein vaccine given IM, will ensure that all pigs are exposed to *L. intracellularis*, and will be protected from clinical signs of PE on later challenge. Intramuscular vaccination may overcome suggested maternal antibody interference and the need to have a seven day antibiotic-free period in newly weaned pigs. However, in some instances vaccination alone does not appear to prevent PE. In these cases, careful monitoring of the herd will help target antibiotic medication treatments.

Finally a better understanding of the costs and production losses associated with both clinical and sub-clinical PE will help producers target treatments to appropriate periods in pig production. In conclusion, it is a combination of management practices that will enable control of PE with significantly reduced reliance on antibiotic medication.

ACKNOWLEDGMENTS

The author wishes to thank both Australian Pork Limited and the Co-operative Research Centre for High Integrity Pork for their financial support for PE research in Australia over the last twenty years. This research was greatly assisted by colleagues, technical staff and students at the Department of Primary Industries, NSW and the University of Sydney.

REFERENCES

1. Lawson, G.H.K.; Gebhart, C.J. Proliferative enteropathy. *J. Comp. Pathol.* **2000**, *122*, 77–100.
2. McOrist, S.; Jasni, S.; Mackie, R.A.; Macintyre, N.; Neef, N.; Lawson, G.H.K. Reproduction of porcine proliferative enteropathy with pure cultures of Ileal symbiont intracellularis. *Infect. Immun.* **1993**, *61*, 4286–4292.
3. Biester, H.E.; Schwarte, L.H. Intestinal adenoma in swine. *Am. J. Pathol.* **1931**, *7*, 175–185.
4. Emsbo, P. Terminal or regional ileitis in swine. *Nord. Vet. Med.* **1951**, *3*, 1–28.
5. Dodd, D.C. Adenomatous intestinal hyperplasia (proliferative ileitis) of swine. *Vet. Pathol.* **1968**, *5*, 333–341.
6. Rowland, A.C.; Rowntree, P.G. A haemorrhagic bowel syndrome associated with intestinal adenomatosis in the pig. *Vet. Rec.* **1972**, *91*, 235–241.
7. O'Hara, P.J. Intestinal haemorrhage syndrome in the pig. *Vet. Rec.* **1972**, *91*, 517–518.
8. Redman Chu, R.M.; Hong, C.B. Haemorrhagic bowel syndrome in pigs in Taiwan. *Vet. Rec.* **1973**, *93*, 562–563.
9. Martinsson, H.; Holmgren, N.; Jonsson, L.; Nordstrom, G. Studies in terminal ileitis in piglets. *Sven Vet.* **1974**, *26*, 347–354.
10. Love, R.J.; Love, D.N.; Edwards, M.J. Proliferative haemorrhagic enteropathy in pigs. *Vet. Rec.* **1977**, *100*, 65–68.
11. Rowland, A.C.; Lawson, G.H.; Maxwell, A. Intestinal adenomatosis in the pig: Occurrence of a bacterium in affected cells. *Nature* **1973**, *243*, 417.
12. Lawson, G.H.K.; McOrist, S.; Jasni, S.; Mackie, R.A. Intracellular bacteria of porcine proliferative enteropathy: Cultivation and maintenance *in vitro*. *J. Clni. Microbiol.* **1993**, *31*, 1136–1142.
13. Pointon, A.M. *Campylobacter* associated intestinal pathology in pigs. *Aust. Vet. J.* **1989**, *66*, 90–91.
14. Christensen, N.H.; Cullinane, L.C. Monitoring the health of pigs in New Zealand abattoirs. *N. Z. Vet. J.* **1990**, *38*, 136–141.
15. Holyoake, P.K.; Cutler, R.S.; Caple, I.W. A diagnostic dilemma: Detecting proliferative enteritis in pigs at slaughter. *Aust. Vet. J.* **1994**, *71*, 308–309.
16. Maes, D.; Verdonck, M.; Castryck, F.; Ducatelle, R.; de Ridder, E.; Weyns, A.; Soenen, K.; de Kruif, A. Porcine proliferative enteropathy: Preliminary epidemiological and pathological observations in Belgian pig herds. *Vlaams Diergeneeskundig Tijdschrift* **1999**, *68*, 231–236.
17. Stege, H.; Jensen, T.K.; Moller, K.; Baekbo, P.; Jorsal, S.E. Prevalence of intestinal pathogens in Danish finishing pig herds. *Prev. Vet. Med.* **2000**, *46*, 279–292.
18. Roberts, L.; Lawson, G.H.; Rowland, A.C.; Laing, A.H. Porcine intestinal adenomatosis and its detection in a closed pig herd. *Vet. Rec.* **1979**, *104*, 366–368.
19. Holyoake, P.K.; Cutler, R.S.; Caple, I.W. Prevalence of proliferative enteritis on pig farms in Australia. *Aust. Vet. J.* **1994**, *71*, 418–422.

20. Smith, S.H.; McOrist, S.; Green, L.E. Questionnaire survey of proliferative enteropathy on British pig farms. *Vet. Rec.* **1998**, *142*, 690–693.
21. Corzo, C.A.; Friendship, R.M.; Dewey, C.E.; Blackwell, T. Seroprevalence of *Lawsonia intracellularis* in Ontario swine herds. *J. Swine Health Prod.* **2005**, *13*, 314–317.
22. Bane, D.; Norby, B.; Gardner, I.; Roof, M.; Knittel, J.P.; Bush, E.J. Prevalence of Management Risk Factors Associated with *Lawsonia intracellularis* Seropositivity in the US Swine Herd. In Proceedings of Allen D Leman Swine Conference, University of Minnesota, Minneapolis, MN, USA, 21 Sepetember 1997; pp. 10–11.
23. Holyoake, P.K.; Emery, D.; Gonsalves, J.; Donahoo, M.; Collins, A. Prevalence of antibodies to *Lawsonia intracellularis* in pig herds in Australia. *Aust. Vet. J.* **2010**, *88*, 186–188.
24. Wilson, T.M.; Chang, K.; Gebhart, C.J.; Kurtz, H.J.; Drake, T.R.; Lintner, V. Porcine proliferative enteritis: Serological, microbiological and pathological studies from three field epizootics. *Can. J. Vet. Res.* **1986**, *50*, 217–220.
25. Winkelman, N.L. Ileitis: An update. *Compend. Contin. Educ. Pract. Vet.* **1996**, *18*, S19–S25.
26. Chang, W.L.; Wu, C.F.; Wu, Y.; Kao, Y.M.; Pan, M.J. Prevalence of *Lawsonia intracellularis* in swine herds in Taiwan. *Vet. Rec.* **1997**, *141*, 103–104.
27. Kim, O.; Kim, B.; Chae, C. Prevalence of *Lawsonia intracellularis* in selected pig herds in Korea as determined by PCR. *Vet. Rec.* **1998**, *143*, 587–589.
28. Pozo, J.; Carvajal, A.; de Arriba, M.L.; Lanza, I.; Rubio, P. Presence of *Lawsonia intracellularis* as an Aetiologic Agent in Enteric Disorders in Spanish Swine Herds. In Proceedings of the 15th International Pig Veterinary Society Congress, Birmingham, UK, 5–9 July 1998; p. 416.
29. Chiriboga, A.E.; Guimaraes, W.V.; Vanetti, M.C.; Araujo, E.F. Detection of *Lawsonia intracellularis* in faeces of swine from the main producing regions in Brazil. *Can. J. Microbiol.* **1999**, *45*, 230–234.
30. McOrist, S.; Boid, R.; Lawson, G.H.K. Antigenic analysis of *Campylobacter* species and an intracellular *Campylobacter*-like organism associated with porcine proliferative enteropathies. *Infect. Immun.* **1989**, *57*, 957–962.
31. McOrist, S.; Boid, R.; Lawson, G.H.K.; McConnell, I. Monoclonal antibodies to intracellular *Campylobacter*-like organisms of the porcine proliferative enteropathies. *Vet. Rec.* **1987**, *121*, 421–422.
32. Holyoake, P.K.; Cutler, R.S.; Caple, I.W.; Monckton, R.P. Enzyme-linked immunosorbent assay for measuring ileal symbiont intracellularis-specific immunoglobulin G response in sera of pigs. *J. Clin. Microbiol.* **1994**, *32*, 1980–1985.
33. Knittel, J.P.; Jordan, D.M.; Schwartz, K.J.; Janke, B.H.; Roof, M.B.; McOrist, S.; Harris, D.L. Evaluation of antemortem polymerase chain reaction and serologic methods for detection of *Lawsonia intracellularis*-exposed pigs. *Am. J. Vet. Res.* **1998**, *59*, 722–726.
34. Guedes, R.M.C.; Gebhart, C.J.; Deen, J.; Winkelman, N.L. Validation of an immunoperoxidase monolayer assay as a serologic test for porcine proliferative enteropathy. *J. Vet. Diagn. Investig.* **2002**, *14*, 528–530.
35. Boesen, H.T.; Jensen, T.K.; Moler, K.; Nielsen, L.H.; Jungersen, G. Evaluation of a novel enzyme-linked immunosorbent assay for serological diagnosis of porcine proliferative enteropathy. *Vet. Microbiol.* **2005**, *109*, 105–112.
36. Kroll, J.J.; Eichmeyer, M.A.; Schaeffer, M.L.; McOrist, S.; Harris, D.L.; Roof, M.B. Lipopolysaccharide-based enzyme-linked immunosorbent assay for experimental use in detection of antibodies to *Lawsonia intracellularis* in pigs. *Clin. Diagn. Lab. Immun.* **2005**, *12*, 693–699.
37. Vu Ngoc, Q.; Nguyen Dang, H.; Cu Huu, P.; Nguyen Ngoc, N.; Collins, A.; Love, R. Epidemiological survey of *Lawsonia intracellularis* infection and the risk factors in industrial pig farms in Vietnam. *Khoa Hoc Ky Thuat Thu Y* **2003**, *10*, 23–31.
38. Collins, A.M.; Barchia, I.; Gonsalves, J.R.; Fell, S.; Eamens, G.J. Reducing Survival and Transmission of *Lawsonia intracellularis* in Conventional Piggeries. In *Manipulating Pig Production XI*; Paterson, J.E., Barker, J.A., Eds.; Australasian Pig Science Association: Werribee, Australia, 2007; p. 125.

39. Collins, A.M.; Fell, S.A.; Barchia, I.M. Cleaning and disinfection with Virkon S significantly reduces *Lawsonia intracellularis* survival and transmission to naive pigs. *J. Swine Health Prod.* **2013**, *21*, 144–147.
40. Collins, A.; Gonsalves, J.; Fell, S.; Barchia, I. Comparison of a commercial ELISA with an indirect fluorescent antibody test to detect antibodies to *Lawsonia intracellularis* in experimentally challenged pigs. *Aust. Vet. J.* **2012**, *90*, 97–99.
41. Jacobson, M.; Wallgren, P.; Nordengrahn, A.; Merza, M.; Emanuelson, U. Evaluation of a blocking ELISA for the detection of antibodies against *Lawsonia intracellularis* in pig sera. *Acta Vet. Scand.* **2011**, *53*, 23.
42. Guedes, R.M.C.; Gebhart, C.J. Onset and duration of fecal shedding, cell-mediated and humoral immune responses in pigs after challenge with a pathogenic isolate or attenuated vaccine strain of *Lawsonia intracellularis. Vet. Microbiol.* **2003**, *91*, 135–145.
43. Collins, A.M.; Love, R.J. Re-challenge of pigs following recovery from proliferative enteropathy. *Vet. Microbiol..* **2007**, *120*, 381–386.
44. Guedes, R.M.C.; Gebhart, C.J. Evidence of cell-mediated immune response and specific local mucosal immunoglobulin (Ig) A production against *Lawsonia intracellularis* in experimentally infected swine. *Can. J. Vet. Res.* **2010**, *74*, 97–101.
45. Cordes, H.; Riber, U.; Jensen, T.K.; Jungersen, G. Cell-mediated and humoral immune responses in pigs following primary and challenge-exposure to *Lawsonia intracellularis. Vet. Res.* **2012**, *43*, 9.
46. Nogueira, M.G.; Collins, A.M.; Donahoo, M.; Emery, D. Immunological responses to vaccination following experimental *Lawsonia intracellularis* virulent challenge in pigs. *Vet. Microbiol.* **2013**, *164*, 131–138.
47. Jones, G.F.; Ward, G.E.; Gebhart, C.J.; Murtaugh, M.P.; Collins, J.E. Use of a DNA probe to detect the intracellular organism of proliferative enteritis in swine feces. *Am. J. Vet. Res.* **1993**, *54*, 1585–1590.
48. Jones, G.F.; Ward, G.E.; Murtaugh, M.P.; Lin, G.; Gebhart, C.J. Enhanced detection of intracellular organism of swine proliferative enteritis, ileal symbiont intracellularis, in feces by polymerase chain reaction. *J. Clin. Microbiol.* **1993**, *31*, 2611–2615.
49. Moller, K.; Jensen, T.K.; Jorsal, S.E.; Leser, T.D.; Carstensen, B. Detection of *Lawsonia intracellularis, Serpulina hyodysenteriae*, weakly beta-haemolytic intestinal *spirochaetes, Salmonella enterica*, and haemolytic *Escherichia coli* from swine herds with and without diarrhoea among growing pigs. *Vet. Microbiol.* **1998**, *62*, 59–72.
50. Collins, A.; Love, R.J.; Pozo, J.; Smith, S.H.; McOrist, S. Studies on the *ex vivo* survival of *Lawsonia intracellularis. Swine Health Prod.* **2000**, *8*, 211–215.
51. Jacobson, M.; Lofstedt, M.G.; Holmgren, N.; Lundeheim, N.; Fellstrom, C. The prevalences of *Brachyspira* spp. and *Lawsonia intracellularis* in Swedish piglet producing herds and wild boar population. *J. Vet. Med.* **2005**, *52*, 386–391.
52. Echeverria, P.; Taylor, D.; Seriwatana, J.; Brown, J.E.; Lexomboon, U. Examination of colonies and stool blots for detection of enteropathogens by DNA hybridization with eight DNA probes. *J. Clin. Microbiol.* **1989**, *27*, 331–334.
53. Millar, D.S.; Withey, S.J.; Tizard, M.L.V.; Ford, J.G.; Hermon-Taylor, J. Solid-phase hybridization capture of low-abundance target DNA sequences: Application to the polymerase chain reaction detection of *Mycobacterium paratuberculosis* and *Mycobacterium avium* subsp. *silvaticum. Anal. Biochem.* **1995**, *226*, 325–330.
54. La, T.; Collins, A.M.; Phillips, N.D.; Oksa, A.; Hampson, D.J. Development of a multiplex-PCR for rapid detection of the enteric pathogens *Lawsonia intracellularis, Brachyspira hyodysenteriae*, and *Brachyspira pilosicoli* in porcine faeces. *Lett. Appl. Microbiol.* **2006**, *42*, 284–288.
55. Nathues, H.; Oliveira, C.J.B.; Wurm, M.; Beilage, E.G.; Givisiez, P.E.N. Simultaneous detection of *Brachyspira hyodysenteriae, Brachyspira pilosicoli* and *Lawsonia intracellularis* in porcine faeces and tissue samples by multiplex-PCR. *Zentralblatt Vet.* **2007**, *54*, 532–538.
56. Nathues, H.; Holthaus, K.; Beilage, E.G. Quantification of *Lawsonia intracellularis* in porcine faeces by real-time PCR. *J. Appl. Microbiol.* **2009**, *107*, 2009–2016.

57. Richter, B.; Ladinig, A.; Nedorost, N.; Weissenbock, H. A TaqMan quantitative polymerase chain reaction assay for the detection of *Lawsonia intracellularis* in fecal and tissue samples from pigs. *J. Vet. Diagn. Investig.* **2010**, *22*, 70–73.
58. Collins, A.M.; Fell, S.; Pearson, H.; Toribio, J.A. Colonisation and shedding of *Lawsonia intracellularis* in experimentally inoculated rodents and in wild rodents on pig farms. *Vet. Microbiol.* **2011**, *150*, 384–388.
59. Stahl, M.; Kokotovic, B.; Hjulsager, C.K.; Breum, S.O.; Angen, O. The use of quantitative PCR for identification and quantification of *Brachyspira pilosicoli*, *Lawsonia intracellularis* and *Escherichia coli* fimbrial types F4 and F18 in pig feces. *Vet. Microbiol.* **2011**, *151*, 307–314.
60. Pedersen, K.S.; Skrubel, R.; Stege, H.; Angen, O.; Stahl, M.; Hjulsager, C.; Larsen, L.E.; Nielsen, J.P. Association between average daily gain, faecal dry matter content and concentration of *Lawsonia intracellularis* in faeces. *Acta Vet. Scand.* **2012**, *54*, 58.
61. Johansen, M.; Nielsen, M.; Dahl, J.; Svensmark, B.; Baekbo, P.; Kristensen, C.S.; Hjulsager, C.K.; Jensen, T.K.; Stahl, M.; Larsen, L.E.; *et al.* Investigation of the association of growth rate in grower-finishing pigs with the quantification of *Lawsonia intracellularis* and porcine circovirus type 2. *Prev. Vet. Med.* **2013**, *108*, 63–72.
62. Collins, A.M.; Barchia, I.M. The critical threshold of *Lawsonia intracellularis* in pig faeces that causes reduced average daily gains. *Vet. Microbiol.* **2013**, in press.
63. Pedersen, K.S.; Angen, O.; Nielsen, J.P. Pooling of Faecal Samples for Quantitative Assessment of *Lawsonia intracellularis* Excretion Levels. In Proceedings of the 22nd International Pig Veterinary Society Congress, Jeju, Korea, 10–13 June 2012; p. 284.
64. Collins, A.M.; Barchia, I.M. Optimal Pooling of Faeces to Quantify *Lawsonia intracellularis* in Clinically and Sub-Clinically Affected Pigs. In *Manipulating Pig Production*; Pluske, J.A.J., Ed.; Australasian Pig Science Association: Werribee, Australia, 2013; *XIV*, in press.
65. Collins, A. Efficacy of Antibiotics to Terminate Faecal Shedding of *Lawsonia intracellularis*. In Proceedings of the 22nd International Pig Veterinary Society Congress, Jeju, Korea, 10–13 June 2012; p. 675.
66. Love, R.J. Gastro Intestinal Disorders of Growers. In *Gastroenteric Conditions Seminar*; Post-Graduate Committee in Veterinary Science, Ed.; The University of Sydney: Sydney, Australia, 1981; pp. 397–401.
67. McOrist, S.; Smith, S.H.; Shearn, M.F.H.; Carr, M.M.; Miller, D.J.S. Treatment and prevention of porcine proliferative enteropathy with oral Tiamulin. *Vet. Rec.* **1996**, *139*, 615–618.
68. Schwartz, K.J.; Knittel, J.P.; Walter, D.; Roof, M.; Anderson, M. Effect of oral tiamulin on the development of porcine proliferative enteropathy in a pure-culture challenge model. *Swine Health Prod.* **1999**, *7*, 5–11.
69. Moore, G.M.; Zimmermann, A.G.; Armbruster, G.A. Efficacy of Tylosin Phosphate Administration in Feed for the Prevention and/or Control of Porcine Proliferative Enteropathies (PPE) in Commercial Swine Herds. In Proceedings of the 28th Annual Meeting of the American Association of Swine Practitioners, Quebec City, Canada, 1–4 March 1997; pp. 57–58.
70. McOrist, S.; Morgan, J. ; Veenhuizen, M.F.; Lawrence, K.; Kroger, H.W. Oral administration of Tylosin phosphate for treatment and prevention of proliferative enteropathy in pigs. *Am. J. Vet. Res.* **1997**, *58*, 136–139.
71. Winkelman, N.L.; Cornell, P.; Bradford, J. Lincomycin Feed Medication and Two Water Medications Against Ileitis Caused by *Lawsonia intracellularis*. In Proceedings of 29th Annual Meeting of the American Association of Swine Practitioners, Des Moines, IA, USA, 7–10 March 1998; pp. 195–199.
72. Winkelman, N.L.; Gebhart, C.; Collins, J.; King, V.; Hannon, M.; Wolff, T. An Evaluation of Aureomycin Chlortetracycline (CTC) Granular Feed Additive for Prevention or Treatment of Swine Ileitis. In Proceedings of the 28th Annual Meeting of the American Association of Swine Practitioners, Quebec City, Canada, 1–4 March 1997; pp. 79–83.
73. Pauling, G.E.; Paradis, M.A.; Dick, C.P.; Brennan, J.; Wilson, J. Evaluation of Tylosin Tartrate in Drinking Water for the Treatment of Porcine Proliferative Enteropathy. In Proceedings of the 30th Annual Meeting of the American Association of Swine Practitioners, St. Louis, MO, USA, 28 Feberary–2 March 1999; pp. 119–121.

74. Vu, N.Q.; Collins, A.M.; van Dijk, M.; Love, R. Effect of Antibiotic Medication on the Development of Immunity to *Lawsonia Intracellularis* in Pigs. In *Manipulating Pig Production*; Cranwell, P.D., Ed.; Australasian Pig Science Association: Werribee, Australia, 2001; p. 148.
75. Collins, A.M.; van Dijk, M.; Vu Ngoc, Q.; Pozo, J.; Love, R.J. Immunity to *Lawsonia intracellularis*. In Proceedings of the Allen D Leman Swine Conference, Minneapolis, MN, USA, 16 September 2001; pp. 67–69.
76. Love, R.J.; Love, D.N. Control of proliferative haemorrhagic enteropathy in pigs. *Vet. Rec.* **1977**, *100*, 473.
77. Riber, U.; Cordes, H.; Boutrup, T.S.; Jensen, T.K.; Heegaard, P.M.H.; Jungersen, G. Primary infection protects pigs against re-infection with *Lawsonia intracellularis* in experimental challenge studies. *Vet. Microbiol.* **2011**, *149*, 406–414.
78. Collins, A.M.; van Dijk, M.; McOrist, S.; Love, R.J. Strategic Medication and Development of Immunity to *Lawsonia intracellularis*. In Proceedings of the 16th International Pig Veterinary Society Congress, Melbourne, Australia, 17–21 September 2000; Cargill, C., McOrist, S., Eds.; IPVS: Melbourne, Australia, 2000; p. 30.
79. Flø, H.; Bock, R.; Oppengaard, O.J.; Bergsjø, B.; Lium, B. An Attempt to Eradicate *Lawsonia Intracellularis and Brachyspira* sp. from Swine Herds. In Proceedings of the 16th International Pig Veterinary Society Congress, Melbourne, Australia, 17–21 Sepetember 2000; p. 66.
80. Collins, A. Eradication of Proliferative Enteropathy, Current Limitations in Our Knowledge. In Proceedings of the Australian Pig Veterinarians, Melbourne, Australia, 22–24 October 2006; pp. 47–52.
81. Johansen, M.; Baekbo, P.; Jensen, T.K.; Møller, K.; Nielsen, V.R. Attempt to Eradicate *Lawsonia intracellularis* by Medication in 9 Sow Herds—Preliminary Results. In Proceedings of the 17th International Pig Veterinary Society Congress, Ames, IA, USA, 2–5 June 2002.
82. Ellegaard, B.; Koerting, K.; Szancer, J. Attempt to Eradicate *Lawsonia intracellularis* from an SPF Farm by Medication. In Proceedings of the 20th International Pig Veterinary Society Congress, Durban, South Africa, 22–25 June 2008; p. 152.
83. Bane, D.P.; Neumann, E.; Gebhart, C.J.; Gardner, I.A.; Norby, B. Porcine proliferative enteropathy: A case-control study in swine herds in the United States. *J. Swine Health Prod.* **2001**, *9*, 155–158.
84. Bronsvoort, M.; Norby, B.; Bane, D.P.; Gardner, I.A. Management factors associated with seropositivity to *Lawsonia intracellularis* in US swine herds. *J. Swine Health Prod.* **2001**, *9*, 285–290.
85. Collins, A.M.; Love, R.J. Risk Factors Associated with *Lawsonia intracellularis* Infection. In *Manipulating Pig Production*; Paterson, J.E., Ed.; Australasian Pig Science Association: Werribee, Australia, 2003; p. 34.
86. Holyoake, P.K.; Thomson, P.; Hamilton, D.; Collins, A.M. Ecology of *Lawsonia intracellularis* infection across rearing systems in Australia. In Proceedings of the 21st International Pig Veterinary Society Congress, Vancouver, Canada, 18–21 July 2010; p. 688.
87. Pearce, G.P. Epidemiology of enteric disease in grower and finisher pigs: A postal survey of pig producers in England. *Vet. Rec.* **1997**, *144*, 338–342.
88. Wattanaphansak, S.; Singer, R.S.; Gebhart, C.J. Evaluation of *in vitro* bactericidal activity of commercial disinfectants against *Lawsonia intracellularis. J. Swine Health Prod.* **2010**, *18*, 11–17.
89. Kroll, J.J.; Roof, M.; McOrist, S. Evaluation of protective immunity in pigs following oral administration of an avirulent live vaccine of *Lawsonia intracellularis. Am. J. Vet. Res.* **2004**, *65*, 559–565.
90. Nogueira, M.G.; Collins, A.M.; Emery, D. The Immune Response to a Ten-Times Oral or Intramuscular Immunisation with *Lawsonia* Vaccine in Weaner Pigs. In *Manipulating Pig Production XIII*; Australian Pig Science Association: Werribee, Australia, 2011; p. 47.
91. Holyoake, P.K.; Collins, A.; Donahoo, M.; Lising, R.; Emery, D. Identifying obstacles to reducing the use of antibiotics to control porcine proliferative enteropathy. *Aust. Vet. J.* **2009**, *87*, 33–34.

92. Lawson, G.H.K.; Rowland, A.C.; Roberts, L.; Fraser, G.; McCartney, E. Proliferative haemorrhagic enteropathy. *Res. Vet. Sci.* **1979**, *27*, 46–51.
93. Charles, S.; Hussain, I.; Choi, C.; Nagaranja, K.; Sivanandan, V. Adjuvanted subunit vaccines for the control of *Salmonella enteritidis* in turkeys. *Am. J. Vet. Res.* **1994**, *55*, 636–642.
94. Dale, C.J.H.; Strugnell, R.A.; Lew, A.M.; Hasse, D.; Sinistaj, M.; Panaccio, M. Vaccination against Proliferative Enteropathy in Pigs. In *Manipulating Pig Production VI*; Cranwell, P.D., Ed.; Australasian Pig Science Association: Werribee, Australia, 1997; p. 182.
95. Jacobs, A.; Vermeij, P.; Sefers, R.; Schrier, C. Vaccine for Protection against *Lawsonia intracellularis, Mycoplasma hyopneumoniae* and Porcine Circo Virus. US Patent 6,921,536, 26 July 2005.
96. Paradis, M.A.; Gebhart, C.J.; Toole, D.; Vessie, G.; Winkelman, N.L.; Bauer, S.A.; Wilson, J.B.; McClure, C.A. Sub-clinical ileitis: Diagnostic and perfromance parameters in a multi-dose mucosal homogenate challenge model. *J. Swine Health Prod.* **2012**, *20*, 137–141.
97. Vannucci, F.A.; Foster, D.N.; Gebhart, C.J. Comparative transcriptional analysis of homologous pathogenic and non-pathogenic *Lawsonia intracellularis* isolates in infected porcine cells. *PLoS One* **2012**, *7*, e46708.
98. Mapother, M.E.; Joens, L.A.; Glock, R.D. Experimental reproduction of porcine proliferative enteritis. *Vet. Rec.* **1987**, *121*, 533–536.
99. Gogolowski, R.P.; Cook, R.W.; Batterham, E.S. Suboptimal growth associated with porcine intestinal adenomatosis in pigs in nutritional studies. *Aust. Vet. J.* **1991**, *68*, 406–408.
100. Holyoake, P.K.; Collins, A.; Mullan, B. Simulation of the Economic Impact of *Lawsonia intracellularis* Infection. In Proceedings of the 21st International Pig Veterinary Society Congress, Vancouver, Canada, 18–21 July 2010; p. 233.
101. Holyoake, P.K.; Mullan, B.P.; Cutler, R.S. Simulation of the economic impact of proliferative enteritis on pig production in Australia. *Aust. Vet. J.* **1996**, *73*, 89–92.
102. McOrist, S.; Smith, S.H.; Green, L.E. Estimate of direct financial losses due to porcine proliferative enteropathy. *Vet. Rec.* **1997**, *140*, 579–581.
103. Collins, A.M.; Fell, S.; van Straaten, J.; Bolsius, N.; Penrose, L. Increased Variation in Pig Weight and Reduced Feed Intake in Pigs Sub-Clinically Affected with Proliferative Enteropathy. In Proceedings of the 21st International Pig Veterinary Society Congress, Vancouver, Canada, 18–21 July 2010; p. 690.
104. Collins, A.M.; Fell, S.; van Straaten, J.; Bolsius, N. Reduced Muscle Growth in Pigs Sub-Clinically Affected with Proliferative Enteropathy. In Proceedings of the 21st International Pig Veterinary Society Congress, Vancouver, Canada, 18–21 July 2010; p. 234.
105. Jordan, D.M.; Knittel, J.P.; Schwartz, K.J.; Roof, M.; Hoffman, L.J. A *Lawsonia intracellularis* transmission study using a pure culture inoculated seeder-pig sentinel model. *Vet. Microbiol.* **2004**, *104*, 83–90.
106. Jordan, D.M.; Knittel, J.P.; Schmoll, E.M.; Schwartz, K.J.; Roof, M.B.; Larson, D.J.; Hoffman, L.J. A *Lawsonia intracellularis* Transmission Study Using a Pure Culture Inoculated Seeder-Pig Sentinel Model. In Proceedings of the American Association of Swine Practitioners, Quebec, Canada, 1–4 March 1997; pp. 243–248.
107. Jones, G.F.; Ward, G.E.; Murtaugh, M.P.; Rose, R.; Gebhart, C.J. Relationship between Ileal symbiont intracellularis and porcine proliferative enteritis. *Infect. Immun.* **1993**, *61*, 5237–5244.
108. Smith, S.H.; McOrist, S. Development of persistent intestinal infection and excretion of *Lawsonia intracellularis* by piglets. *Res. Vet. Science* **1997**, *62*, 6–10.
109. Stabel, T.J.; Fedorka-Cray, P.J. Effect of 2-deoxy-d-glucose induced stress on *Salmonella choleraesuis* shedding and persistence in swine. *Res. Vet. Sci.* **2004**, *76*, 187–194.
110. Jensen, T.K.; Møller, K.; Lindecrona, R.; Jorsal, S.E. Detection of *Lawsonia intracellularis* in the tonsils of pigs with proliferative enteropathy. *Res. Vet. Sci.* **2000**, *68*, 23–26.
111. Collins, A.M.; Fell, S.; Gonsalves, J.; Barchia, I. Immune Suppression does not Lead to a Carrier State in Pigs Previously Infected with *Lawsonia intracellularis.* In *Manipulating Pig Production XII*; Barneveld, R.J., Ed.; Australasian Pig Science Association: Werribee, Australia, 2009; p. 42.

112. Starek, M.; Bilkei, G. Sows seropositive to *Lawsonia intracellularis* (LI) influence performance and seropositivity of their offspring. *ACTA Vet. Brno* **2004**, *73*, 341–345.
113. Jensen, T.K.; Vigre, H.; Sorensen, V.; Moller, K. Naturally acquired *Lawsonia intracellularis* infection in pigs studied from weaning to slaughter by indirect immunofluorescence antibody test and polymerase chain reaction on faeces. *Res. Vet. Sci.* **2005**, *79*, 93–98.
114. Farmer, C.; Sorensen, M.T. Factors affecting mammary development in gilts. *Livest. Prod. Sci.* **2001**, *70*, 141–148.
115. Rooke, J.A.; Bland, I.M. The acquisition of passive immunity in the new-born piglet. *Livest. Prod. Sci.* **2002**, *78*, 13–23.
116. Bandrick, M.; Pieters, M.; Pijoan, C.; Molitor, T.W. Passive transfer of maternal *Mycoplasma hyopneumoniae*—Specific cellular immunity to piglets. *Clin. Vaccine Immunol.* **2008**, *15*, 540–543.
117. Pozo, J.; Collins, A.M.; Rubio, P.; Love, R.J. Maternal Immunity to *Lawsonia intracellularis* Infection. In Proceedings of the 16th International Pig Veterinary Society Congress, Melbourne, Australia, 17–21 Sepetember 2000; Cargill, C., McOrist, S., Eds.; Causal Productions Pty Limited: Melbourne, Australia, 2000; p. 108.
118. Chang, K.; Kurtz, H.J.; Ward, G.E.; Gebhart, C.J. The Use of Weaned Mouse as a Model to Study the Role of *Campylobacter* Species in Proliferative Enteritis of Swine. In Campylobacter *III: The Third International Workshop on* Campylobacter *Infections*; Pearson, A.D., Ed.; Public Health Laboratory Service: Ottawa, Canada, 1985; p. 102.
119. Vandenberghe, J.; Verheyen, A.; Lauwers, S.; Geboes, K. Spontaneous adenocarcinoma of the ascending colon in Wistar rats: The intracytoplasmic presence of a *Campylobacter*-like bacterium. *J. Comp. Pathol.* **1985**, *95*, 45–55.
120. Eriksen, K.; Landsverk, T.; Bratberg, B. Morphology and immunoperoxidase studies of intestinal adenomatosis in the blue fox, *Alopex lagopus*. *J. Comp. Pathol.* **1990**, *102*, 265–278.
121. Hotchkiss, C.E.; Shames, B.; Perkins, S.E.; Fox, J.G. Proliferative enteropathy of rabbits—The intracellular *Campylobacter*-like organism is closely related to *Lawsonia intracellularis*. *Lab. Anim. Sci.* **1996**, *46*, 623–627.
122. Williams, N.M.; Harrison, L.R.; Gebhart, C.J. Proliferative enteropathy in a foal caused by *Lawsonia intracellularis*-like bacterium. *J. Vet. Diagn. Investig.* **1996**, *8*, 254–256.
123. Cooper, D.M.; Swanson, D.L.; Gebhart, C.J. Diagnosis of proliferative enteritis in frozen and formalin-fixed, paraffin-embedded tissues from a hamster, horse, deer and ostrich using a *Lawsonia intracellularis*-specific multiplex PCR assay. *Vet. Microbiol.* **1997**, *54*, 47–62.
124. LeMarchand, T.X.; Tully, T.N.J.; Shane, S.M.; Duncan, D.E. Intracellular *Campylobacter*-like organisms associated with rectal prolapse and proliferative enteroproctitis in emus (*Dromaius novaehollandiae*). *Vet. Pathol.* **1997**, *34*, 152–156.
125. Tomanova, K.; Bartak, P.; Smola, J. Detection of *Lawsonia intracellularis* in wild pigs in the Czech Republic. *Vet. Rec.* **2002**, *151*, 765–767.
126. McOrist, S.; Blunt, R.; Gebhart, C.J. Pig-associated *Lawsonia intracellularis* in various on-farm dipterous fly stages. *J. Swine Health Prod.* **2011**, *19*, 277–283.
127. Collins, A.M.; Love, R.J.; Jasni, S.; McOrist, S. Attempted infection of mice, rats and chickens by porcine strains of *Lawsonia intracellularis*. *Aust. Vet. J.* **1999**, *77*, 120–122.
128. Go, Y.Y.; Lee, J.K.; Ye, J.Y.; Lee, J.B.; Park, S.Y.; Song, C.S.; Kim, S.K.; Choi, I.S. Experimental reproduction of proliferative enteropathy and the role of IFN-gramma in protective immunity against *Lawsonia intracellularis* in mice. *J. Vet. Sci.* **2005**, *6*, 357–359.
129. Murakata, K.; Sato, A.; Yoshiya, M.; Kim, S.; Watarai, M.; Omata, Y.; Furuoka, H. Infection of different strains of mice with *Lawsonia intracellularis* derived from rabbit or porcine proliferative enteropathy. *J. Comp. Pathol.* **2008**, *139*, 8–15.
130. Friedman, M.; Bednar, V.; Klimes, J.; Smola, J.; Mrlik, V. *Lawsonia intracellularis* in rodents from pig farms with the occurrence of porcine proliferative enteropathy. *Lett. Appl. Microbiol.* **2008**, *47*, 117–121.
131. De Viott, A.M.; Franca, S.A.; Vannucci, F.A.; Cruz Junior, E.C.C.; Costa, M.C.; Gebhart, C.J.; Guedes, R.M.C. Infection of sparrows (*Passer domesticus*) and different mice strains with *Lawsonia intracellularis*. *Pesqui. Vet. Bras.* **2013**, *33*, 372–378.

CHAPTER 5

Effect of spineless cactus intake (*Opuntia ficus-indica*) on blood glucose levels in lactating sows and its impact on feed intake, body weight loss, and weaning-estrus interval

Gerardo Ordaz-Ochoa[1], Aureliano Juárez-Caratachea[1], Rosa Elena Pérez-Sánchez[2], Rafael María Román[3], BravoRuy Ortiz-Rodríguez[4]

ABSTRACT

The effect of spineless cactus intake (*Opuntia ficus-indica*) on blood glucose (BG) levels in lactating sows and its impact on daily and total feed intake (dFI^{-1} and TFI, respectively), body weight loss (BWL), and weaning-estrus interval length (WEI) were evaluated. Thirty-four hybrid (Yorkshire × Landrace × Pietrain) sows in lactation phase were used. Sows were divided into two groups: G1 (n = 17) where they received commercial feed and G2 (n = 17) provided with commercial feed plus an average of 2.0 ± 0.5 kg spineless cactus, based on a sow's body weight. The variables evaluated were BG, dFI^{-1}, TFI, BWL, and WEI. Statistical analysis was performed by using a fixed and mixed model methodology, under a repeated measurements experiment. Group effects were found on all analyzed variables ($P < 0.05$). The BG was lower in G2 (55.2 and 64.5 mg/dL pre- and post-prandial, respectively), compared to that in G1 (70.9 and 80.1 mg/dL pre- and post-prandial, respectively) ($P < 0.05$). G2 showed better performance than G1 for dFI^{-1}, BWL, and WEI ($P < 0.05$) whose averages were 5.5 ± 1.8 kg, 7.4 ± 4.5%, and 5.3 ± 1.2 days, respectively. Averages for these variables in G1 were 4.7 ± 1.5 kg, 16.8 ± 4.6%, and 6.1 ± 1.6 days, respectively. Intake of spineless cactus reduced BG levels in lactating sows, generating greater dFI^{-1}, lower BWL at the end of lactation, and a lower WEI.

Keywords: Feeding, Hypophagia, Lactation, Glycaemia

1. INTRODUCTION

The feeding of the sow during lactation is among the most important indicators to maintain the productivity of the swine production systems (Segura et al., 2013; Xie et al., 2015). During this stage of production, the sows experience lactational physiologic hypophagia, which is associated with a gradual development of insulin resistance (Koketzu et al., 1998; Schenkel et al., 2010), and therefore, to an increase in blood glucose (Pére and Etienne, 2007; Klosterbuer et al., 2012). The increase of blood glucose in lactating sows causes alterations in appetite, reflected in a decrease in voluntary feed intake during the first week post-farrowing (Rigón et al., 2008), which is manifested in body weight loss of the sows and in the decrease of reproductive and productive post-lactation indicators (weaning) (Schenkel et al., 2010).

A sows' body weight loss caused by the lactational physiologic hypophagia is not only associated with the reduced voluntary feed intake but is also the result of energy imbalance and the removal of corporal reserves during lactation (Moreira et al., 2013). However, only if the loss of body weight is greater than 10% at the end of lactation, the reproductive processes subsequent to weaning are negatively affected (Cools et al., 2014); it mainly generates a delayed response in the resumption of ovarian activity and increases the weaning-estrus interval, reducing the fertility and prolificacy of the sows (Schenkel et al., 2010).

Given the above reasoning, nutritional alternatives for lactating sows are required to solve or minimize the effects of the lactational physiologic hypophagia and at the same time preserve animal health (Quesnel et al. 2009), production performance, and product quality (Tikabo et al., 2006).

Thus, the forage spineless cactus (*Opuntia* spp), for its hypoglycemic quality, high fiber content, and digestible energy (Alarcon et al., 2003; Pinos et al., 2010), can be a nonconventional alternative in the nutritional strategy of lactating sows, to counter the lactational physiologic hypophagia effect and to improve reproductive and productive activity after weaning. Therefore, the objective of this research was to evaluate the effect of spineless cactus intake (*Opuntia ficus-indica*) on blood glucose levels in lactating sows and its impact on daily and total feed intake, body weight loss, and weaning-estrus interval.

2. MATERIALS AND METHODS

This research was carried out at the Swine Unit of "La Posta Zootécnica" belonging to the Veterinary Medicine and Husbandry Faculty of Universidad Michoacana de San Nicolás de Hidalgo (FMVZ-UMSNH), Tarímbaro, Michoacán, México, located at km 9.5 Morelia-Zinapécuaro Road; 19° 46′ N, 101° 08′ W, and altitude of 1855 m (INEGI, 2010).

2.1 Animals, diets, and housing

Thirty-four hybrid sows (Yorkshire × Landrace × Pietrain), with 2.6 ± 1.5 average farrowing (1–7 farrowing range), were selected at random from the reproductive herd on the evaluated swine unit. The sows were served by natural mating with hybrid boars (Yorkshire × Pietrain) when presenting post-weaning estrus and were housed by groups (n = 7) in 16-m^2 pens during 108 days of gestation. All sows were fed 2.0 kg day^{-1} of commercial feed during the first two thirds of gestation (89.9% dry matter, 12.5% crude protein, 3.7% crude fat, 3.1% crude fiber, 10% crude ash, all on as-fed basis); of the last third, until the 108th day of gestation, sows were fed 2.5 kg day^{-1} divided in two rations: 08:00 and 14:00 h. The feed was provided in individual concrete feeders. The diet composition is shown in Table 1. The water supply was ad libitum through an automatic nipple drinker.

One week before the probable farrowing date (day 109 of gestation), the sows were randomly selected to form each one of the two groups (G) established in the experimental design: G1 (n = 17), which was offered commercial feed, and G2 (n = 17), sows with commercial feed plus spineless cactus. The sows were then transferred to the farrowing house where they were fed a lactation diet until the farrowing day (the ingredients and nutritional composition are shown in Table 1).

Table 1 Ingredients and nutrient composition of conventional gestation diet, conventional lactation diet (group 1), and experimental lactation diet (group 2)

Item	Gestation diet	Lactation diet	
Ingredient, g/kg		Group 1	Group 2
Sorghum	824.0	649.7	649.7
Soybean paste	60.0	100.0	100.0
Canola paste	61.5	185.3	185.3
Orthophosphate	11.8	5.4	5.4
Calcium carbonate	14.0	12.4	12.4
Soy oil	22.0	38.5	38.5
Lysine	1.2	2.5	2.5
Salt	4.0	4.0	4.0
Vitamin and mineral premix [a]	2.0	2.5	2.5
Nutrient composition for spineless cactus (*O. ficus-indica*)[b]			
Crude protein, %			5.6
Crude fat, %			0.2
Fiber, %			28.8
Humidity, %			88.6
Ash, %			24.5
Nitrogen-free elements, %			40.8
Mucilage, g 300 g^{-1} dry base			2.6
Nutrient composition[c]			
Metabolizable energy, Mcal/kg [d]	2.3	2.3	2.3
Crude protein, %	12.5	17.5	17.3
Crude fat, %	3.7	4.5	4.4
Fiber, %	3.1	4.3	4.7
Humidity, %	12.0	12.0	13.8
Ash, %	10.0	10.0	12.9

Calcium, % [d]	0.75	0.75	0.76
Phosphorus, % [d]	0.60	0.60	0.59
Lysine, % [d]	0.52	0.95	0.94
Met-Cist, % [d]	0.43	0.59	0.59

[a]Provided per kilogram of diet: Cu 30 mg; Fe 160 mg; Zn 160 mg; Mn 55 mg; Se 0.5; Cr 0.2 mg; vitamin A 14,200 IU; vitamin D_3 2800 IU; vitamin E 125 mg; vitamin K_3 5 mg; vitamin B_1 2.4 mg; vitamin B_2 8.7 mg; vitamin B_6 4.5 mg; vitamin B_{12} 0.05 mg; pantothenic acid 35 mg; folic acid 6 mg

[b] Supply of spineless cactus in fresh base was given once a day in the morning throughout the lactation phase. Quantity was 1% of the sow's pre-parity body weight

[c] In order to determine the nutritional composition of the diet supplemented with spineless cactus, 1% of spineless cactus was added in dry base to the conventional feed sample before bromatological analysis

[d] Chemical concentrations calculated using feed ingredient values from NRC (1998)

After farrowing, all sows in both groups were fed ad libitum during the 21 days of lactation. The only difference in the feeding of the post-farrowing sows was the addition of the spineless cactus (*O. ficus-indica*), as-fed basis (AF), to the diet of G2 (Table 1): commercial feed +1% of spineless cactus (according to the weight of the pre-partum sow). The corresponding spineless cactus quantity, added to the sow^{-1} diet, was on average 2.0 ± 0.5 kg spineless cactus. Due to the variety of the spineless cactus (no spikes), the cladodes were only fragmented into 3 × 2 cm pieces approximately and added to the corresponding diet at 08:00 h.

The sows in both groups were monitored and had the same husbandry practices during lactation, which was done in the farrowing house (maternity). This area has an installed capacity for six cages for farrowing and lactation; each cage has a bucket-type stainless steel feeder (44.5 cm/wide, 37.0 cm/height, and 33.0 cm/deep) and an automatic nipple drinker. The temperature in the maternity area remained constant (18 °C) during the experimental period (temperature for the sows and their litters); an automatic infrared heater Holme® brand with power from 750 to 1500 W, which was regulated at 18 °C. The ventilation was through curtains.

2.2 Experimental procedures

The variables evaluated sow^{-1} group^{-1} were blood glucose (BG), daily and total feed intake (dFI^{-1} and TFI, respectively), body weight loss (BWL), and weaning estrus-interval (WEI). BG was determined by a glucometer for human use (ACCU-CHEK Performan®), according to the methodology described by Perez et al. (2015b). Blood samples were taken at 8:00 h by right ear vein puncture on gestation days 85, 100, and 110 and lactation days 3, 7, 10, 14, 17, and 21, prior to a fasting period of 8 h. Feed intake was determined with a portable digital scale (model CRS5000, TOR REY®, Mexico) with capacity of 0.100 to 500 kg. Rejected feed sow^{-1} group^{-1}was weighed daily in the morning, prior to feeding, with the aim of determining the dFI^{-1} from the day before and TFI. The live weight of the sows when entering the farrowing house (7 days pre-farrowing) and at weaning (21 days post-farrowing) was obtained using a fixed electronic scale (STG-1500-T1500SL, OCONY®, México) with capacity from 1 to 1500 kg. The BWL (kg and %) was determined at the end of lactation through the following formulas:

$$\mathrm{BWL}_{\%} = 100 - \left(\frac{\text{Weight of the sow at weaning} \times 100}{\text{Weight of the sow pre farrowing}}\right).$$

2.3 Statistical analysis

The data was analyzed by ANOVA using the methodology of the fixed and mixed effects models (SAS, 2010). Data for sows BG and dFI^{-1} were analyzed using ANOVA for repeated measures experiment, with sow nested within group as a random effect and group, day, and group × day interaction, and the adjustment for the linear effect of the body weight pre-farrowing as fixed effects. The model for TFI, LBW, and WEI included group as a fixed effect and taking pre-farrowing sow body weight and litter size at weaning as covariates. Differences between means were estimated by using the least square means SAS statement setting $\alpha = 0.05$. Thus, the explicitly models used were:

$$Y_{\mathrm{ijkl}} = \mu + G_i + S(G)_{j(i)} + D_k + G \times D_{ik} + \beta_1 X_{1j} + \varepsilon_{\mathrm{ijkl}}$$

Where:

Y_{ijkl} = response variable: BG and dFI^{-1}; μ = constant common to the population; G_i = fixed effect of i-th group, with i = 1, 2; $S(G)_{j(i)}$ = random effect of the j-th sow, nested within the i-th group, with j = 1, 2,..., 17 and i = 1, 2; D_k = fixed effects of the k-th lactation day, with k = 1, 2, 3,..., 21; $G{\times}D_{ik}$ = fixed effects of interaction of i-th group in k-th lactation day; X_{1j} = effect of covariate pre-farrowing body weight of j-sow; ε_{ijkl} = random effect associated with each observation (~NID = 0, σ^2_e).

$$Y_{ij} = \mu + G_i + \beta_1 X_{1j} + \beta_2 X_{2j} + \varepsilon_{ij}$$

Where:

Y_{ij} = response variable: TFI, BWL, and WEI; μ = constant common to the population; G_i = fixed effects of the i-th group, with i = 1, 2; X_{1j} = effect of the covariate pre-parity body weight for j-sow; X_{2j} = effect of the covariate litter weight at weaning for j-sow; ε_{ij} = random effect associated with each observation (~NID = 0, σ^2_e).

3. RESULTS

There was no group effect on BG levels pre- and post-prandial during the last third of gestation (P = 0.085): 70.3 ± 7.2 vs 75.2 ± 7.9 mg/dL for the control group (G1) and experimental group (G2), respectively. The post-prandial BG levels were 79.8 ± 8.2 mg/dL for G1 and 83.2 ± 6.7 mg/dL for G2. However, during lactation, there were group effects on BG levels pre- and post-prandial (P < 0.001). Pre-prandial BG levels in G1 were greater during lactation (P < 0.05): 70.9 ± 8.2 mg/dL in comparison with 55.2 ± 8.5 mg/dL from G2. The post-prandial BG levels showed the same effect; it was greater in G1 than in G2 (P < 0.05; Table 2).

BG level reduction in G2 sows did not show immediately; interaction group × day was highly significant (P < 0.001). At 24-h post-administration of the spineless cactus, a 7.9% reduction of the pre-prandial BG was observed with respect to the pre-prandial BG levels of the sows in G1 (66.4 mg/dL); at 168 h, the reduction was 24.7%, and at 336 h, the greatest reduction of pre-prandial BG was observed (35.3%) in comparison to that of G1 (Fig. 1). The analysis by orthogonal polynomials indicated that the response in both groups can be described by a second-degree polynomial, and the estimated prediction equations for each group were G1 Yˆi=68.5152+ 1.0139X−0.0527X2Y^i=68.5152+1.0139X−0.0527X2 and G2 Yˆi=66.5247−2.7426X+0.1118X2Y^i=66.5247−2.7426X+0.1118X2. Deriving both equations and equaling to 0, the critical points for each group were 9.6 for G1 and 12.2 for G2; this implies that the sows in G1 show the greatest level of pre-prandial BG around day 9.6 and the sows in G2 show the lowest level of pre-prandial BG around day 12.2.

Table 2 Least squares mean for blood glucose levels pre- and post-prandial according to lactation interaction group × day

Day	Control group (G1)		Group fed spineless cactus (G2)		Contrast *P* value*	
	BG pre-pandrial	BG post-pandrial	BG pre-pandrial	BG post-pandrial	1	2
85 to 110[&]	$70.3^{1} \pm 7.2$	$79.8^{1} \pm 8.2$	$75.2^{1} \pm 7.9$	$83.2^{1} \pm 6.7$	<0.0001	<0.0001
1 to 7[†]	$72.0^{a1} \pm 8.3$	$81.5^{a1} \pm 9.0$	$59.7^{a2} \pm 5.4$	$67.2^{a2} \pm 12.1$	<0.0001	<0.0001
8 to 14[†]	$71.1^{a1} \pm 8.5$	$81.4^{a1} \pm 10.6$	$46.4^{a2} \pm 7.5$	$58.7^{b2} \pm 5.7$	<0.0001	<0.0001
15 to 21[†]	$69.2^{a1} \pm 7.7$	$76.9^{a1} \pm 12.9$	$57.1^{a2} \pm 7.1$	$66.1^{a2} \pm 9.5$	<0.0001	<0.0001

[&]Gestation phase

[†] Lactation phase

[a,b]Different letters indicate statistical difference ($P < 0.05$) within the column

[1,2]Different numerals indicate statistical difference ($P < 0.05$) between groups for BG pre-prandial and post-prandial, respectively

*Contrast: 1, BG pre-prandial group 1 vs BG pre-prandial group 2; 2, BG post-prandial group 1 vs BG post-prandial group 2

There was no group effect regarding the dFI^{-1} per sow during the last third of gestation ($P = 0.090$). During lactation, the feed intake increased ($P < 0.05$) in sows of G2 (4.4 ± 1.7 kg day^{-1}) in the first week post-farrowing in comparison with the dFI^{-1} from the sows of G1 (Table 3). The average dFI^{-1} during the 21 days of lactation were 4.7 ± 1.5 and 5.5 ± 1.8 kg for sows in G1 and G2, respectively ($P < 0.05$). Regarding the highest dFI^{-1} for each group, their averages were 5.2 and 6.6 kg for G1 and G2 groups, respectively (Fig. 2). Moreover, increase of dFI^{-1} was not linear during the 21-day lactation period, since there was a decline in voluntary feed intake at the beginning of the third lactation week (Fig. 2).

The increase of dFI^{-1} sow^{-1} $week^{-1}$ observed in G2 impacted TFI during the whole lactation phase, with a total of 126.6 kg of commercial feed in 21 days, resulting in 28.2 kg more of commercial feed compared to sows of G1 (Fig. 3). Moreover, G2 sows had 31.5 ± 7.4 kg of spineless cactus in fresh base in addition to the commercial feed consumed during the 21-day lactation phase, resulting in a TFI 158.1 kg sow^{-1}.

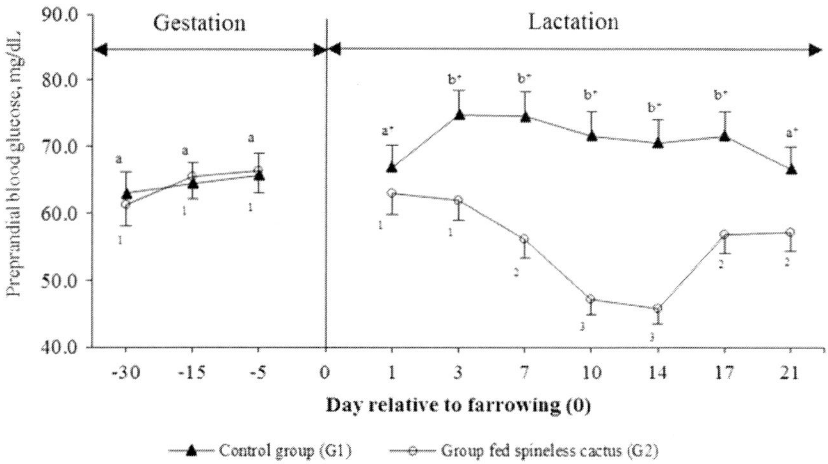

Figure. 1 Least squares means for pre-prandial blood glucose levels (mean ± SD) according to the lactation day in sows fed a conventional diet (*G1*, *n* = 17) and sows fed a conventional diet supplemented with spineless cactus (*G2*, *n* = 17). *Different letters* indicate statistical difference ($P < 0.05$) between day of lactation for *G1*. *Different numerals* indicate statistical difference ($P < 0.05$) between days of lactation for *G2*. †Statistical difference ($P < 0.05$) between groups

Table 3 Least squares means for average feed intake day^{-1} for lactating sows according to group

	Control group (G1)	Group fed spineless cactus (G2)					*P* value*
	CF	CF	CF + SC *FB*	CF + SC *DB*	SC *FB*	SC *DB*	
Day 85 to 110&	2.5^{a}	2.5^{a}	–	–	–	–	0.087
Day 1 to 7†	3.8 ^{a1} ± 1.5	4.4 ^{b1} ± 1.7	6.2^{1} ± 2.2	4.6^{1} ± 1.8	1.7^{1} ± 0.6	0.20^{1} ± 0.06	<0.001
Day 8 to 14†	5.1 ^{a2} ± 1.3	5.9 ^{b2} ± 1.9	7.5^{2} ± 2.1	6.1^{2} ± 1.9	1.9^{1} ± 0.11	0.19^{1} ± 0.07	<0.001
Day 15 to 21†	5.4 ^{a2} ± 1.4	6.2 ^{b2} ± 1.6	7.7^{2} ± 1.6	6.3^{2} ± 1.6	1.5^{1} ± 0.7	0.17^{1} ± 0.10	<0.001
General average	4.7 ^{a} ± 1.5	5.5 ^{b} ± 1.8	7.1 ± 1.9	5.7 ± 1.8	1.6 ± 0.6	0.19 ± 0.09	<0.001

CF commercial fed, *SC* spineless cactus, *BH* fresh base, *BS* dry base

&Gestation phase

† Lactation phase

[a,b]Different letters indicate statistical differences ($P < 0.05$) for the intake CF/group

[1,2]Different numerals indicate statistical differences ($P < 0.05$) within the column

*Probability for the group effect on CF only

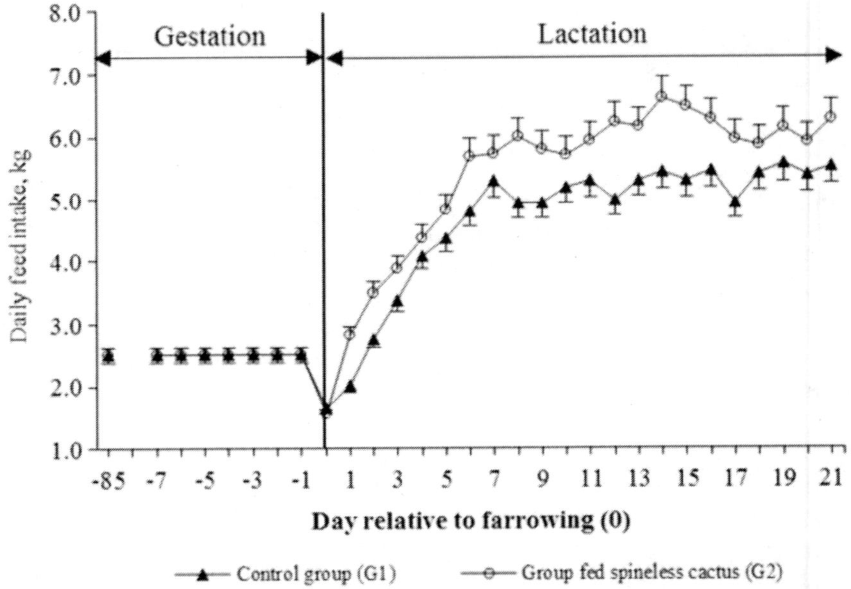

Figure. 2 Least squares means for voluntary feed intake (mean ± SD) during the last third of gestation (30 days pre-farrowing) and lactation (21 days post-farrowing) in sows fed with conventional diet (*G1*, $n = 17$) and sows fed a diet supplemented with spineless cactus (*G2*, $n = 17$)

The sows that consumed spineless cactus during lactation (G2) showed a lower percentage of BWL in comparison to sows of G1 ($P < 0.05$; Table 4). It could also be established that the sows with a lower BWL (G2) showed a lower WEI compared to those in G1 ($P < 0.05$): 5.3 ± 1.2 and 6.1 ± 1.6 days, respectively. Finally, the association between TFI and WEI was $r = 0.36$; ($P < 0.001$), and that of spineless cactus intake and WEI was $r = -0.40$; ($P < 0.001$). And the regression estimator of spineless cactus intake on TFI was $\beta_1 = 1.1$ kg; ($P < 0.05$); whereas that of spineless cactus intake on WEI was $\beta_1 = -0.03$ days; ($P < 0.05$), and the linear regression of TFI on WEI was $\beta_1 = -0.06$ days; ($P < 0.05$). It can be inferred that for each kilogram of spineless cactus intake during lactation, the TFI increased in 1.1 kg, which originates in a reduction of the WEI in 0.3 days for each kilogram of feed consumed and of 0.6 days per each kilogram of spineless cactus intake. Therefore, the sows that consumed spineless cactus not only showed a lower WEI ($P < 0.05$), but also the optimal WEI is achieved (7 days) with less TFI due to the extra energy intake provided by the spineless cactus (Fig. 4).

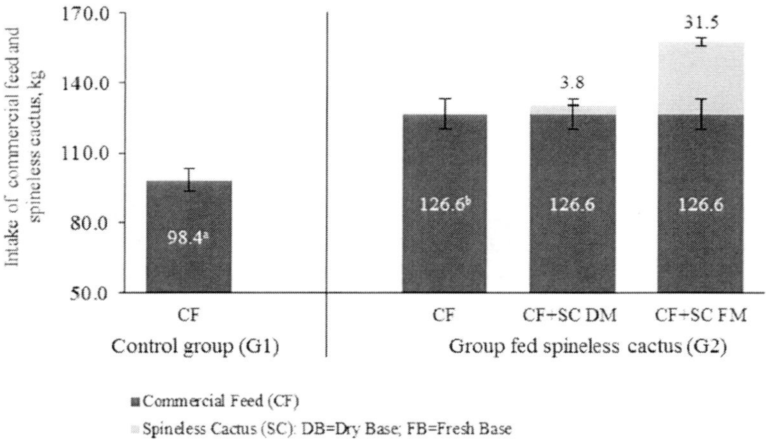

Figure. 3 Least squares means for commercial feed (*CF*) total intake by sows fed conventionally (*G1*, $n = 17$) and total intake of commercial feed plus spineless cactus (*CF + SC*), in dry and fresh base (*DM* and *FM*, respectively) for sows fed a commercial diet plus spineless cactus (*G2*, $n = 17$), during lactation phase (21 days). Mean ± SD. *Different letters* indicate statistical difference ($P < 0.05$) between groups regarding *CF* total intake

Table 4 Least squares means for sows' body weight loss during lactation phase and weaning estrus-interval according to the group

	Control group (G1)	Group fed spineless cactus (G2)	*P* value
Body weight loss, kg	$25.7^{a} \pm 11.3$	$16.8^{b} \pm 10.0$	<0.0001
Body weight loss, %	$16.8^{a} \pm 4.6$	$7.4^{b} \pm 4.5$	<0.0001
Weaning-estrus interval, days	$6.1^{a} \pm 1.6$	$5.3^{b} \pm 1.2$	<0.0001

[a,b]Different letters indicate statistical difference ($P < 0.05$) within the row

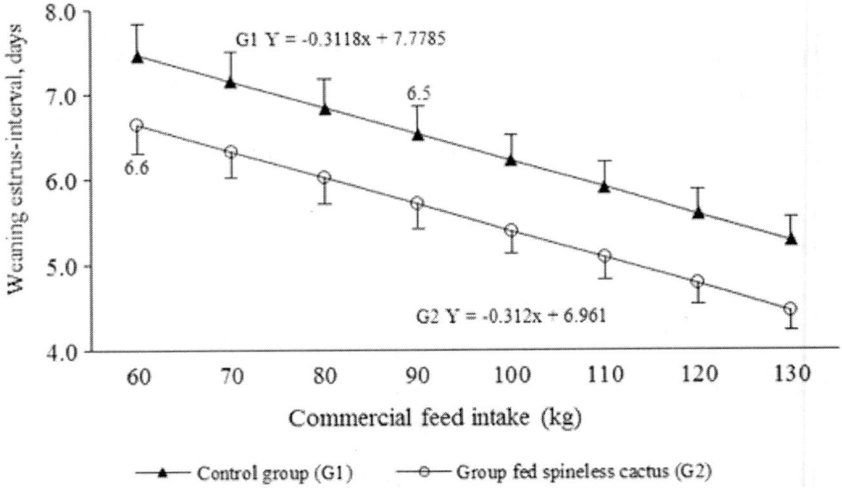

Figure. 4 Commercial feed intake (kg) linear effect during lactation phase (21 days) on weaning estrus-interval (mean ± SD) in sows fed conventionally (*G1*, n = 17) and sows whose diet was supplemented with spineless cactus (*G2*, n = 17)

4. DISCUSSION

The reduced BG levels in lactating sows fed the diet supplemented with spineless cactus (Table 2) is consistent with the effects observed in humans and other animal species after consuming spineless cactus (Alarcón et al., 2003; Halmi et al., 2013). In rabbits, it has been documented that spineless cactus intake reduces glucose levels up to 21.2% ($P < 0.05$) (Halmi et al., 2013); apparently, the hypoglycemic effect of the cactus is due to the dietary fiber, especially pectin, which causes a decrease in carbohydrate absorption by the pectin-gel formation (Sánchez et al., 2008). In addition, diets high in dietary fiber increase microbial enzymatic activity, affecting the activity of α-glucosidase and β-galactosidase, resulting in the inhibition of the glycosidic bond hydrolysis (Igho et al., 2015). Nevertheless, maximum reduction (35.3%) in pre-prandial BG levels in G2 sows was not observed until 336-h post-initiation of the treatment with spineless cactus (Fig. 1); these results show the possibility of modifying the lactational physiological hypophagia effect and increased feed intake in lactating sows by decreasing BG levels within the first 12 days of post-parity through the addition of spineless cactus to the diet.

A decrease in BG levels, due to spineless cactus intake, may be a consequence of an increase in insulin sensitivity due to the suppression of hepatic glucose production (Andrade y Wiedenfeld, 2011). Possibly, the nonfermentable digestible dietary fiber of the spineless cactus increases the release of intestinal peptides similar to glucagon-1 (GLP-1) (Chen et al., 2014), causing an increase of insulin production, inhibiting the release of glucagon, and influencing the renewal of intestinal epithelial cells (Jasleen et al., 2002). Additionally, the high calcium content (0.59 mg/100 g) of the cactus (Pinos et al., 2010) could stimulate insulin secretion by closing the K^+/ATP channels and depolymerize the membrane because of the increase of Ca^{2+} in the plasma membrane of the cells: the main key to insulin release (Pari y Latha, 2005).

The metabolic route of the lactational physiologic hypophagia is associated with the increase of the hepatic gluconeogenesis and the increase, from lipolysis, in the formation of nonsteroidal fatty acid precursors of leptin (Paspala et al., 2012). However, the increase of leptin post-parity originates hypophagia due to the following: (1) blocking of the ghrelin receptor at the gastric level and (2) blocking the orexigenic action of the neuropeptides-*Y* and agouti protein type at the hypothalamic level (Ylonem et al., 2003), whose effect is more evident during the first week of post-parity (Pére y Etienne, 2007; Segura et al., 2013), which is consistent with the results found in this research (Table3).

In addition to the hypoglycemic effect of the spineless cactus, its high fiber content (300 g/kg) (Pinos et al., 2010) could have possibly caused (i) the capturing of cholesterol and triglycerides (appetite regulators) and (ii) greater gastric distension on the sows, which allowed greater daily feed intake (Fig. 2) and increased ($P < 0.05$) TFI during lactation (Fig. 3).

On the contrary, in G1, the absence of a factor that decreases the BG levels, cholesterol, low-density lipoproteins, and triglycerides in the blood during lactation maintained active metabolic mechanisms that cause lactational hypophagia (Paspala et al., 2012) and its negative effects on voluntary feed intake (Rigón et al., 2008). The fact that the increase in dFI^{-1} of sows in G2 was not lineal during the 3 weeks of lactation (Fig. 2) is that on the 15th day of lactation, post-farrowing ovarian activity initiates, which is characterized by the increase of reproductive hormones: follicle stimulating hormone (FSH), luteinizing hormone (LH), and estrogens (E2); these hormones reduce dFI^{-1} by blocking the chemical mediators (neuropeptides-*Y*) with orexigenic effect at the hypothalamic level (Pére y Etienne, 2007; Barb et al., 2008). Thus, the decrease in dFI^{-1} at the beginning of the third week of lactation can be considered physiologically as normal.

The dFI^{-1} of the lactating sow has a direct repercussion on the TFI at the end of lactation. However, there are differences regarding the TFI of the sows during lactation, since previous researches report a TFI ranging between 92.8 (Pérez et al., 2015a) and 103.3 kg (Cools et al., 2014) of feed in 21-day lactations. This variable (TFI) is affected primarily by the sows' genotype, dFI^{-1} during gestation, corporal condition at farrowing, feeding frequency, environmental temperature, water availability, age, and metabolic physiology of the sow during lactation (Pére y Etienne, 2007; Olsson et al.,2011).

The total spineless cactus intake from the sows in G2 is an important aspect to consider (31.5 ± 7.4 kg AF) and can establish the gastric capacity during lactation (Table 3). In addition, this spineless cactus intake (1.6 ± 0.5 kg day^{-1} sow^{-1}) was able to counteract the negative effects of lactational hypophagia by stimulating a greater feed intake (12.4%) and decreasing the BWL in sows of G2 during lactation (Table4).

It is possible that the spineless cactus, besides its hypoglycemic effects, can improve the digestive processes of lactating sows, since it has been observed that the greater live weight gain in animals fed diets complemented with cactus is due to the high soluble carbohydrate content (53.9%) (Tikabo et al.,2006). Additionally, the nonstarch-soluble polysaccharides, present in the cactus, increase the viscosity of the alimentary bolus (Chen et al., 2014) and inhibit the phosphodiesterase in the intestinal smooth muscle (Baldassano et al., 2010), causing the reduction of gastric emptiness and the transit rate of the nutritional content through the gastrointestinal tract (Wang et al., 2003), generating greater nutrient absorption (Le Goff et al., 2002). Also the nonstarch polysaccharides, when undergoing fermentation by the microbiota of the colon, lead to a greater production of volatile fatty acids, including acetate, propionate, butyrate, and valeric acid (Cani et al.,2006), metabolites intended to supply energy to the organism (Chen et al.,2014). This longer time of the feed in the gastrointestinal tract, triggered by the spineless cactus intake, could generate lower BWL of the sows during lactation (Table 4). The incorporation of fermentable nonstarch polysaccharides modifies the activity of the intestinal microbiota and makes digestibility more efficient, improving with the weight and age of the swine (Rajesh y Berrocoso,2016).

In this regard, it has been determined that a BWL greater than 10% during lactation affects subsequent productive indicators, like the increase in the WEI, repeat-service percentage, and litter size reduction in the next farrowing (Cools et al., 2014). This agrees with the results from this research, especially of the WEI (Table 4), where the sows with lower BWL (G2) showed lower WEI. The reduction of the WEI of the sows from G2 was possibly due to the increased carbohydrates in the diet provided by the spineless cactus; as it has been observed that the implementation of glycogenic treatments trigger estrus (Viñoles et al., 2008); and combined with the energy content of the cactus (13.0 MJ/kg) (Nefzaoui y Ben Salem,

2002), the fiber content and its hypoglycemic effect could have positively affected the follicular dynamic of the ovaries (Sakly et al.,2013).

In the reproductive processes of the sows, the energy balance during lactation is essential for the resumption of reproductive activity post-weaning and the insulin appears to be the main modulator post-partum between the nutritional balance and the energy removed for ovarian reactivation (Sakly et al., 2013). Therefore, the intake of spineless cactus could be responsible for the increase in feed intake (Table 2), due to the suppression of hepatic glucose production (Viñoles et al., 2008) contributing to improve energy balance and minimize the BWL (Table 4) whose effect was reflected in a lower WEI (Fig.4).

5. CONCLUSION

The addition of 1% of spineless cactus (according to the live weight of the sows) to the diet of lactating sows reduces blood glucose levels during the first 2 weeks of lactation, generating increased voluntary feed intake and reduction in body weight loss in lactating sows, and its consequence is reflected in a reduction of the weaning-estrus interval.

ACKNOWLEDGEMENTS

The authors want to thank the Facultad de Medicina Veterinaria y Zootecnia-UMSNH for their support in conducting this research and CONACYT for the funding provided.

REFERENCES

1. Alarcón, A.F.J., Valdez, A.A., Xolalpa, M.S., Banderas, D.T., Jiménez, E.M., Hernández, GE., Román, R.R. 2003. Hypoglycemic activity of two polysaccharides isolated from *Opuntia ficus-indica* and *Opuntia streptacantha.* Proc West Pharmacol Soc. 46, 139–142.
2. Andrade, C.A., Wiedenfeld, H. 2011. Anti-hyperglycemic effect of *Opuntia streptacantha Lem.* J. Ethnopharmacol. 133, 940–943.
3. Baldassano, S., Tesoriere, L., Rotondo, A., Livrea, M.A., Serio, R., Mula, R. 2010. The inhibition of the mechanical activity of the Cactus mouse ileum (*Opuntia ficus indica, L, Mill.*) Extract of the fruit and its pigment indicaxanthin. J Agric Food Chem, 58, 7565–7571.
4. Barb, C.R., Hausman, G.H., Lents, C.A. 2008. Energy Metabolism and Leptin: Effects on Neuroendocrine Regulation of Reproduction in the Gilt and Sow. Reprod. Dom. Anim. 43(2), 324–330.

5. Cani, P.D., Knauf, C., Iglesias, M.A., Drucker, D.J., Delzenne, N.M., Burcelin, R. 2006. Improvement of glucose tolerance and hepatic insulin sensitivity by oligofructose requires a functional glucagon-like peptide 1 receptor. Diabetes. 55, 1484–1490.
6. Chen, X.B., Mao, L.Q., Che, B., Yu, J., He, J., Yu, G.Q., Han, Z.Q., Huang, P., Zheng, D.W., 2014. Impact of fiber types on gut microbiota, gut environment and gut function in fattening pigs. Anim. Feed Sci. Technol. 195, 101–111.
7. Cools, A., Maesb, D., Decaluwéa, R., Buysec, J., Kempend, T.A.G.T., Liesegange, A., Janssens, G.P.J. 2014. Ad libitum feeding during the peripartal period affects body condition, reproduction results and metabolism of sows. Anim. Feed Sci. Technol. 145, 130–140.
8. Halmi, B.S., Benlaksira, B., Bechtarzi, K., Berouel, K., Serakta, M., Richi, F., Djaalab, H., Maameri, Z., Djerrou, Z., Hamdipacha, Y. 2013. Pharmaco-toxicological study of *Opuntia ficus indica* L. aqueous extract in experimental animal. Int. J. Med. Arom Plants. 3(3), 375–381.
9. Igho, J., Onakpoya, M.D., Jack, O.S., Carl, J.H. 2015. The effect of cactus pear (*Opuntia ficus-indica*) on body weight and cardiovascular risk factors: A systemic review and meta-analysis of randomized clinical trials. Nutrition. 31, 640–646.
10. INEGI: Instituto Nacional de Estadística Geografía e Informática. Anuario Estadístico del Estado de Michoacán. 2010; 127–142.
11. Jasleen, J., Ashley, S.W., Shimoda, N., Zinner, M.J., Whang, E.E., 2002. Glucagon-like peptide 2 stimulates intestinal epithelial proliferation in vitro. Digest. Dis. Sci. 47, 1135–1140.
12. Klosterbuer, A.S., Thomas, W., Slavin, J.L. 2012. Resistant starch and pullulan reduce postprandial glucose, insulin, and GLP-1, but have no effect on satiety in healthy humans. J. Agric. Food Chem. 60, 11928–11934.
13. Koketsu, Y., Dial, G., Pettigrew, J., Xue, J., Yang, H., Lucia, T. 1998. Influence of lactation length and feed intake on reproductive performance and blood concentrations of glucose, insulin and luteinizing hormone in primiparous sows. Anim. Repro. Sci. 52, 153–163.
14. Le Goff, G., Van Milgen, J., Noblet, J. 2002. Influence of dietary fiber on digestive utilization rate of passage in growing pigs, finishing pigs and adult sows. Anim Sci. 74, 503–515.
15. Moreira, F., Corcino, C.D., Mondadori, R.G., Gevehr, F.G., Mendes, F.F., Araújp, E.G., Lucia, T. 2013. Leptin and mitogen-activated protein kinase (MAPK) in oocytes of sows and gilts. Anim. Repro. Sci. 139, 89–94.
16. Nefzaoui, A., Ben Salem, H. 2002. Supplementation of *Acacia cyanophylla Lindl.* diets based foliage with barley or shrubs in arid zones (*Opuntia ficus-indica var. intermis* and *Atriplex nummularia* l.) on growth and digestibility in sheep. J. Anim. Alimentación. Sci. Tecnología. 96, 15–30.
17. Olsson, A. Ch., Andersson, M., Botermans, J., Rantzer, D., Svendsen, J. 2011. Animal interaction and response to electronic sow feeding (ESF) in 3 different herds and effects of function settings to increase capacity. Livest. Sci. 137, 268–272.
18. Pari, L., Latha, M. 2005. Antidiabetic effect of *Scoparia dulcis:* effect on lipid peroxidation in streptozotocin diabetes. Gen Physiol Biophys. 24, 13–26.
19. Paspala, I., Katsiki, N., Kapoukranidou, D., Mikhailidis, D.P., Tsiligiroglou, F.A. 2012. The Role of Psychobiological and Neuroendocrine Mechanisms in Appetite Regulation and Obesity. The Open Cardiovascular Medicine Journal. 6, 147–155.
20. Pére, M.C., Etienne, M. 2007. Insulin sensitivity during pregnancy, lactation, and postweaning in primiparous gilts. J Anim Sci. 85(1), 101–10.
21. Pérez, S.R.E., Ordaz, O.G., Juárez, C.A., García, V.A., Ortiz, R.R. 2015b. Efecto del número de parto sobre el consumo de alimento voluntario de las cerdas durante la fase de lactancia y su repercusión en el intervalo destete-estro. Nota técnica. Rev. Cient. FCV-LUZ. XXV(2), 145–152.
22. Pérez, SRE, Ortiz, RR., Ordaz, OG. 2015b. 2015a. Validación de un glucómetro comercial de uso humano para determinar glucosa en cerdos. LI Reunión Nacional de Investigación Pecuaria Memoria (RNIP). Toluca de Lerdo, Edo. de México. Sección: Nutrición animal. Año 1. Núm.1. Vol. 1. Nov. 258.
23. Pinos, R.J.M., Velázquez, J.C., González, S.S., Aguirre, J.R., García, J.C., Álvarez, G., Jasso, Y. 2010. Effects of cladode age on biomass yield and nutritional value of intensively

produced spineless cactus for ruminants. South African Journal of Animal Science 40(3), 245–250.

24. Quesnel, H., Meunier-Salaün, M.C., Hamard, A., Guillemet, R., Etienne, M., Farmer, C., Dourmad, J.Y., Pére, M.C. 2009. Dietary fiber for pregnant sows: Influence on sow physiology and performance during lactation. J Anim Sci. 87, 532–543.
25. Rajesh, J.H.A., Berrocoso, J.F.D. 2016. Dietary fiber and protein fermentation in the intestine of swine and their interactive effects on gut health and on the environment: A review. Anim. Feed Sci. Technol. 2012, 18–26.
26. Rigón, RC,A,, Lovattom PA,, Weschenfelderm VA,, Lehnen, Ch. R., Bruno, N.F., Andretta, I., Speroni, C.M. 2008. Metanálise da relação entre espessura de toicinho e variáveis nutricionais de porcas gestantes e lactantes. Cien. Rur. 38(4), 1085–1091.
27. Sakly, C., Rekik, M., Ben Salem, I., Lassoued, N., Gonzalez, A.B., Ben Salem, H. 2013 Reproductive response of fat-tailed Barbarine ewes subjected to short-term nutritional treatments including spineless cactus (*Opuntia ficus-indica F. inermis*) cladodes. J Anim Phys and Anim Nutr. 98, 43–49.
28. Sánchez, D., Muguerza, B., Moulay, L., Hernández, R., Miguel, M., Aleixandre, A. 2008. Highly methoxylated pectin improves insulin resistance and other cardiometabolic risk factors in zucker fatty rats. J. Agric. Food Chem. 56, 3574–3581.
29. SAS: Statistical Analysis System Institute. Guide for personal computers. Version 8. 2010.
30. Schenkel, A.C., Bernardi, M.L., Bortolozzo, F.P., Wentz, I., 2010. Body reserve mobilization during lactation in first parity sows and its effect on second litter size. Livest. Sci. 132, 165–172.
31. Segura, C.J.C., Alzina, L.A., Santos, R.R.H. 2013. Risk factors associated with the occurrence of the second-litter syndrome in sows in southeastern Mexico. The Scientific World Journal. 1-4.
32. Tikabo, G., Solomon, M., Alemu. 2006. Effect of different levels of cactus (*Opuntia ficus-indica*) inclusion on feed intake, digestibility and body weight gain in tef (*Eragrostis tef*) straw-based feeding of sheep. Anim Feed Sci and Tec. 131, 42–51.
33. Viñoles, C., Meikle, A., Martin, G.B. 2008. Short-term nutritional treatments grazing legumes or feeding concentrates increase prolificacy in Corriedale ewes. Anim Rep Sci. 113, 82–92.
34. Wang, J.F., Li, D.F., Jensen, B.B., Jakobsen, K., Xing, J.J., Gonga, L.M., Zhub, Y.H. 2003. Effect of type and level of fibre on gastric microbial activity and short-chain fatty acid concentrations in gestating sows. Anim. Feed Sci. Technol. 104, 95–110.
35. Xie, Ch., Guoa, X., Longa, C., Fana, Z., Xiao, C., Ruanc, Z., Dengc, Z., Wua, J., Yina, Y. 2015. Supplementation of the sow diet with chitosan oligosaccharide during late gestation and lactation affects hepatic gluconeogenesis of suckling piglets. Anim. Rep. Sci. 159, 109–117.
36. Ylonem, K., Saloranta, C., Kronberg-Kippila, C., Goop, L., Aro, A., Virtanen, S.M. 2003. Association of dietary fiber with glucose metabolism in nondiabetic relatives of subject with type 2 diabetes. The Botnia Dietary Study. 26, 1979–1985.

CHAPTER 6

Intestinal IL-17 Expression in Canine Inflammatory Bowel Disease

Kinjo T[1], Azuma YT[2*], Nishiyama K[2], Fujimoto Y[2], Miki M[2], Kuramoto N[1]

[1]*Laboratory of Molecular Pharmacology, Setsunan University Faculty of Pharmaceutical Sciences, Hirakata, Osaka, Japan.*

[2]*Laboratory of Veterinary Pharmacology, Division of Veterinary Science, Osaka Prefecture University Graduate School of Life and Environmental Science, Izumisano, Osaka, Japan.*

ABSTRACT

IL-17 plays important roles in human inflammatory bowel disease (IBD). However, little is known about the possible relation of IL-17 in canine IBD. For comparative medicine and canine disease, it is useful to demonstrate the role of IL-17 in canine IBD. In this study, 7 cases of the dog with IBD typed of lymphocytic-plasmacytic enteritis without Intestinal lymphoma were examined immunohistochemically for possible relation with IL-17. Here, we found that IL-17 is expressed in intestinal mucosa of the dog with IBD. Among 7 cases, there are strong expressions of IL-17 in cases 2, moderate expression of it in cases 2, weak expression of it in cases 2, and very low expression of it in case 1. The score of IL-17 expression is not correlated with sex, age, and the breed. Using biopsy specimen, our results revealed the immunopathological relevance of IL-17 in the dog with IBD.

1. INTRODUCTION

The gastrointestinal tract is crucial for the immunological system in host defense in addition to food intake, food digestion, and nutrient absorption within humans and animals. Epithelium barrier in the intestine plays to block pathogens from the lumen. Inflammatory bowel diseases (IBDs) in dogs, as well as humans, are characterized by dysregulated intestinal inflammation and mucosal tissue damage in parts of the gastrointestinal tract [9, 11-13, 15, 18, 24, 28]. Regarding of histologic criteria, ulcerative colitis (UC) and Crohn's disease (CD) are the two types of human IBD. Despite having a common basis in over responsiveness to mucosal antigens, UC and CD have considerably different pathophysiologies. In contrast, IBD in dogs divided into the several types by histological criteria of inflammatory infiltrate such as neutrophilic, eosinophilic, lymphocytic, plasmacytic and granulomatous. However, little is known whether these types reveal the common basis or different basis. Some reports show the pathology is different between humans and dogs, whereas other reports show common mucosal immune system between human and dogs [1,6, 8, 19].

The maintenance of intestinal homeostasis and the development of IBD involve interactions among the intestinal microflora, epithelium, and host mucosal immune system [25, 28]. Especially, the host mucosal immune system was controlled by the balance among inflammatory mediators, anti-inflammatory mediators and regulatory cytokines then contributed to the intestinal homeostasis. Cytokines have important roles in the regulation of the mucosal immune system of dogs as well as humans. In human, CD is associated with a Th1 response mainly driven by IL-12 and IFNγ [21], while UC is driven by Th2 cytokines, such as IL-4, IL-5, and IL-13 [27]. In contrast, dogs with IBD express with a mixture of both Th1 and Th2 cytokines, including IL2, IL4, IL5, IL10, IL12, IFNγ, TNFα, and TGFβ, in intestinal mucosa [17]. However, little is known about the expression of IL-17 in the dog with IBD. Data implicating mucosal IL-17 in the pathogenesis of dog with IBD are limited. Th17 is responsible for the IL-17 production, a key inflammatory cytokine and increased in human IBD [29]. The aim of the present study was to report the expression and localization of IL-17 in 7 cases of dogs with IBD typed of lymphocytic-plasmacytic gastroenteritis.

2. MATERIALS AND METHODS

2.1 The Case Animals

The details of 7 cases of dogs with IBD typed of lymphocyticplasmacytic gastroenteritis are given in Table 1. In all cases, a biopsy was submitted for microscopical evaluation. A sample of biopsy was fixed in 10% neutral buffered formalin, processed routinely and embedded in paraffin.

Table 1. Details of Dogs with Lymphocytic - Plasmacytic Gastroenteritis.

Case no.	Age (years)	Breed	Sex
1	11	Dachshund	Male
2	10	Dachshund	Female neutered
3	11	Dachshund	Male neutered
4	7	Papillon	Male neutered
5	2	Toy Poodle	Female
6	8	Shiba	Female neutered
7	10	Labrador retriever	Female neutered

Table 2. Histological Score of IL-17 Expression.

Case no.	IL-17 expression score
1	++
2	+++
3	+
4	+++
5	++
6	+
7	±

2.2 Immunohistochemistry (IHC)

IHC of 3-μm-thick paraffin-embedded sections was performed as previously described [20], with minor modifications. We used primary antibodies against IL-17 (ab79056; Abcam, Cambridge, MA, USA). To absorption the antibody, we used IL17 peptide (ab190773; Abcam). Immunoreactivity was detected using peroxidase-conjugated anti-rabbit polymers (Histofine, Nichirei, Tokyo, Japan) and a DAB substrate kit (Vector Laboratories Inc., Burlingame, CA, USA).

2.3 Histology

The histological score of IL-17 expression was graded by ranging from 0 to 3 as follows: 0, no expression; 1, low expression; 2, moderate expression; 3, high expression. Scores were obtained by a pathologist.

2.4 Statistical Analysis

Histological score was statistically analyzed using the Mann-Whitney U test. Differences with P values of less than 0.05 were considered significant.

3. RESULTS

All of 7 cases investigated in this study (Table 1) were diagnosed IBD typed of lymphocytic- plasmacytic enteritis by histological criteria of inflammatory infiltrate. In addition, these 7 cases did not include intestinal lymphoma. IHC was performed to investigate the expression and localization of IL-17-positive cells in the intestinal mucosa of dogs with IBD. The treatment of recombinant IL-17 in addition to IL-17 antibody indicated as negative controls (each right panel). In Figure 1, the breed was all Dachshund. In case 1, the IL-17 was expressed in lamina propria (Figure 1). The expression of IL-17 in case 2 is stronger than it in case 1 (Figure 1). In contrast, the IL-17 was weakly expressed in lamina propria of case 3 (Figure 1). These results indicate that IL-17-positive cells are present in the lamina propria, and the expression level of IL-17 varies by case.

In case 4, there are dense IL-17-positive cells in lamina propria (Figure 2). In case 5, the IL-17 was expressed in lamina propria (Figure 2). In case 6, there is a little expression of IL-17 in lamina propria (Figure 2). In case 7, there is no marked expression of IL-17 in lamina propria (Figure 2). These results also indicate that the expression level of IL-17 varies by case.

A histological score of the IL-17 expression in all 7 cases is shown in Table 2. Regarding of the score of IL-17 expression in lamina propria, there are no significant differences in sex of dogs with IBD (Figure 3 upper). Furthermore, no significant differences in age were observed in the score of IL-17 expression in lamina propria (Figure 3 lower). These results suggest that the expression level of IL-17 may vary by the case rather than the breed.

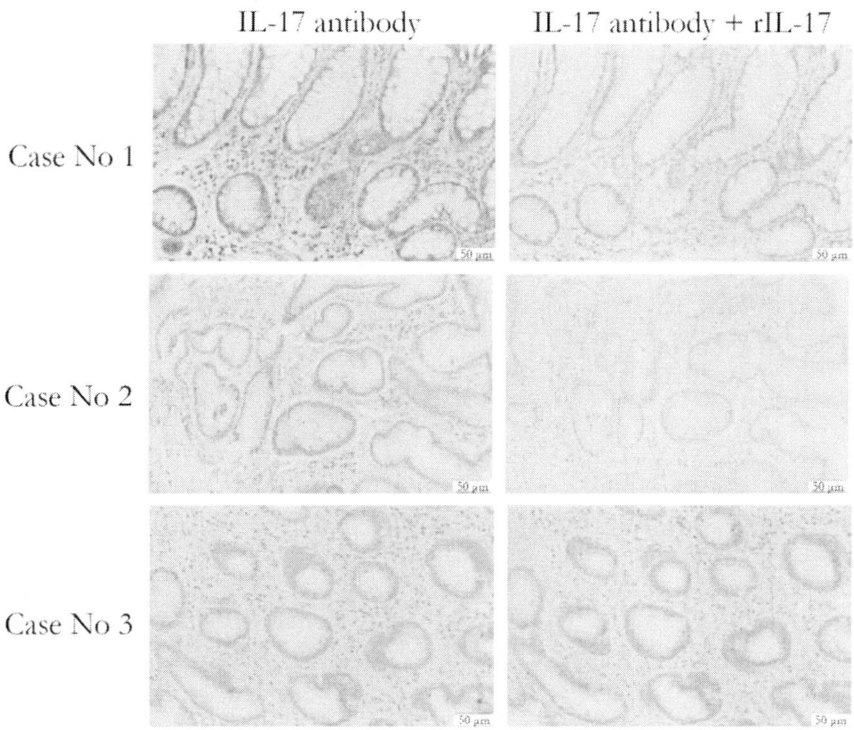

Figure 1. Cases 1-3. Expression of IL-17 in intestinal mucosa by biopsy specimen. Expressions of IL-17 are present in the lamina propria (left panels), which are the loss by the absorption of IL-17 antibody by recombinant IL-17 (right panels). Scale bar, 50 μm.

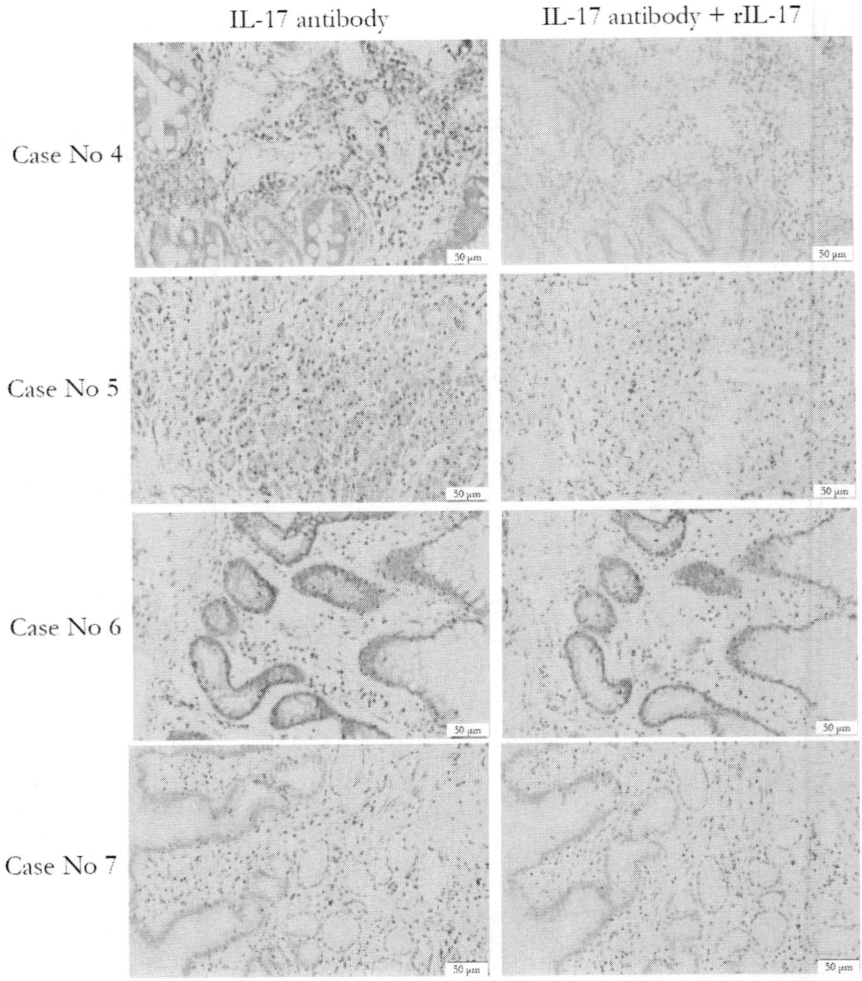

Figure 2. Cases 4-7. Expression of IL-17 in intestinal mucosa by biopsy specimen.Expressions of IL-17 are present in the lamina propria (left panels), which are the loss by the absorption of IL-17 antibody by recombinant IL-17 (right panels). Scale bar, 50 µm.

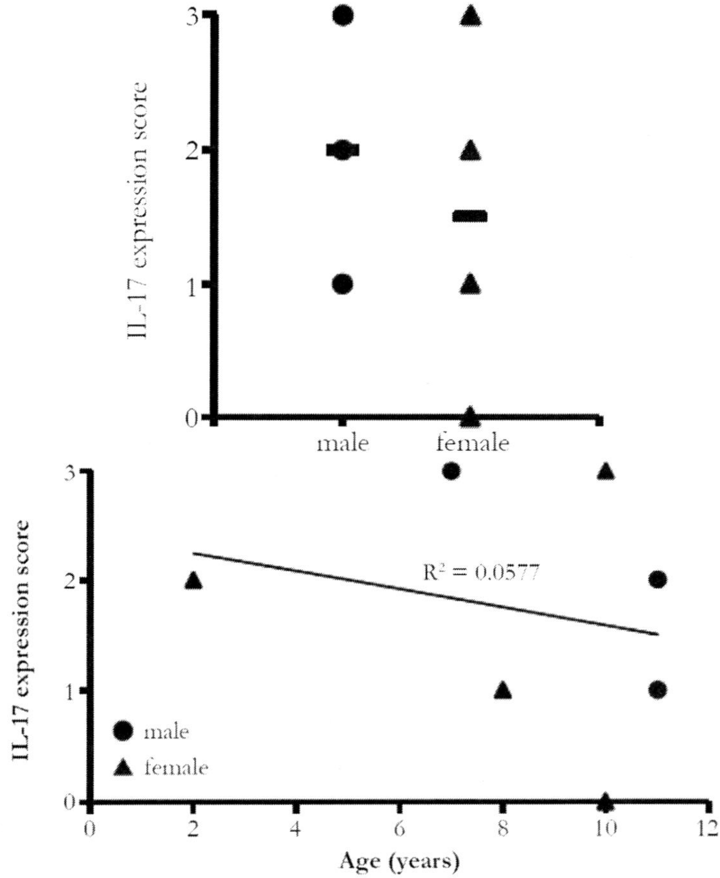

Figure 3. The relation analysis. (Upper) The relation of IL-17 expression score with sex. The horizontal bar means the mean. (Lower) The relation of IL-17 expression score with age (years).

4. DISCUSSION

In this study, we defined the mucosal IL-17 expression in intestinal biopsies from dogs with IBD. Namely, we observed marked expression of IL-17 and the different pattern of localization of IL-17 even the expressed in the intestinal mucosa in dogs with IBD. We press that the expression pattern of intestinal mucosa IL-17 did not depend on the breed type of dogs with IBD. These indicate that IL17-mediated immune responses are related to the development of dogs with IBD, similar to human UC, at large. The classified etiology of dogs with IBD is not well understood although dogs with IBD are compared with human IBD such as UC and CD [5, 15]. Similar to human IBD, histologic features of dogs with IBD show infiltration of mononuclear cells into the lamina propria, goblet cell depletion, crypt hyperplasia and epithelial erosion [15]. T-cells also are related to onset and development of dogs with IBD, because CD^{3+}, CD^{4+}, and CD^{8+} T cells are elevated in the mucosa [10, 14, 16, 26].

Our strength of this study is that dog with IBD expressed IL-17 in duodenal mucosa. Up to the present, little has been reported on possible relation of IL-17 in canine diseases including IBD. These previous reports and our result show that IL-17 plays important role in canine as well as human. Our other strength is that a well-normalized and validated methodology of IHC is useful to evaluate the expression of interleukins in endoscopic biopsy [8, 23].

Cases 1, 2, 4 and 5 show high expression of IL-17, indicating increased infiltration of Th17 into the intestinal mucosa. In contrast, cases 3, 6 and 7 indicate low expression of IL-17, suggesting two possibilities. The first possibility is that Th17 population is low in spite of increased infiltration of the T-cells. The second possibility is that infiltration of the T-cells is not increased into the mucosa.

Unbalance among inflammatory mediators, anti-inflammatory mediators and regulatory cytokines result in IBD. Dysregulated balance between Th1/Th17-mediated inflammatory cytokines and Th2/Treg-mediated immunoregulatory cytokines results in patients with IBD and mice models with colonic inflammation like human IBD [2-4, 7, 10, 20, 22, 23].

A limitation of our results is that dogs with IBD in each case were not almost same by age (years). Aging dogs make a spontaneous progress of terrible and prolonged enterocolitis with diarrhea and body weight loss. Further studies will be needed to assess the relation of IL-17 expression level and age of dogs. Like a human with IBD, genetic factors may contribute to the pathogenesis of dogs with IBD, because an enhanced risk for the development of IBD is related to especial breeds including Basenjis, French bulldogs, German shepherd dogs and soft-coated Wheaton terriers [10, 15]. Further studies are also needed for the expression levels of IL-17 in these breed types.

Our results indicate the increased expression of IL-17 as an inflammatory cytokine in dogs with IBD, and constitute one of the several potential immune mechanisms that can contribute to canine IBD pathology. In conclusion, our results provide the strong insight into the important roles of IL-17 in dogs with IBD. Our results are also able to contribute the relationship among IL-17 expression, canine IBD pathology, drug therapeutics, and clinical outcomes, due to the reduced number of cases. IL-

17 and its receptor will be an important consideration in the design of therapeutic agents in dogs with IBD.

REFERENCES

1. Allenspach K, Bergman PJ, Sauter S, Gröne A, Doherr MG, et al., (2006) P-glycoprotein expression in lamina propria lymphocytes of duodenal biopsy samples in dogs with chronic idiopathic enteropathies. J Comp Pathol. 134(1): 1-7.
2. Azuma YT, Hagi K, Shintani N, Kuwamura M, Nakajima H, et al., (2008) PACAP provides colonic protection against dextran sodium sulfate induced colitis. J Cell Physiol. 216(1): 111-119.
3. Azuma YT, Matsuo Y, Kuwamura M, Yancopoulos GD, Valenzuela DM, et al., (2010) Interleukin-19 protects mice from innate-mediated colonic inflammation. Inflamm Bowel Dis. 16(6): 1017-1028.
4. Azuma YT, Nishiyama K, Matsuo Y, Kuwamura M, Morioka A, et al., (2010) PPARα contributes to colonic protection in mice with DSS-induced colitis. Int Immunopharmacol. 10(10): 1261-1267.
5. Baumgart M, Dogan B, Rishniw M, Weitzman G, Bosworth B, et al., (2007) Culture independent analysis of ileal mucosa reveals a selective increase in invasive Escherichia coli of novel phylogeny relative to depletion of Clostridiales in Crohn's disease involving the ileum. ISME J. 1(5): 403-418.
6. Cerquetella M, Spaterna A, Laus F, Tesei B, Rossi G, et al., (2010) Inflammatory bowel disease in the dog: differences and similarities with humans. World J Gastroenterol. 16(9): 1050-1056.
7. Fujimoto Y, Azuma YT, Matsuo Y, Kuwamura M, Kuramoto N, et al., (2017) Interleukin-19 contributes as a protective factor in experimental Th2-mediated colitis. Naunyn Schmiedebergs Arch Pharmacolin press.
8. German AJ, Hall EJ, Day MJ (2001) Immune cell populations within the duodenal mucosa of dogs with enteropathies. J Vet Intern Med. 15(1): 14- 25.
9. German AJ, Hall EJ, Day MJ (2003) Chronic intestinal inflammation and intestinal disease in dogs. J Vet Intern Med. 17(1): 8-20.
10. German AJ, Helps CR, Hall EJ, Day MJ (2000) Cytokine mRNA expression in mucosal biopsies from German shepherd dogs with small intestinal enteropathies. Dig Dis Sci. 45(1): 7-17.
11. Grimble RF (2003) Inflammatory response in the elderly. Curr Opin Clin Nutr Metab Care. 6(1): 21-29.
12. Jacobs G, Collins-Kelly L, Lappin M, Tyler D (1990) Lymphocytic-plasmacytic enteritis in 24 dogs. J Vet Intern Med. 4(2): 45-53.
13. Jergens AE (1999) Inflammatory bowel disease. Current perspectives. Vet Clin North Am Small Anim Pract. 29(2): 501-521.
14. Jergens AE, Gamet Y, Moore FM, Niyo Y, Tsao C, et al., (1999) Colonic lymphocyte and plasma cell populations in dogs with lymphocytic-plasmacytic colitis. Am J Vet Res. 60(4): 515-520.
15. Jergens AE, Moore FM, Haynes JS, Miles KG (1992) Idiopathic inflammatory bowel disease in dogs and cats: 84 cases (1987-1990). J Am Vet Med Assoc. 201(10): 1603-1608.
16. Jergens AE, Moore FM, Kaiser MS, Haynes JS, Kinyon JM (1996) Morphometric evaluation of immunoglobulin A-containing and immunoglobulin G-containing cells and T cells in duodenal mucosa from healthy dogs and from dogs with inflammatory bowel disease or nonspecific gastroenteritis. Am J Vet Res. 57(5): 697-704.
17. Jergens AE, Sonea IM, O'Connor AM, Kauffman LK, Grozdanic SD, et al., (2009) Intestinal cytokine mRNA expression in canine inflammatory bowel disease: a meta-analysis with critical appraisal. Comp Med. 59(2): 153-162.

18. Kleinschmidt S, Meneses F, Nolte I, Hewicker-Trautwein M (2007) Characterization of mast cell numbers and subtypes in biopsies from the gastrointestinal tract of dogs with lymphocytic-plasmacytic or eosinophilic gastroenterocolitis. Vet Immunol Immunopathol. 120(3-4): 80-92.
19. Locher C, Tipold A, Welle M, Busato A, Zurbriggen A, et al., (2001) Quantitative assessment of mast cells and expression of IgE protein and mRNA for IgE and interleukin 4 in the gastrointestinal tract of healthy dogs and dogs with inflammatory bowel disease. Am J Vet Res. 62(2): 211-216.
20. Matsuo Y, Azuma YT, Kuwamura M, Kuramoto N, Nishiyama K, et al., (2015) Interleukin 19 reduces inflammation in chemically induced experimental colitis. Int Immunopharmacol. 29(2): 468-475.
21. Monteleone G, Monteleone I, Fina D, Vavassori P, Del Vecchio Blanco G, et al., (2005) Interleukin-21 enhances T-helper cell type I signaling and interferon-gamma production in Crohn's disease. Gastroenterol. 128(3): 687-694.
22. Niessner M, Volk BA (1995) Altered Th1/Th2 cytokine profiles in the intestinal mucosa of patients with inflammatory bowel disease as assessed by quantitative reversed transcribed polymerase chain reaction (RT-PCR). Clin Exp Immunol. 101(3): 428-435.
23. Ridyard AE, Nuttall TJ, Else RW, Simpson JW, Miller HR (2002) Evaluation of Th1, Th2 and immunosuppressive cytokine mRNA expression within the colonic mucosa of dogs with idiopathic lymphocytic-plasmacytic colitis. Vet Immunol Immunopathol. 86(3-4): 205-214.
24. Roth L, Walton AM, Leib MS, Burrows CF (1990) A grading system for lymphocytic plasmacytic colitis in dogs. J Vet Diagn Invest. 2(4): 257-262.
25. Sartor RB (2006) Mechanisms of disease: pathogenesis of Crohn's disease and ulcerative colitis. Nat Clin Pract Gastroenterol Hepatol. 3(7): 390-407.
26. Stonehewer J, Simpson JW, Else RW, Macintyre N (1998) Evaluation of B and T lymphocytes and plasma cells in colonic mucosa from healthy dogs and from dogs with inflammatory bowel disease. Res Vet Sci. 65(1): 59-63.
27. Tanaka J, Saga K, Kido M, Nishiura H, Akamatsu T, et al., (2010) Proinflammatory Th2 cytokines induce production of thymic stromal lymphopoietin in human colonic epithelial cells. Dig Dis Sci. 55(7): 1896-1904.
28. Xavier RJ, Podolsky DK (2007) Unravelling the pathogenesis of inflammatory bowel disease. Nature. 448(7152): 427-434.
29. Zhu J, Wang Y, Yang F, Sang L, Zhai J, et al., (2015) IL-33 alleviates DSSinduced chronic colitis in C57BL/6 mice colon lamina propria by suppressing Th17 cell response as well as Th1 cell response. Int Immunopharmacol. 29(2): 846-853.

CHAPTER 7

Global Change and Helminth Infections in Grazing Ruminants in Europe: Impacts, Trends and Sustainable Solutions

Eric R. Morgan[1,*], Johannes Charlier[2], Guy Hendrickx[3], Annibale Biggeri[4], Dolores Catalan[4], Georg von Samson-Himmelstjerna[5], Janina Demeler[5], Elizabeth Müller[6], Jan van Dijk[7], Fiona Kenyon[8], Philip Skuce[8], Johan Höglund[9], Padraig O'Kiely[10], Bonny van Ranst[11], Theo de Waal[12], Laura Rinaldi[13], Giuseppe Cringoli[13], Hubertus Hertzberg[14], Paul Torgerson[15], Adrian Wolstenholme[16] and Jozef Vercruysse[2]

[1] *School of Veterinary Sciences, University of Bristol, Langford House, Langford, North Somerset BS40 5DU, UK*

[2] *Laboratory of Parasitology, Faculty of Veterinary Medicine, Ghent University, Merelbeke B-9820, Belgium*

[3] *Avia-GIS, Zoersel 2980, Belgium*

[4] *Cooperativa Epidemiologia e Prevenzione "Giulio Alfredo Maccacaro", Milan 20148, Italy*

[5] *Institute for Parasitology and Tropical Veterinary Medicine, Freie Universität Berlin, Berlin 14195, Germany*

[6] *Laboklin, Bad Kissingen D-97668, Germany*

[7] *Institute of Infection and Global Health, University of Liverpool, Liverpool L69 3BX, UK*

[8] *Moredun Research Institute, Edinburgh EH26 0PZ, Scotland, UK*

[9] Department of Biomedicine and Vet Public Health, Swedish University of Agricultural Sciences, Uppsala 753 12, Sweden

[10] Animal & Grassland Research and Innovation Centre, Teagasc, Grange, Dunsany, Co. Meath, Irelan

[11] UNIFORM-AGRI BV, Assen 9401, The Netherlands

[12] School of Veterinary Medicine, University College Dublin, Dublin 4, Ireland

[13] Department of Pathology and Animal Health, Faculty of Veterinary Medicine, University of Naples "Federico II", Naples 80137, Italy

[14] Institute of Parasitology, Vetsuisse Faculty, University of Zürich, Zürich 8057, Switzerland

[15] Section of Epidemiology, Vetsuisse Faculty, University of Zürich, Zürich 8057, Switzerland

[16] Department of Infectious Diseases, University of Georgia College of Veterinary Medicine, Athens, GA 30602, US

ABSTRACT

Infections with parasitic helminths (nematodes and trematodes) represent a significant economic and welfare burden to the global ruminant livestock industry. The increasing prevalence of anthelmintic resistance means that current control programmes are costly and unsustainable in the long term. Recent changes in the epidemiology, seasonality and geographic distribution of helminth infections have been attributed to climate change. However, other changes in environment (e.g., land use) and in livestock farming, such as intensification and altered management practices, will also have an impact on helminth infections. Sustainable control of helminth infections in a changing world requires detailed knowledge of these interactions. In particular, there is a need to devise new, sustainable strategies for the effective control of ruminant helminthoses in the face of global change. In this paper, we consider the impact of helminth infections in grazing ruminants, taking a European perspective, and identify scientific and applied priorities to mitigate these impacts. These include the development and deployment of efficient, high-throughput diagnostic tests to support targeted intervention, modelling of geographic and seasonal trends in infection, more thorough economic data and analysis of the impact of helminth infections and greater translation and involvement of end-users in devising and disseminating best practices. Complex changes in helminth epidemiology will require innovative solutions. By developing and using new technologies and models, the use of anthelmintics can be optimised to limit the development and spread of drug resistance and to reduce the overall economic impact of helminth infections. This will be essential to the continued productivity and profitability of livestock farming in Europe and its contribution to regional and global food security.

Keywords: helminthoses; ruminants; diagnosis; control; infection risk; global change; climate change; anthelmintic resistance; risk management; spatio-temporal modelling; epidemiology; food security

1. IMPACT OF HELMINTH INFECTION ON THE SUSTAINABILITY AND EFFICIENCY OF LIVESTOCK FARMING

1.1. Livestock Farming As a Cornerstone of Society

Livestock farming is central to the sustainability of rural communities around the world, as well as being socially, economically and politically highly significant at national and international levels. In the European Union (EU), for instance, there are currently around 88 million cattle, 101 million sheep and 12 million goats [1]. Ongoing socioeconomic and climatic changes will increasingly emphasise the need for food security, obtained from sustainable intensification of agriculture [2]. It is inevitable that the production of meat and dairy products will also have to expand to meet the demands of increasing world population. Efficient ruminant livestock production is crucial to achieving this goal, especially in areas in which land is unsuitable for growing crops [3]. In this context, the competitiveness of livestock farming will largely depend on the degree to which this industry can achieve sustainable optimal production levels under changing environmental and socioeconomic pressures.

1.2. Costs to Economies and Animals

All grazing animals are exposed to helminth infections at pasture and any future intensification of pasture-based systems will likely increase the risk of helminth disease. Gastrointestinal nematodes (GIN) and liver fluke are the two major causes of lost productivity in ruminants, with lungworms also important in some situations. The economic costs of parasitic disease are currently difficult to quantify; however, some estimates exist within the scientific literature. For example, studies in the United Kingdom estimated the cost of parasitic nematodoses of sheep to be on the order of 99 million € per year [4], and in Switzerland, the cost of liver fluke disease has been estimated at 52 million € per year in cattle alone [5]. Within the EU as a whole, annual sales of anthelmintic drugs used to control these infections in ruminants have been estimated to be on the order of 400 million € [6]. It is likely that these figures only represent the tip of the iceberg when it comes to calculating the true cost of livestock helminthoses endemic within the EU [7].

Parasitic gastroenteritis (PGE) in European cattle results principally from infections with *Ostertagia ostertagi* and *Cooperia oncophora* [8]. Although *Cooperia* is less pathogenic than *Ostertagia*, these parasite species usually occur together in the same host, with one contributing significantly to the pathogenic effect of the other. In European sheep and goats, *Teladorsagia circumcincta*, *Haemonchus contortus*, *Trichostrongylus* spp. and *Nematodirus* spp., parasitizing the intestine, are the most pathogenic GIN species, contributing significantly to PGE. Fasciolosis (liver fluke disease), caused by the trematode, *Fasciola hepatica*, is a worldwide infection of livestock, especially sheep and cattle, while lungworms in the genus *Dictyocaulus* cause significant disease in cattle and to a lesser extent in small ruminants. The main effect of all these infections is to reduce production efficiency, although under certain circumstances, mortalities can also be high: up to 20% with genera, such as *Haemonchus*, *Nematodirus* and *Fasciola* [9]. The major economic impact of parasitism is due to sub-clinical infections causing production losses. These costs have become increasingly important in the current economic climate with the low profit margins in the livestock sector.

At present, there is only a limited and fragmented understanding of the true costs of helminth parasitism, including costs associated with its control. This must be rectified if future control strategies are to be economically sensible and well integrated into farming systems. Recent advances in diagnostics, such as quantification of bulk milk anti-nematode antibodies, have enabled improved estimates of the production impacts of parasites in dairy herds and underpin optimized strategies for their control [10,11,12]. Further gains in understanding are likely to be made by comparing animal production in high and low intervention situations [13] and where anthelmintic efficacy fails. It is only by more fully characterizing and understanding the production impacts of parasites that future control can be optimized effectively.

2. INCREASED RISKS OF HELMINTH INFECTION IN LIVESTOCK DUE TO GLOBAL CHANGE

2.1. The Changing Environment

In recent years, sharp increases in helminth-associated disease frequency and intensity have been reported within the ruminant sector in some regions [14]. Climate warming, which in temperate regions tends to increase the developmental success of parasites, might be expected to increase pasture contamination with infective stages and may be one driver behind this trend. For example, there have already been reports of altered seasonal patterns of nematode and liver fluke infections in northern parts of the UK [15,16]. In Switzerland, unpublished data suggest that *H. contortus* transmission is occurring at higher altitudes than previously recorded, and in Sweden, transmission occurs near the Arctic Circle [17]. If these trends continue as predicted, European farmers may be faced with new and unfamiliar parasitological challenges that they are ill equipped to meet. However, the study of the effects of climate change on the endemic diseases of livestock is still in its infancy [18,19], and the effects of additional factors, such as altered land use, the emergence of anthelmintic resistance (AR) and farm management practices, have received little detailed attention [19]. Thus, attempts have been made to integrate the abundant knowledge of environmental, especially climatic, effects on parasite stages outside the livestock hosts into predictive mathematical models, especially focusing on selection for AR [20,21,22,23,24], and some have tried to predict future infection patterns [25]. However, the abundance and distribution of helminth infections of livestock is a complex and dynamic issue affected by a whole range of parameters, including those that could be classed as global changes (Figure 1). The situation is further complicated by interacting regional, seasonal and host-specific factors that influence disease and the fact that helminthoses are usually seen in animals that have concurrent multi-species infections. The fact that global change is much more than climate change alone must be acknowledged and incorporated more fully into future research approaches in this area.

2.2. The Importance of Anthelmintic Resistance

An equally important driver of increased disease and production loss due to helminths is likely to be treatment failure, which is being reported ever more frequently. The increasing occurrence of AR in helminth populations threatens the sustainability, as well as the efficiency of livestock production. Although there are a number of different approaches to the control of helminth parasitism in livestock, including nutritional, immunological and biological interventions [26], at present, effective control relies almost exclusively on effective anthelmintic drugs. Most of these compounds have broad-spectrum activity, *i.e.*, they kill all the common roundworm species (there are specific products effective against fluke), but effective control ultimately relies on multiple treatments each year. When first introduced, all these drugs were highly efficacious, but frequent and widespread use and misuse has resulted in the emergence of resistant parasite populations, such that AR is now a major global problem, especially (but not exclusively) in parasites of small ruminants [27], and is the greatest threat to the sustainable control of helminthoses in the short to medium term. Anthelmintic inefficacy and resistance has also become apparent in cattle, reported initially in the southern hemisphere [28,29,30], but more recently also within Europe [31,32].

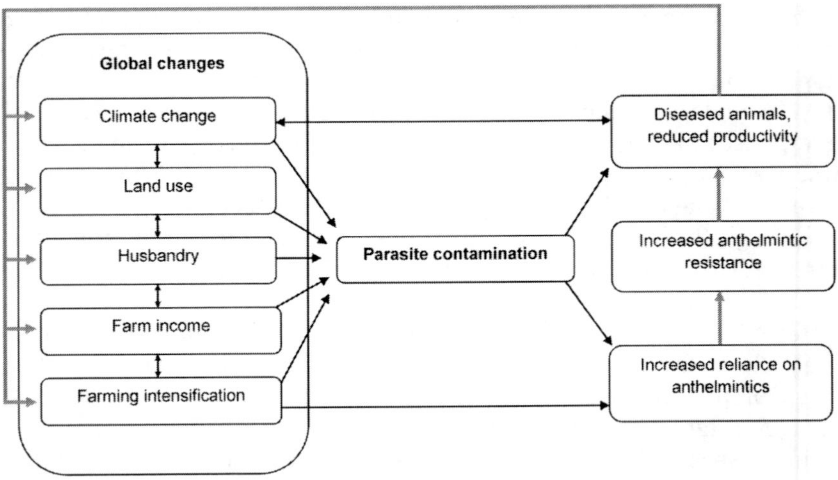

Figure 1. An association network illustrating how global changes may influence pasture contamination with helminth parasites. Black arrows represent positive enforcement, and those in red (bold) show potential associations arising from global change.

The problem of AR is compounded by the fact that many of these parasite populations are resistant to more than one class of anthelmintic [33,34] and that the prevalence of resistance is increasing inexorably in many regions [35]. Despite the recent introduction of two novel anthelmintic drug classes for sheep [36,37], the situation regarding the development and transmission of AR remains largely unaltered. Accordingly, there is an urgent need to reduce anthelmintic usage, while timing treatments optimally, taking into account global change-driven alterations in parasite seasonality. To maintain or, in some regions, regain control over these parasites, an improved understanding and quantification of the key mechanisms involved in their spatial and temporal prevalence is also paramount.

2.3. The Need for Increased Carbon Efficiency in Ruminant Farming

The increased prevalence of helminths, especially drug-resistant parasites, poses an even more pressing problem against a background of increasing pressure towards lowering the carbon footprint of ruminant livestock. Worldwide, ruminant livestock farming accounts for 70% of agricultural land use, occupies 30% of the ice-free surface of the planet and produces some 40% of the global agricultural Gross Domestic Product (GDP). Set against this, it is estimated to account for up to 18% of worldwide greenhouse gas emissions [38]. There is a new requirement for European livestock farming to become carbon emission efficiency-driven [39]. Animals emitting greenhouse gasses, while not growing or producing, because of illness, add significantly to a farm's carbon footprint; so, combating infectious and parasitic diseases on livestock farms is essential for any improvements in the carbon efficiency of production [40].

A particular issue regarding AR is that a response to increased infection pressure as a result of climate and other environmental change that relies on increased use of anthelmintic drugs will undermine the sustainability of control, as increased drug use selects more strongly for AR. As drug efficacy declines, production efficiency would be expected to decrease and greenhouse gas emissions per unit production, increase. This could trigger a positive feedback cycle, fuelling ongoing climate change and further increases in infection and its effects.

3. SCIENTIFIC PRIORITIES TO SUPPORT SUSTAINABLE HELMINTH CONTROL

The development and implementation of innovative, refined approaches to worm control, targeted at the appropriate regional scale, is a prerequisite for reducing the enormous burden helminth parasitism imposes upon ruminant livestock production. This goal can be supported by exploiting developments in high-throughput technologies coupled to novel diagnostics to identify sub-clinical infections within an affordable and flexible diagnostic capacity. However, the provision of innovative diagnostic tools to identify the species implicated in infections, and individual and/or herd/flock markers of helminthoses, are only the first steps in the process. To have real impact, new diagnostic capacity needs to integrate with decision support tools, such as measures of economic impact [11], predictive disease risk forecasting models and livestock management software. Statistically-based risk models of helminthoses that incorporate Geographical Information System (GIS)-based surveillance, as well as temporally explicit models predicting periods of high risk, could inform farm management responses to regional environmental, climatological, parasitological and socioeconomic changes, and themselves depend on data generated from enhanced diagnostic capacity. In particular, optimising intervention strategies, including extent and intensity of drug use to control worm infections, must be based on evidence. This requires accurate and efficiently collected information on levels of challenge, infection and production loss and integration of this information in such a way that optimal strategies can be devised. Scientific research can support this endeavour in four main fields: (1) diagnosis of helminthoses in livestock at the individual and herd level with specific attention to multi-species infections and the distribution of anthelmintic resistance; (2) prediction of the impact of global changes on the epidemiology of parasitic infections, as well as the distribution and spread of anthelmintic resistance; (3) explanation of current, and future predictions of, seasonal trends in helminth infections of grazing livestock; (4) strategies for the sustainable management of helminth infections in a changing landscape. The rationale for prioritising these areas, and the scientific and technical challenges that they present, are discussed below.

3.1. Diagnosis

Many methods are already available for the diagnosis of helminth infections, clinical and sub-clinical disease and AR, but all have their weaknesses. The principal clinical signs, such as diarrhoea, anorexia and ill-thriftiness are common sequelae of many other infectious diseases and syndromes and, consequently, lack sensitivity and specificity. Many of the methods used to diagnose helminthoses can only be applied to individual animals and/or to hosts infected with one helminth species, with little or no capacity to deal with infections where multiple species are present. Current parasitological, immunological and molecular diagnostic technologies are often labour-intensive and, therefore, expensive, with limited application either on-farm or for large-scale surveillance.

Because helminthoses inevitably involve multiple parasite species, in order to ensure that the appropriate treatment is applied, it is important to identify the key species responsible for the signs of disease and/or production losses. Most routine helminth diagnoses rely on the detection of parasite stages (usually eggs) in faecal samples. Parasitological diagnoses offer the advantage of relatively low cost and can be conducted in non-specialised laboratories; however, the morphological identification of nematode eggs and larvae to the species (or even genus) level is difficult and requires experienced personnel [38]. In some circumstances, for example, treatment decisions based on increasing faecal nematode egg counts in grazing ruminants, specific identification is not a priority. However, control strategies that are truly directed at the major parasitic threats to health and production should be based on more detailed knowledge of these threats, including their economic impact.

Although species specificity is often more easily achievable through immunological detection, a major drawback with these approaches is that they do not necessarily indicate the presence of a currently active infection. A further complication is in obtaining suitable samples, since some tests require invasive procedures, such as blood sampling, which are not routine practices in sheep and cattle. Molecular genetic (Polymerase chain reaction, PCR-based) approaches have potential for this application, because they are sensitive and specific and can be applied to helminth parasite material [41], most notably, eggs and larvae extracted from faecal samples. To date, however, these approaches have tended to concentrate on single species and work best on material derived from single parasites [42,43], which requires considerable upstream processing. Real-time PCR and pyrosequencing approaches have also been developed in an attempt to provide quantitative information on species composition and show much promise, but more development work is required before validated molecular species identification tests can be made available [44,45].

AR is an additional complicating factor with regard to the diagnosis of helminthoses. Currently, the principal means of identifying AR *in vivo* is the faecal egg count reduction test (FECRT), based on a series of pre- and post-treatment individual egg counts. This test has limitations, not only because it is labour intensive, but also regarding the optimum sampling and counting criteria, especially in cattle, and can be confounded by over-dispersed parasite populations [46]. However, the FECRT is still a useful indicator of drug efficacy in the field, since samples can be submitted to diagnostic laboratories by post, but the test does still require simplification and further optimisation to encourage uptake by end-users. There are several *in vitro* bioassays available that examine the effects of differing drug concentrations on parasite behaviour and development, e.g., the egg hatch test (EHT), larval migration inhibition assay (LMIA), larval feeding inhibition assay (LFIA) and larval development test (LDT) [47]. These tests are useful for characterising single-species parasite isolates in the laboratory, but are not yet sufficiently standardised to be of use as routine diagnostic tests for resistance in the field. Moreover, they do not all work equally well with all available drug classes, e.g., the EHT can only be used to detect benzimidazole (BZ) resistance. Their utility is further compromised by the fact that different parasite species have their own inherent sensitivities to certain drugs, making the interpretation of test results from natural multi-species infections extremely difficult. Since AR has a genetic component, molecular (DNA-based) methods have great potential as putative diagnostic test systems here, too. However, this development requires validated genetic markers for resistance, and to date, this is only available for one drug class (the BZs) and not yet in all species [48]. Nonetheless, sensitive and accurate pyrosequencing assays have been published for the detection of BZ resistance-associated single nucleotide polymorphisms (SNPs) in *Haemonchus* and *Teladorsagia* [49,50] and have shown that BZ resistance allele frequencies correlate well with the outputs from FECRTs conducted in the field [51]. The search for resistance markers for the other drug classes has largely focused on candidate resistance genes, but has not been proven, to date, to be a fruitful exercise [52]. Although whole genome sequencing has the potential to reveal the genetic determinants of resistance, this approach is still very much in its infancy [53].

Cost and speed of reporting remain important barriers to the support of control with accurate diagnosis of helminth infection and disease. The use of emerging high-throughput immunological and molecular-based technologies offers the potential to reduce costs and offer diagnosis at a scale suitable for large-scale monitoring within Europe. The advent of microbead-based technologies has led to the development of a number of multiplex assay platforms, e.g., LUMINEX®, which permit multiple assays to be performed on the same samples and provide a range of versatile assay designs, including antibody/antigen, primer/probe and enzyme/substrate interactions [54,55,56,57,58]. The use of this platform allows the simultaneous determination of several parameters (in theory, up to 100; in practice, probably 10–20) from a single serum sample in a streamlined highly automatable workflow. This greatly enhances test/sample throughput and the efficiency of sample utilisation, both prerequisites for effective surveillance and monitoring activities. Such methodologies are widely used in human diagnostic facilities, but, to date, have not been widely applied in veterinary parasitology. This platform is ideally suited for high-throughput Enzyme-linked immunosorbent assay (ELISA)-based test systems. Possible relevant parameters to be measured include not only circulating parasite antigen or anti-parasite specific antibodies [59,60,61], but also health indicators, such as markers of inflammation, e.g., acute phase proteins [62], and tissue damage, e.g., pepsinogen [63], for which validated assays already exist.

Multivalent platforms are suited to the incorporation of serological- and molecular-based tests. Prototype multiplex PCR-based molecular assays already exist for the detection of individual GIN species infecting sheep and cattle [43,44]. Such multiplex LUMINEX® assays have already been described for the detection of the protozoan pathogens, *Giardia* and *Cryptosporidium* spp., from faecal samples [64,65,66], but have not yet been developed or validated for helminth parasites. These high-throughput diagnostic platforms require considerable capital investment and experienced personnel to maintain the instruments and to run assays. There is still a need for relatively simple diagnostic tests that require little specialised equipment and that might, ultimately, have "pen-side" applications. Innovative approaches based on the recently developed loop-mediated isothermal amplification (LAMP) method show promise in this regard. This is a commercially available detection method to amplify nucleic acids and offers a rapid, accurate and cost-effective diagnosis of infectious diseases [67,68,69,70]. LAMP technology has the added advantage that it is not prone to inhibition by contaminants within faecal samples, as is the case with traditional PCR-based methods. These novel analytical platforms also have the potential to detect and quantify anthelmintic resistance-associated DNA polymorphisms in key parasite genera.

3.2. Spatial Epidemiology and Forecasting of Helminth Disease

More efficient targeting of anthelmintic-based control strategies can benefit not only from improved diagnosis, but also prediction of disease threats in space and time. In the ~50 years since the first epidemiologically-based helminth parasite model was published [71], a number of simplified models have been developed for both nematodes and trematodes [72]. However, simplified models often neglect crucial interactions, while those with a more realistic level of detail are usually over-parameterised, making it hard to interpret their output. Most existing models of disease and/or parasite ecology were not constructed against a background of global change and do not, therefore, include crucial factors, such as changing land use and farm management practices. Exclusion of such important drivers leads to a non-realistic output, which might help explain why such models have had a limited impact on parasite control in practice. The climate-based forecasting models previously developed for animal fasciolosis in different areas of Europe, Africa and the USA [73] also require further development to incorporate global change factors, including farm management practices, animal movements and climate change. Therefore, the more realistic models become by incorporating such factors, the less reliably their predictions can be extrapolated to other areas, circumstances and research or control questions. Efforts to model the dynamics and control of helminths in ruminants in other parts of the world [20,23] have value in highlighting general aspects of system dynamics, but ultimately, models for prediction and decision support in Europe should be constructed and validated in relation to local conditions [74]. Models will only be plausible to animal health advisors if shown to work well on the ground, and proper validation is vital if their use in decision support tools is to gain traction. An approach that builds local detail and variation onto a common, general framework for helminth-livestock interactions might, therefore, be a useful way forward.

Ultimately, both spatial and temporal models of parasite occurrence and abundance rely on solid data for their development and validation [75] and can, therefore, benefit from improved high-throughput diagnostics. The prohibitive costs and logistics of conducting extensive and repeated surveys of disease for purely descriptive information on distribution have led to an increased interest in computer-based geospatial technologies. Human and veterinary medicine has seen an increasing reliance on the application of geospatial tools—*i.e.*, GIS, Global Positioning System (GPS), satellite-based remote sensing (RS) and Virtual Globes (e.g., Google Earth™)—to study the spatial and temporal distribution of infectious and parasitic diseases and their vectors [76,77]. Health research based on geospatial tools is considered a timely approach in a changing environment to understand climatic-environmental-health linkages [78]. The application of spatial sampling strategies to animal diseases is relatively new, and the study of pathogen distribution and abundance at a geospatial scale has focused mainly on vector-borne diseases (VBD), mainly due to their direct link with the environment. As an example, 51% of the papers published by the journal, Geospatial Health [79], between 2005 and 2010 dealt with VBD, only 6% with non-vector transmitted helminths and 19% with trematode infections; however, the latter mainly pertain to schistosomes and, to a lesser extent, *Fasciola*. There therefore seems to be considerable scope for applying existing, as well as new geospatial approaches to understanding and predicting parasite distribution and disease risk in livestock. Important parameters, such as environmental factors and farm management, are also rarely taken into consideration when planning cross-sectional or longitudinal surveys for spatial modelling.

3.3. Predicting the Timing of Parasite Risk

Spatial variation in parasite disease risk enables the design of monitoring and intervention strategies that are locally appropriate, while also providing a systematic basis for assessing trends in risk as a result of global change. However, effective intervention against parasitic disease is temporally sensitive, relying on the coincidence of animal grazing with infective stage availability [80]. The timing of treatment and other actions, including grazing practice and diagnostic monitoring in support of targeted treatment strategies, is therefore very important to the outcome in terms of the reduction of disease challenge and selection for anthelmintic resistance. Moreover, one of the main effects of climate change on parasite epidemiology is likely to be the altered seasonal availability of infective stages [9,81]. For example, warmer temperatures appear to be reducing overwinter survival of GIN of sheep in the UK, but accelerating build-up of infection in summer, leading to decreases in recorded disease in spring, but increases in autumn [82]. Therefore, comprehensive modelling approaches should also consider temporal patterns of infection [83], for example, by building and populating mechanistic representations of the life-cycles of the major helminth species, with components that explicitly incorporate climatic stochasticity, forage availability and utilisation and variation in farm management at the regional level. These can be extended to consider global change scenarios through variation in climate, farm systems and parasite biology, as well as, potentially, parasite adaptation. The extent to which parasite adaptation to changes in climate and management will affect their control has barely been considered in trichostrongyloids beyond drug resistance, with some exceptions [84]. The breadth of this approach and the specific consideration of the spatial scale over which important determinants of epidemiology act are crucial, not only to advancing existing scientific understanding of the processes involved [72], but also to practical application of modelling in support of disease control, in that explicit consideration of the timing of helminth transmission is needed to optimise control strategies. A future vision of mathematical and geospatial modelling for parasite control would combine temporal and spatial aspects, to provide forecasting and scenario analysis tools that are integrated at local, regional and global levels. These could be applied to on-farm decision making through decision support tools, to horizon-scanning for disease threats by industry and animal health authorities and to policy making at national and supra-national levels.

Validation of models with experimental and field data will be crucial to model plausibility. However, model exploration through scenario analysis is also essential to consider the possible effects of global change and, often, is impossible to validate fully, because scenarios will often go beyond the range of past or present experience. For the same reason, purely empirical approaches will be limited in their ability to demonstrate and predict the effects of global change: there is no controlled experiment possible to emulate the scope of global change on parasitism within farming systems as a whole. At the same time, limited standardisation and centralisation of parasitic disease data, and the many confounding factors in endemic disease surveillance, make it difficult to validate predictions of global change impacts on parasitic disease [82]. This explains the lack of well-documented, proven effects of global, including climate, change on disease incidence in livestock, which should not be taken to indicate the absence of such effects.

3.4. Sustainable Parasite Control

Ruminant production systems vary considerably, depending on social, economic and other environmental factors, such as soil, climate and farm structure. Predictable (and unpredictable) global changes will inevitably change the way these systems are operated [3]. These changes will also have both direct and indirect effects on helminth lifecycles (see above) and, thus, on the most appropriate preventative and remedial strategies required on farms.

Current chemically-based approaches to helminth control generally utilise frequent whole flock/herd treatments, even though these are known to lead to an increased rate of the development of anthelmintic resistance [85]. Additionally, there are societal concerns about the levels of chemicals present in food products. Although anthelmintic residue concentrations have been found to be very low in beef in Europe [86], concerns persist and underpin restrictions on anthelmintic use in livestock, especially in dairy cows [87]. In any event, the risk of AR is sufficiently serious that current chemical-based helminth control strategies can reasonably be deemed unsustainable unless considerably modified. Previous studies, reviewed in [85], have established proof-of-concept for *refugia*-based treatment strategies and have demonstrated the benefits of optimised anthelmintic usage in maintaining animal performance and drug efficacy, though mostly in small ruminants, rather than cattle. The ability to optimise treatments will likely change the way in which anthelmintics are used; they will only be given to those animals that actually need treatment, instead of, as is common practice now, treatment being given to all animals in a flock/herd simultaneously. Optimising drug treatments will slow the development of AR [88] and, thus, maximise the life of those anthelmintic families where resistance is known to be an issue and prolong the life expectancy of the two new sheep anthelmintic families, currently represented by monepantel [36] and derquantel [37]. Proof-of-concept studies so far published suggest that targeted whole-flock treatment or individual animal treatments are effective and pragmatic strategies for optimising anthelmintic use in Europe, as in a range of production systems across the world [63,85,89,90,91,92,93,94].

Targeted treatment (TT), *i.e.*, optimised whole flock/herd treatment, has been shown to be beneficial in controlling nematode infections, in both large and small ruminants. Targeted selective treatment (TST), *i.e.*, individual animal treatment, has been shown to reduce anthelmintic usage in small ruminants, whilst maintaining animal production and drug efficacy. However, to date, TST approaches have rarely been studied in cattle, and neither approach has been applied to liver fluke infections. The wide-scale uptake of these strategies can only be achieved with a full understanding of their potential costs and benefits; this has also not yet been properly evaluated, and the optimum balance between worm control and maintenance of efficacy in TST and Integrated Pest Control (IPC) programmes still needs to be identified. Current TT and TST strategies for small ruminants, for example, are most applicable for large-scale producers, because of the investment needed in efficient animal handling and performance monitoring systems. Indicators of the need to treat at the group or individual level, which are easily applicable within smaller scale enterprises, are needed.

In order to make it possible to integrate these new strategies into routine farm management practices, farmers (and their advisors) will need to fully understand the costs and benefits of these novel treatment strategies. Previous studies [85] showed that TT and TST strategies can reduce anthelmintic use and maintain drug efficacy; however, there may be some increased costs associated with the use of the new strategies (such as increased labour and the costs of new technology and diagnostics). The economic costs and benefits of these more sustainable treatment strategies, in the short and long term, have yet to be rigorously analysed, with few specific studies [92], and the resulting uncertainty is perhaps the major obstacle to their adoption by livestock producers. This is compounded by the scarcity of hard data on the costs of anthelmintic resistance, as well as uncertainty regarding future access to new treatment compounds and their likely longevity in the face of selection for resistance.

Finally, to encourage the implementation of new treatment strategies into routine farm practices, decision support tools are needed. This is because targeted strategies are inherently more complex than universal protocol-based systems and must be flexible to differences in farm systems, climate and other context-specific factors [95]. Integration of decision support tools into pre-existing herd management software would help farmers to incorporate the ideas and approaches discussed above.

4. CONCLUSIONS

Anthelmintic resistance and global change are dominant factors underpinning current and future trends in parasitic disease in grazing ruminants. Climate warming acts on immature parasite stages outside the main host and could alter the level and timing of peak infection pressure. The way in which this translates to altered disease patterns is modified by many factors, including host immunity, grazing patterns and other farm management practices [96]. Meeting increased infection pressure with more frequent administration of anthelmintic drugs is unsustainable, due to rapid development of resistance in nematode and, probably, trematode populations. Therefore new approaches are required. Increasingly, the targeting of treatment at the group and individual level appears to be the only practical way forward for sustainable helminth control on farms. Much remains to be learned regarding the optimal design and implementation of such strategies in different contexts. We have outlined some of the challenges in this regard and identified key areas in which advances in science and technology can help to support effective and efficient strategies for maintaining productivity in the face of major future challenges. The adoption of improved parasite control practices is crucial for sustainable and efficient production from ruminants at pasture.

ACKNOWLEDGEMENTS

The authors are grateful for funding from EU FP7 project 288975 GLOWORM.

REFERENCES

1. FAOSTAT (2009). Available online: http://faostat.fao.org/ (accessed on 15 July 2013).
2. Baulcombe, D.; Crute, I.; Dunwell, J.; Gale, M.; Jones, J.; Pretty, J.; Sutherland, W.; Toulmin, C. *Reaping the Benefits—Science and the Sustainable Intensification of Global Agriculture*; Royal Society Policy Document: London, UK, 2009.
3. Chiotti, Q.P.; Johnston, T. Extending the boundaries of climate change research—A discussion on agriculture. *J. Rural Stud.* **1995**, *11*, 335–350.
4. Nieuwhof, G.J.; Bishop, S.C. Costs of the major endemic diseases in Great Britain and the potential benefits of reduction in disease impact. *Anim. Sci.* **2005**, *81*, 23–29.
5. Schweizer, G.; Braun, W.; Deplazes, D.; Torgerson, P.R. The economic effects of bovine fasciolosis in Switzerland. *Vet. Rec.* **2005**, *157*, 188–193.
6. Selzer, P.M. Preface. In *Antiparasitic and Antibacterial Drug Discovery. From Molecular Targets to Drug Candidates*; Wiley-Blackwell: Hoboken, USA, 2009; pp. 11–12.
7. Charlier, J.; Höglund, J.; von Samson-Himmelstjerna, G.; Dorny, P.; Vercruysse, J. Gastrointestinal nematode infections in adult dairy cattle: Impact on production, diagnosis and control. *Vet. Parasitol.* **2009**, *164*, 70–79.
8. Rehbein, S.; Baggott, D.G.; Johnson, E.G.; Kunkle, B.N.; Yazwinski, T.A.; Yoon, S.; Cramer, L.G.; Soll, M.D. Nematode burdens of pastured cattle treated once at turnout with eprinomectin extended-release injection. *Vet. Parasitol.* **2013**, *192*, 321–331.
9. Van Dijk, J.; Morgan, E.R. The influence of temperature on the development, hatching and survival of *Nematodirus battus* larvae. *Parasitology* **2008**, *135*, 269–283.
10. Mejia, M.E.; Perri, A.F.; Licoff, N.; Miglierina, M.M.; Cseh, S.; Ornstein, A.M.; Becu-Villalobos, D.; Lacau-Mengido, I.M. Comparison of three methods for gastrointestinal nematode diagnosis determination in grazing dairy cattle in relation to milk production. *Vet. Parasitol.* **2011**, *183*, 174–177.
11. Charlier, J.; van der Voort, M.; Hogeveen, H.; Vercruysse, J. ParaCalc®—A novel tool to evaluate the economic importance of worm infections on the dairy farm. *Vet. Parasitol.* **2012**, *184*, 204–211.
12. Charlier, J.; Levecke, B.; Devleesschauwer, B.; Vercruysse, J.; Hogeveen, H. The economic effects of whole-herd versus selective anthelmintic treatment strategies in dairy cows. *J. Dairy Sci.* **2012**, *95*, 2977–2987.
13. Mason, W.A.; Pomroy, W.E.; Lawrence, K.E.; Scott, I. The effect of repeated, four-weekly eprinomectin treatment on milk production in pasture-based, seasonally-calving dairy cattle. *Vet. Parasitol.* **2012**, *189*, 250–259.
14. Van Dijk, J.; Sargison, N.D.; Kenyon, F.; Skuce, P. Climate change and infectious disease: Helminthological challenge to farmed ruminants in temperate regions. Invited Review. *Animal* **2010**, *4*, 377–392.
15. Sargison, N.D.; Wilson, D.J.; Bartley, D.J.; Penny, C.D.; Jackson, F. Haemonchosis and teladorsagiosis in a Scottish sheep flock putatively associated with the over-wintering of hypobiotic fourth stage larvae. *Vet. Parasitol.* **2007**, *147*, 326–331.
16. Kenyon, F.; Sargison, N.D.; Skuce, P.J.; Jackson, F. Sheep helminth parasitic disease in South-Eastern Scotland arising as a possible consequence of climate change. *Vet. Parasitol.* **2009**, *163*, 293–297.

17. Lindqvist, A.; Ljungström, B.L.; Nilsson, O.; Waller, P.J. The dynamics, prevalence and impact of nematode infections in organically raised sheep in Sweden. *Acta Vet. Scand.* **2001**, *42*, 377–389.
18. De la Rocque, S.; Rioux, J.A.; Slingenbergh, J. Climate change: Effects on animal disease systems and implications for surveillance and control. *Rev. Sci. Tech. Int. Off. of Epizoot.* **2008**, *27*, 339–354.
19. Morgan, E.R.; Wall, R. Climate change and parasitic disease: Farmer mitigation? *Trends Parasitol.* **2009**, *7*, 308–313.
20. Leathwick, D.M.; Barlow, N.D.; Vlassoff, A. A model for nematodiasis in New Zealand lambs. *Int. J. Parasitol.* **1992**, *22*, 789–799.
21. Reynecke, D.P.; Waghorn, T.S.; Oliver, A.-M.B.; Miller, C.M.; Vlassoff, A.; Leathwick, D.M. Dynamics of the free-living stages of sheep intestinal parasites on pasture in the North Island of New Zealand. 2. Weather variables associated with development. *Vet. J.* **2011**, *59*, 287–292.
22. Leathwick, D.M.; Waghorn, T.S.; Miller, C.M.; Candy, P.M.; Oliver, A.-M.B. Managing anthelmintic resistance—Use of a combination anthelmintic and leaving some lambs untreated to slow the development of resistance to ivermectin. *Parasitology* **2012**, *187*, 285–294.
23. Dobson, R.J.; Barnes, E.H.; Tyrrell, K.L.; Hosking, B.C.; Larsen, J.W.A.; Besier, R.B.; Love, S.; Rolfe, P.F.; Bailey, J.N. A multi-species model to assess the effect of refugia on worm control and anthelmintic resistance in sheep grazing systems. *Aust. Vet. J.* **2011**, *89*, 200–208.
24. Laurenson, Y.C.S.M.; Bishop, S.C.; Forbes, A.B.; Kyriazakis, I. Modelling the short- and long-term impacts of drenching frequency and targeted selective treatment on the performance of grazing lambs and the emergence of anthelmintic resistance. *Parasitology* **2013**, *140*, 780–791.
25. Fox, N.J.; White, P.C.L.; McClean, C.J.; Marion, G.; Evans, A.; Hutchings, M.R. Predicting impacts of climate change on *Fasciola hepatica* risk. *PLoS One* **2011**, *6*, e16126.
26. Jackson, F.; Miller, J. Alternative approaches to control—Quo vadit? *Vet. Parasitol.* **2006**, *139*, 371–384.
27. Kaplan, R.M. Drug resistance in nematodes of veterinary importance: A status report. *Trends Parasitol.* **2004**, *20*, 477–481.
28. Familton, A.S.; Mason, P.; Coles, G.C. Anthelmintic-resistant *Cooperia* species in cattle. *Vet. Rec.* **2001**, *149*, 719–720.
29. Sangster, N.C.; Dobson, R.J. Anthelmintic Resistance. In *The Biology of Nematodes*; Lee, D.L., Ed.; Taylor and Francis: London, UK, 2002; pp. 531–567.
30. Sutherland, I.A.; Leathwick, D.M. Anthelmintic resistance in nematode parasites of cattle—A global issue? *Trends Parasitol.* **2010**, *27*, 176–181.
31. Demeler, J.; van Zeveren, A.M.; Kleinschmidt, N.; Vercruysse, J.; Höglund, J.; Koopmann, R.; Cabaret, J.; Claerebout, E.; Areskog, M.; von Samson-Himmelstjerna, G. Monitoring the efficacy of ivermectin and albendazole against gastro intestinal nematodes of cattle in Northern Europe. *Vet. Parasitol.* **2009**, *160*, 109–115.
32. El-Abdellati, A.; Charlier, J.; Geldhof, P.; Levecke, B.; Demeler, J.; von Samson-Himmelstjerna, G.; Claerebout, E.; Vercruysse, J. The use of a simplified faecal egg count reduction test for assessing anthelmintic efficacy on Belgian and German cattle farms. *Vet. Parasitol.* **2010**, *169*, 352–357.
33. Bartley, D.J.; Jackson, F.; Jackson, E.; Sargison, N. Characterisation of two triple resistant field isolates of *Teladorsagia* from Scottish lowland sheep farms. *Vet. Parasitol.* **2004**, *123*, 189–199.
34. Sargison, N.D.; Jackson, F.; Bartley, D.J.; Wilson, D.J.; Stenhouse, L.J.; Penny, C.D. Observations on the emergence of multiple anthelmintic resistance in sheep flocks in the south-east of Scotland. *Vet. Parasitol.* **2007**, *45*, 65–76.
35. Wolstenholme, A.J.; Fairweather, I.; Prichard, R.; von Samson-Himmelstjerna, G.; Sangster, N.; Nicholas, C. Drug resistance in veterinary helminths. *Trends Parasitol.* **2004**, *20*, 469–476.

36. Kaminsky, R.; Ducray, P.; Jung, M.; Clover, R.; Rufener, L.; Bouvier, J.; Weber, S.S.; Wenger, A.; Wieland-Berghausen, S.; Goebel, T.; *et al.* A new class of anthelmintics effective against drug-resistant nematodes. *Nature* **2008**, *452*, 176–180.
37. Little, P.R.; Hodges, A.; Watson, T.G.; Seed, J.A.; Maeder, S.J. Field efficacy and safety of an oral formulation of the novel combination anthelmintic, derquantel-abamectin, in sheep in New Zealand. *N. Z. Vet. J.* **2010**, *58*, 121–129.
38. Steinfeld, H.; Gerber, P.; Wassenaar, T.; Castel, V.; Rosales, M.; de Haan, C. *Livestock's Long Shadow: Environmental Issues and Options*; Report presented to the Food and Agricultural Organisation of the United Nations (FAO): Rome, Italy, 2006; p. 284.
39. Gill, M.; Smith, P.; Wilkinson, J.M. Mitigating climate change: The role of domestic livestock. *Animal* **2010**, *4*, 323–333.
40. Thornton, P.K. Livestock production: Recent trends, future prospects. *Philos. Trans. R. Soc. B Biol. Sci.* **2010**, *365*, 2853–2867.
41. Taylor, M.A.; Hunt, K.R.; Goodyear, K.L. Anthelmintic resistance detection methods. *Vet. Parasitol.* **2002**, *103*, 183–194.
42. Gasser, R.B.; Bott, N.J.; Chilton, N.B.; Hunt, P.; Beveridge, I. Toward practical, DNA-based diagnostic methods for parasitic nematodes of livestock—Bionomic and biotechnical implications. *Biotechnol. Adv.* **2008**, *26*, 325–334.
43. Zarlenga, D.S.; Chute, M.; Gasbarre, L.C.; Boyd, P.C. A multiplex PCR assay for differentiating economically important gastrointestinal nematodes of cattle. *Vet. Parasitol.* **2001**, *97*, 199–209.
44. Wimmer, B.; Craig, B.H.; Pilkington, J.G.; Pemberton, J.M. Non-invasive assessment of parasitic nematode species diversity in wild Soay sheep using molecular markers. *Int. J. Parasitol.* **2004**, *34*, 625–631.
45. Learmount, J.; Conyers, C.; Hird, H.; Morgan, C.; Craig, B.H.; von Samson-Himmelstjerna, G.; Taylor, M. Development and validation of real-time PCR methods for diagnosis of *Teladorsagia circumcincta* and *Haemonchus contortus* in sheep. *Vet. Parasitol.* **2009**, *166*, 268–274.
46. Dobson, R.J.; Sangster, N.C.; Besier, R.B.; Woodgate, R.G. Geometric means provide a biased efficacy result when conducting a faecal egg count reduction test (FECRT). *Vet. Parasitol.* **2009**, *161*, 162–167.
47. Coles, G.C.; Jackson, F.; Pomroy, W.E.; Prichard, R.K.; von Samson-Himmelstjerna, G.; Silvestre, A.; Taylor, M.A.; Vercruysse, J. The detection of anthelmintic resistance in nematodes of veterinary importance. *Vet. Parasitol.* **2006**, *136*, 167–185.
48. Von Samson-Himmelstjerna, G.; Blackhall, W.J.; McCarthy, J.S.; Skuce, P.J. Single nucleotide polymorphism (SNP) markers for benzimidazole resistance in veterinary nematodes. *Parasitology* **2007**, *134*, 1077–1086.
49. Von Samson-Himmelstjerna, G.; Walsh, T.K.; Donnan, A.A.; Carriere, S.; Jackson, F.; Skuce, P.J.; Rohn, K.; Wolstenholme, A.J. Molecular detection of benzimidazole resistance in *Haemonchus contortus* using real-time PCR and pyrosequencing. *Parasitology* **2009**, *136*, 349–358.
50. Skuce, P.J.; Stenhouse, L.; Jackson, F.; Hypsa, V.; Gilleard, J.S. Benzimidazole resistance allele haplotype diversity in United Kingdom isolates of *Teladorsagia circumcincta* supports a hypothesis of multiple origins of resistance by recurrent mutation. *Int. J. Parasitol.* **2010**, *40*, 1247–1255.
51. Höglund, J.; Gustaffson, K.; Ljungstrom, B.L.; Engstrom, A.; Donnan, A.A.; Skuce, P.J. Anthelmintic resistance in Swedish sheep flocks based on a comparison of the results from the faecal egg count reduction test and resistant allele frequencies of the beta-tubulin gene. *Vet. Parasitol.* **2009**, *161*, 60–68.
52. Beech, R.N.; Skuce, P.; Bartley, D.J.; Martin, R.J.; Prichard, R.K.; Gilleard, J.S. Anthelmintic resistance: Markers for resistance, or susceptibility? *Parasitology* **2010**, *138*, 160–174.
53. Gilleard, J.S. Understanding anthelmintic resistance: The need for genomics and genetics. *Int. J. Parasitol.* **2006**, *36*, 1227–1239.
54. Vignali, D.A.A. Multiplexed particle-based flow cytometric assays. *J. Immunol. Methods* **2000**, *243*, 243–255.

55. Pickering, J.W.; Martin, T.B.; Schroder, M.C.; Hill, H.R. Comparison of a multiplex flow cytometric assay with Enzyme-Linked Immunosorbent Assay for quantification of antibodies to Tetanus, Diphtheria and *Haemophilus influenza* Type b. *Clin. Diagn. Lab. Immunol.* **2002**, *9*, 872–876.
56. Dunbar, S.A.; Vander Zee, C.A.; Oliver, K.G.; Karem, K.L.; Jacobson, J.W. Quantitative, multiplexed detection of bacterial pathogens: DNA and protein applications of the Luminex LabMAPTM system. *J. Microbiol. Methods* **2003**, *53*, 245–252.
57. Morgan, E.R.; Varro, R.; Sepulveda, H.; Ember, J.A.; Apgar, J.; Wislon, J.; Lowe, L.; Chen, R.; Shivraj, L.; Agadir, A.; *et al.* Cytometric bead array: A multiplexed assay platform with applications in various areas of biology. *Clin. Immunol.* **2004**, *110*, 252–266.
58. Dunbar, S.A. Application of Luminex xMAPTM technology for rapid, high-throughput multiplexed nucleic acid detection. *Clinia Chim. Acta* **2006**, *363*, 71–82.
59. Charlier, J.; Vercruysse, J.; Smith, J.; Vanderstichel, R.; Stryhn, H.; Claerebout, E.; Dohoo, I. Evaluation of anti-*Ostertagia ostertagi* antibodies in individual milk samples as decision parameter for selective anthelmintic treatment in dairy cows. *Prev. Vet. Med.* **2009**, *93*, 147–152.
60. McCann, C.M.; Baylis, M.; Williams, D.J.L. Seroprevalence and spatial distribution of *Fasciola hepatica*-infected dairy herds in England and Wales. *Vet. Rec.* **2010**, *166*, 612–617.
61. Andersen, U.V.; Howe, D.K.; Dangoudoubiyam, S.; Toft, N.; Reinemeyer, C.R.; Lyons, E.T.; Olsen, S.N.; Monrad, J.; Nejsum, P.; Nielsen, M.K. SvSXP: A *Strongylus vulgaris* antigen with potential for prepatent diagnosis. *Parasites Vectors* **2013**, *6*, 84; doi:10.1186/1756-3305-6-84.
62. Ganheim, C.; Höglund, J.; Waller, P. Acute phase proteins in response to *Dictyocaulus viviparus* infection in calves. *Acta Vet. Scand.* **2004**, *45*, 79–86.
63. Charlier, J.; Dorny, P.; Levecke, B.; Demeler, J.; von Samson-Himmelstjerna, G.; Höglund, J.; Vercruysse, J. Serum pepsinogen levels to monitor gastrointestinal nematode infections in cattle revisited. *Res. Vet. Sci.* **2011**, *90*, 451–456.
64. Bandyopadhyay, K.; Kellar, K.L.; Moura, I.; Casaqui Carollo, M.C.; Graczyk, T.K.; Slemenda, S.; Johnston, S.P.; da Silva, A.J. Rapid microsphere assay for identification of *Cryptosporidium hominis* and *Cryptosporidium parvum* in stool and environmental samples. *J. Clin. Microbiol.* **2007**, *45*, 2835–2840.
65. Li, W.; Zhang, N.; Gong, P.; Cao, L.; Li, J.; Su, L.; Li, S.; Diao, Y.; Wu, K.; Li, L.; *et al.* A novel multiplex PCR coupled with Luminex assay for the simultaneous detection of *Cryptosporidium* spp., *Cryptosporidium parvum* and *Giardia duodenalis*. *Vet. Parasitol.* **2010**, *173*, 11–18.
66. Stark, D.; Al-Qassab, S.E.; Barratt, J.L.; Stanley, K.; Roberts, T.; Marriott, D.; Harkness, J.; Ellis, J. An evaluation of a multiplex tandem real-time PCR for the detection of *Cryptosporidium* spp, *Dientamoeba fragilis*, *Entamoeba histolytica*, and *Giardia intestinalis* from clinical stool samples. *J. Clin. Microbiol.* **2010**, *49*, 257–262.
67. Iseki, H.; Alhassan, A.; Ohta, N.; Thekisoe, O.M.M.; Yokoyama, N.; Inoue, N.; Nambota, A.; Yasuda, J.; Igarashi, I. Development of a multiplex loop-mediated isothermal amplification (mLAMP) method for the simultaneous detection of bovine *Babesia* parasites. *J. Microbiol. Methods* **2007**, *71*, 281–287.
68. Parida, M.; Sannarangaiah, S.; Dash, P.K.; Rao, P.V.; Morita, K. Loop mediated isothermal amplification (LAMP): A new generation of innovative gene amplification technique; perspectives in clinical diagnosis of infectious diseases. *Rev. Med. Virol.* **2008**, *18*, 407–421.
69. Aonuma, H.; Yoshimura, A.; Perera, N.; Shinzawa, N.; Bando, H.; Oshiro, S.; Nelson, B.; Fukumoto, S.; Kanuka, H. Loop-mediated isothermal amplification applied to filarial parasites detection in the mosquito vectors: *Dirofilaria immitis* as a study model. *Parasites Vectors* **2009**, *2*, 15–21.
70. Mori, Y.; Notomi, T. Loop-mediated isothermal amplification (LAMP): A rapid, accurate and cost-effective diagnostic method for infectious diseases. *J. Infect. Chemother.* **2009**, *15*, 62–69.

71. Ollerenshaw, C.B.; Rowlands, W.T. A method of forecasting the incidence of fascioliasis in Anglesey. *Vet. Rec.* **1959**, *71*, 591–598.
72. Cornell, S. Modelling nematode populations: 20 years of progress. *Trends Parasitol.* **2005**, *21*, 542–545.
73. Fuentes, M.V. Remote sensing and climate data as a key for understanding fasciolosis transmission in the Andes: Review and update of an ongoing interdisciplinary project. *Geospat. Health* **2006**, *1*, 59–70.
74. Rose, H.; Wall, R. Modelling the impact of climate change on spatial patterns of disease risk: Sheep blowfly strike by *Lucilia sericata* in Great Britain. *Int. J. Parasitol.* **2011**, *41*, 739–746.
75. Biggeri, A.; Catelan, D.; Dreassi, E.; Rinaldi, L.; Musella, V.; Veneziano, V.; Cringoli, G. Multivariate spatially-structured variability of ovine helminth infections. *Geospat. Health* **2007**, *2*, 97–104.
76. Hendrickx, G.; Biesemans, J.; de Deken, R. The Use of GIS in Veterinary Parasitology. In *GIS and Spatial Analysis in Vet. Science*; Durr, P., Gatrell, A., Eds.; CABI Publishing: Wallingford, UK, 2004; pp. 145–176.
77. Rinaldi, L.; Musella, V.; Biggeri, A.; Cringoli, G. New insights into the application of geographical information systems and remote sensing in veterinary parasitology. *Geospat. Health* **2006**, *1*, 33–47.
78. Bergquist, R.; Rinaldi, L. Health research based on geospatial tools: A timely approach in a changing environment. *J. Helminthol.* **2010**, *84*, 1–11.
79. Geospatial health: Health applications in geospatial science. Available online: http://www. geospatialhealth.unina.it (accessed on 11 August 2013).
80. Morgan, E.R.; Milner-Gulland, E.J.; Torgerson, P.R.; Medley, G.F. Ruminating on complexity: Macroparasites of wildlife and livestock. *Trends Ecol. Evol.* **2004**, *19*, 181–188.
81. Morgan, E.R.; van Dijk, J. Climate and the epidemiology of gastrointestinal nematode infections of sheep in Europe. *Vet. Parasitol.* **2012**, *189*, 8–14.
82. Van Dijk, J.; David, G.P.; Baird, G.; Morgan, E.R. Back to the future: Developing hypotheses on the effects of climate change on ovine parasitic gastroenteritis from historical data. *Vet. Parasitol.* **2008**, *158*, 73–84.
83. Smith, G. Modeling of parasite populations—Gastrointestinal nematode models. *Vet. Parasitol.* **1994**, *54*, 127–143.
84. Van Dijk, J.; Morgan, E.R. Variation in the hatching behaviour of *Nematodirus battus*: Polymorphic bet hedging? *Int. J. Parasitol.* **2010**, *40*, 675–681.
85. Kenyon, F.; Greer, A.W.; Coles, G.C.; Cringoli, G.; Papadopoulos, E.; Cabaret, J.; Berrag, B.; Varady, M.; van Wyk, J.A.; Thomas, E.; *et al.* The role of targeted selective treatments in the development of refugia-based approaches to the control of gastrointestinal nematodes of small ruminants. *Vet. Parasitol.* **2009**, *164*, 3–11.
86. Cooper, K.M.; Whelan, M.; Kennedy, D.G.; Trigueros, G.; Cannavan, A.; Boon, P.E.; Wapperom, D.; Danaher, M. ProSafeBeef and anthelmintic drug residues-a case study in collaborative application of multi-analyte mass spectrometry to enhance consumer safety. *Anal. Bioanal. Chem.* **2012**, *404*, 1623–1630.
87. Imperiale, F.; Ortiz, P.; Cabrera, M.; Farias, C.; Sallovitz, J.M.; Iezzi, S.; Perez, J.; Alvarez, L.; Lanusse, C. Residual concentrations of the flukicidal compound triclabendazole in dairy cows' milk and cheese. *Food Addit. Contam. Part A Chem. Anal. Control Expo. Risk Assess.* **2011**, *28*, 438–445.
88. Greer, A.W.; Kenyon, F.; Bartley, D.J.; Jackson, E.B.; Gordon, Y.; Donnan, A.A.; McBean, D.W.; Jackson, F. Development and field evaluation of a decision support model for anthelmintic treatments as part of a targeted selective treatment (TST) regime in lambs. *Vet. Parasitol.* **2009**, *164*, 12–20.
89. Cringoli, G.; Rinaldi, L.; Veneziano, V.; Mezzino, L.; Vercruysse, J.; Jackson, F. Evaluation of targeted selective treatments in sheep in Italy: Effects on faecal worm egg count and milk production in four case studies. *Vet. Parasitol.* **2009**, *164*, 36–43.

90. Gallidis, E.; Papadopoulos, E.; Ptoches, S.; Arsenos, G. The use of targeted selective treatments against gastrointestinal nematodes in milking sheep and goats in Greece based on parasitological and performance criteria. *Vet. Parasitol.* **2009**, *164*, 53–58.
91. Höglund, J.; Morrison, D.A.; Charlier, J.; Dimander, S.; Larrson, A. Targeted selective treatments for gastrointestinal nematodes in first-season grazing cattle based on mid-season daily weight gains. *Vet. Parasitol.* **2009**, *164*, 80–88.
92. Leask, R.; van Wyk, J.A.; Thompson, P.N.; Bath, G.F. The effect of application of the FAMACHA© system on selected production parameters in sheep. *Small Rumin. Res.* **2013**, *110*, 1–8.
93. Gaba, S.; Cabaret, J.; Chylinski, C.; Sauve, C.; Cortet, J.; Silvestre, A. Can efficient management of sheep gastro-intestinal nematodes be based on random treatment? *Vet. Parasitol.* **2012**, *190*, 178–184.
94. Terrill, T.H.; Miller, J.E.; Burke, J.M.; Mosjidis, J.A.; Kaplan, R.M. Experiences with integrated concepts for the control of *Haemonchus contortus* in sheep and goats in the United States. *Vet. Parasitol.* **2012**, *186*, 28–37.
95. Van Wyk, J.A.; Reynecke, D.P. Blueprint for an automated specific decision support system for countering anthelmintic resistance in *Haemonchus* spp. at farm level. *Vet. Parasitol.* **2011**, *177*, 212–223.
96. Gauly, M.; Bollwein, H.; Breves, G.; Bruegemann, K.; Daenicke, S.; Das, G.; Demeler, J.; Hansen, H.; Isselstein, J.; Koenig, S.; *et al.* Future consequences and challenges for dairy cow production systems arising from climate change in Central Europe—A review. *Animal* **2013**, *7*, 843–859.

CHAPTER 8

Screening food-borne and zoonotic pathogens associated with livestock practices in the Sumapaz region, Cundinamarca, Colombia

Nelson E. Arenas[1], Diego A. Abril[1], Paola Valencia[1], Surabhi Khandige[1], Carlos Yesid Soto[1], Vilma Moreno-Melo[1]

1 Faculty of Agricultural Sciences, Universidad de Cundinamarca, Diagonal 18 No. 20-29, Fusagasugá, Cundinamarca 252211, Colombia

2 Department of Biochemistry and Molecular Biology, University of Southern Denmark, Campusvej 55, 5230 Odense M, Denmark

3 Department of Chemistry, Faculty of Sciences, Universidad Nacional de Colombia, Carrera 30, No. 45-03, Ciudad Universitaria, Bogotá, Colombia

ABSTRACT

Hazardous practices regarding antibiotics misuse, unsanitary milking procedures, and the commercial sales of raw milk and unpasteurized dairy products are currently being practiced by livestock farmers in the Sumapaz region (Colombia). The purpose of this study was to screen for food-borne and zoonotic pathogens associated with local livestock practices. We evaluated 1098 cows from 46 livestock farms in the Sumapaz region that were selected by random. Of the total population of cattle, 962 animals (88%) were tested for bovine TB using a caudal-fold tuberculin test and 546 (50%) for brucellosis by a competitive ELISA. In the population tested, 23 cows were positive for *Brucella* sp. representing a 4.2% seroprevalence and no cases of bovine tuberculosis were found. In addition, food-borne contamination with *Escherichia coli* and *Staphylococcus aureus* was assessed together with antibiotic susceptibility for ten different antibiotics in milk samples from 16 livestock farms. We found that 12 of the farms (75%) were contaminated with these food-borne pathogens. Noteworthy, all of the isolated pathogenic strains were resistant to multiple antibiotics, primarily to oxytetracycline and erythromycin. Our findings suggest that livestock products could be a source of exposure to *Brucella* and multidrug-resistant *E. coli* and *S. aureus* strains as a result of unhygienic livestock practices in the Sumapaz region. Training in good farming practices is the key to improving safety in food production.

Keywords: Bovine brucellosis, Bovine tuberculosis, Escherichia coli, Staphylococcus aureus, Zoonoses, Antibiotic resistance

1. INTRODUCTION

Prevailing food regulations demand high-quality assurance and safety in the production chain (Grace 2015). Meeting these global safety standards mandates significant changes in farming practices and animal husbandry (Van Boeckel et al. 2015). The Colombian Agriculture and Livestock Institute (Spanish abbreviation: ICA) is the local authority that assesses compliance with international guidelines, infectious disease control, and compulsory epidemiological registration. The ICA's programs include the screening and epidemiological surveillance of zoonotic diseases associated with livestock and agricultural farming. These programs are based on good farming practices (GFPs) for animal production food safety which are procedures formulated to contribute to sustainable livestock production in the country (Roess et al. 2015). In fact, GFPs have demonstrated excellent results in farm productivity and have contributed to local development, the maintenance of good animal health and increasing economic profits for farmers (Crowder and Reganold 2015). GFPs have been focused on the control of brucellosis and tuberculosis (TB), which are considered to have the greatest impact among zoonotic diseases in Colombia and South America.

Zoonotic disease eradication includes screening programs and the epidemiological surveillance of food-borne pathogens as key strategic activities (Grace 2015). In the Sumapaz region, livestock farming is the central economic activity; it is performed empirically by local farmers without an in-depth understanding of GFPs. Epidemiological surveillance is important because the commercialization of unpasteurized dairy products can pose a threat to human public health. In addition, zoonotic infections, such as brucellosis and TB, are common, insidious, and persistent problems for livestock farmers in general in developing countries (Aznar et al. 2014; Roess et al. 2015). Despite control measures and sanitation efforts, zoonotic disease outbreaks have been reported in Colombia (Morales and Combariza 2004; González et al. 2007; Tique et al. 2009, 2010). In 2013, brucellosis was the most common zoonotic infection in Colombia; 12,000 cases are reported annually and 343 cases of these occur in the Sumapaz region and neighboring areas. Consequently, a decrease in milk production from 168,216 L/day in 2007 to 117,538 L/day in 2009 (30% decrease) has been reported in the Sumapaz region. In contrast, the bovine TB prevalence has been estimated to be less than 1% in Colombia (De Waard 2010). Along with its economic impact, the morbidity and mortality associated with such zoonotic infections such as brucellosis and bovine TB are also relatively unknown or underestimated by livestock farmers and government institutions. Our purpose was to develop a screening and sanitation program in accordance with GFPs that would involve local farmers in the control of zoonotic and food-borne diseases such as brucellosis and TB to improve animal health.

2. MATERIAL AND METHODS

2.1 Study area and sampling

A cross-sectional study was performed from January 2014 to June 2015 to detect and identify zoonotic and food-borne pathogens associated with livestock practices in the Sumapaz region. With an area of 1808 km^2, the Sumapaz region is located in the south of Cundinamarca (representing 8% of the province) and is delimited by Tequendama and Soacha in the north, the province of Tolima in the south, Bogotá in the east, and Alto Magdalena in the west (Fig. 1). Livestock farming and animal husbandry is an important economic activity in this region; thus, 46 livestock farms were randomly selected from Fusagasugá, Silvania, Tibacuy, Arbeláez, Cabrera, Pandi, and Pasca municipalities. Paradoxically, livestock farming is performed empirically without quality assurance schemes and marketing of raw milk and unpasteurized dairy products is currently performed in these municipalities. Five breeds of cattle have been recorded including a local breed (named "Creole"), Holstein, Pardo, Normande, and Cebu (called "Mestizo"). One thousand ninety-eight animals (spread over 46 livestock farms randomly selected in the Sumapaz area) were screened for bovine TB and brucellosis. Of the total population of cattle, 962 animals (88%) were tested for bovine TB and 546 (50%) for brucellosis. Out of 46 livestock farms, 16 were based on milk production. In each of them, ten milk samples were aseptically collected from different random selected cows and were pooled in a bulk tank for further microbiological analyses. In addition, nutrition analysis in raw milk pools was determined using an ultrasonic milk analyzer (LactoScan system, Milkotronic, Ltd., Nova Zagora, Bulgaria) testing: the protein percentage, fat percentage, milk solid nonfat, density, pH, lactose percentage, and total solid percentage.

Figure. 1 Geographical location of the Sumapaz region in Cundinamarca, Colombia. *Shadow* area inside the *circle* represents the location of the Sumapaz area

2.2 Bovine tuberculosis and brucellosis detection

For *Brucella* diagnosis, 10 mL of blood was collected in sterile vacutainer tubes and stored at 4 °C until required. Samples were centrifuged at 3000×*g* for 10 min and processed according to guidelines (ICA 2010). A competitive enzyme-linked immunosorbent assay (ELISA) (Boehringer Ingelheim, Uppsala, Sweden) was performed consistently with the previously standardized technique (Rivera et al. 2003). Results were expressed as percent inhibition by each serum considering a value ≥30% was interpreted as a positive and <30% as negative test. On the other hand, caudal-fold tuberculin tests (CFTs; Thermofisher Scientific, Schlieren-Zurich, Switzerland) were performed by ICA's certified veterinarians for TB screening. Briefly, the animals were injected intradermally with purified protein derivative (PPD) tuberculin, and the injection site was measured 72 h later. Hypersensitivity reactions, such as swelling or discoloration, were interpreted according to ICA standard procedures (ICA 2004).

2.3 Bacterial identification

For microbiological tests, approximately 15 mL of pooled milk was aseptically collected in sterile tubes and stored at 4 °C. Milk dilutions were streaked in LB plates and incubated at 37 °C to obtain single colonies. Colonies were isolated and classified according to morphological features, shape, and Gram staining. Bacterial identification was performed by using phenotypic methods based on biochemical reactions and bacterial metabolism (sugar fermentation, indole, methyl-red, Voges-Proskauer, motility, gas production, and citrate tests) for Gram-positive and Gram-negative bacteria. Bacterial isolates were tested for antibiotic susceptibility to ten drugs that are commonly prescribed for veterinary use. Antimicrobial susceptibility tests were performed by Kirby-Bauer disk diffusion method in Mueller-Hinton agar plates and interpreted according to the guidelines of the National Committee for Clinical Laboratory Standards (NCCLS) and national regulations for antibiotic use in livestock practices (Ministerio de Salud y Proteccion Social 2013). Eighty percent of the veterinary antibiotics selected (8/10) have broad antibacterial activity against Gram-negative and Gram-positive bacteria involving different drug mechanisms. Data about current livestock practices were assessed with regard to milking routines, tool sterilization, and drug records gathered from livestock smallholders participating in this study.

2.4 Ethical considerations

All cattle tests followed the laws and regulations recommended by the European Communities Council Directive of November 24, 1986 (86/609/EEC) and were in agreement with the ICA's regulations. Furthermore, in all cases, the livestock farmers were informed verbally about the procedures, risks, and benefits associated with participation in the study.

3. RESULTS

To evaluate zoonotic disease control in the livestock farms ($n = 46$), bovine TB and brucellosis were screened through CFTs and competitive ELISA, respectively.

No cases of bovine TB were detected and 23 (4.2%) sera from clinically normal cattle were positive for brucellosis. Nineteen of the brucellosis cases are presented in Table 1. The animals represented were all female and belonged to the following breeds: Brahman ($n = 9$), Cross-bred ($n = 3$), Bon ($n = 3$), Holstein ($n = 1$), Normande ($n = 1$), Creole ($n = 1$), and unknown ($n = 1$). All sera showed inhibition levels between 47 and 96% in the competitive ELISA tests. Furthermore, 30 livestock farmers who currently performed milking routines on the farms with index cases (brucellosis) were also tested, but were found to be negative for brucellosis and bovine TB.

Table 1: Characteristics of brucellosis infection in cattle. Inhibition ≥30% indicated a positive test, and inhibition <30% indicated a negative test. The results were obtained with competitive ELISA

Location	Municipality	Breed	% inhibition	Age (months)
1	Arbeláez	Brahman	54	25
2	Arbeláez	Brahman	64	27
3	Arbeláez	Brahman	47	45
4	Arbeláez	Brahman	47	43
5	Arbeláez	Brahman	60	45
6	Arbeláez	Brahman	48	41
7	Arbeláez	Brahman	50	29

8	Arbeláez	Brahman	48	24
9	Arbeláez	Brahman	56	47
10	Arbeláez	Cross-breed	89	64
11	Arbeláez	nd	>30	nd
12	Fusagasugá	Cross-breed	96	104
13	Fusagasugá	Cross-breed	96	30
14	Fusagasugá	Bon Cebu	75	37
15	Fusagasugá	Bon	47	58
16	Fusagasugá	Bon Cebu	50	30
17	Pasca	Holstein	96	55
18	Pasca	Normand	>30	nd
19	Pandi	Creole	53	72

nd data non-determined

To assess milk quality and safety before GFP achievement, nutritional composition and microbiological tests were performed in the 16 milk pooled samples. Nutrition analysis did not show any significant differences in nutritional content between samples (data not shown). Nevertheless, the majority of milk pooled samples showed microbial contamination. Thus, cultures of four pooled samples were negative, while 12 of the milk samples (75%) indicated bacterial contamination with *Staphylococcus aureus* ($n = 1$) and *Escherichia coli* ($n = 11$). None of the bacterial isolates displayed sensitivity to any of the tested antibiotics (other than enrofloxacin), instead most of them showed extended drug resistance. All *E. coli* isolates presented multiple resistance to at least four different antibiotics in 90% ($n = 11$) of all cases. The most concerning findings were that four of the examined *E. coli* isolates (33.3%) were resistant to seven of the tested antibiotics, while two other *E. coli* isolates (16.7%) showed resistance to five to six antimicrobial drugs (Table 2). A single *S. aureus* isolate was found to be resistant to oxytetracycline (OXY) and erythromycin (ERY). Antibiotic resistance involving DOX and/or ERY was often detected in 11 (92%) strains; ten strains showed amoxicillin (AMX) resistance (83%) and eight of the bacterial isolates (67%) showed resistance to fosfomycin (FOF), OXY, and gentamicin (GEN) (Fig. 2). Overall, this study reports a very high prevalence of multidrug-resistant strains of *E. coli* involving livestock farming in the Sumapaz region. The milk contamination observed in the course of our study could be explained by the fact that 91% of livestock farmers performed milking routines manually without gloves and missing knowledge about possible sources of bacterial contamination. These practices did not include the sterilization and cleaning of milking tools, and 55% of the farmers included in our study lacked proper records regarding the administration of drugs and medicines to their livestock herds.

Table 2 Detection of multidrug antibiotic resistance in pathogens (*E. coli* and *S. aureus*) associated with livestock farming in the Sumapaz region. The thresholds for minimum inhibitory concentrations were assessed using local regulations for livestock farming

Number of isolates (%)	Number of antibiotics showing resistance	Antimicrobial resistance pattern (number of isolates)
1 (8.3)	2	OXY, ERY (1)
3 (25)	4	AMX, GEN, ERY, DOX (1)
		FOF, AMX, ERY, DOX (1)
		FOF, SXT, NOR, DOX (1)
2 (16.7)	5	SXT, AMX, GEN, ERY, DOX (1)
		FOF, AMX, OXY, ERY, DOX (1)
2 (16.7)	6	FOF, AMX, OXY, GEN, ERY, DOX (2)
4 (33.3)	7	SXT, AMX, CIP, OXY, GEN, ERY, DOX (1)
		FOF, SXT, AMX, OXY, GEN, ERY, DOX (3)

FOF fosfomycin, *SXT* trimethoprim-sulfamethoxazole, *AMX* amoxicillin, *CIP* ciprofloxacin, *OXY* oxytetracycline, *NOR* norfloxacin, *EFX* enrofloxacin, *GEN* gentamicin, *ERY* erythromycin, *DOX* doxycycline

4. DISCUSSION

The adoption of GFP guidelines represents an important step in the development of sustainable agriculture in the Sumapaz region (Cundinamarca, Colombia). Many livestock farmers have a low income and lack required knowledge about GFPs, which leads to low milk productivity and poor quality together with commercial value loss. Therefore, GFP training is essential for helping farmers achieve high-quality standards and develop sustainable livestock practices (Leibler et al. 2016; Abbasi and Abbasi 2016). In Colombia, antibiotic use in livestock farming has not been considered a major public health problem in terms of emerging antibiotic-resistant strains. The lack of adequate data and thorough surveillance make the true impact of antibiotic use in livestock challenging to assess (Landers et al. 2012). Thus, multidrug antibiotic resistance in farming practices might be associated with antibiotic misuse and unsanitary farming practices, as has been reported in Monteria, Colombia (Máttar et al. 2009). In this study, *E. coli* and *S. aureus* were detected in raw milk; their presence was associated with poor milking practices and the cattle's exposure to the indiscriminate use of veterinary drugs. Thus, the milking routine is a key determinant of quality milk production. Unsanitary handling of the cow's udder during milking is a frequent source of contamination with *S. aureus*, which is recognized as a microbial contaminant (Kateete et al. 2013). GFP guidelines address those problems through the implementation of protocols through which livestock farmers can control environmental bacteria such as *E. coli* and *S. aureus*, which primarily tend to colonize the teat skin (Azevedo et al. 2016).

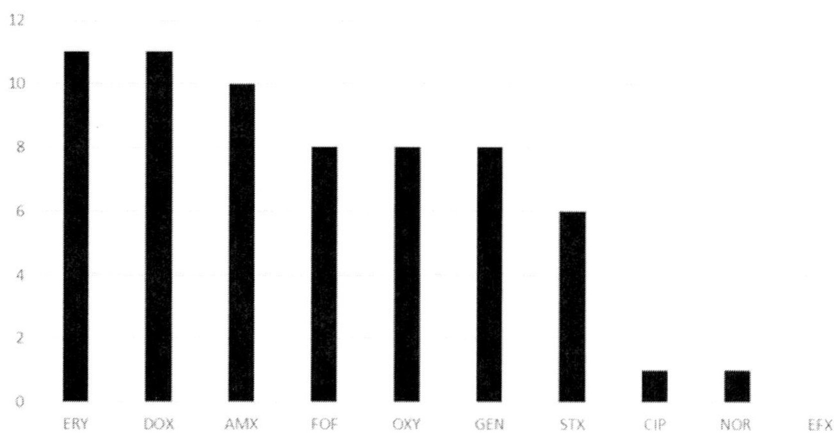

Figure. 2 Frequency of antibiotic resistance of each drug tested in *E. coli* and *S. aureus* isolates. *FOF* fosfomycin, *SXT* trimethoprim-sulfamethoxazole, *AMX* amoxicillin, *CIP* ciprofloxacin, *OXY* oxytetracycline, *NOR* norfloxacin, *EFX* enrofloxacin, *GEN* gentamicin, *ERY* erythromycin, *DOX* doxycycline

The emergence of antibiotic-resistant and multidrug-resistant bacterial contaminants in milk is further complicated by the sale of raw milk to local consumers or small-scale traders in the Sumapaz region. In other settings, the marketing of raw milk has been associated with major health problems, such as the emergence of *E. coli* O157:H7 strains (Addo et al.

2011; Mhone et al. 2011; Rasheed et al. 2014; Amagliani et al. 2016). Concerning antibiotic susceptibility, OXY, ERY, and AMX do not seem to play a significant selective role against microbial contamination in milk; rather, they contribute to increased bacterial resistance or create selective pressure for microorganisms that are already resistant to those drugs. Therefore, livestock farmers should be trained in sustainable livestock production, particularly with regard to antibiotic stewardship (Chang et al. 2015). Contaminated milk might lead to the emergence of strains with extended antibiotic resistance and the potential to spread among human and animal populations, which could influence the farm production environment (Haran et al. 2012; Iweriebor et al. 2015).

Screening for other food-borne pathogens, such as *Brucella abortus* and *Mycobacterium bovis*—the causative agents of brucellosis and bovine TB, respectively—was also performed in this study. Twenty-three cases of brucellosis and none of bovine TB were detected. Brucellosis is a zoonotic infection that is contagious to humans and requires an expensive and prolonged treatment regimen. In Colombia, brucellosis has been recognized as a considerable challenge faced by livestock farmers. The bovine brucellosis incidence in the Caribbean region has been estimated to range from 3.7 to 12.7% in animals and farms (Aznar et al. 2014). Approximately 4% of humans working in slaughterhouses in Tolima have been diagnosed with brucellosis (Morales and Combariza 2004). Additionally, the real prevalence of brucellosis might be underestimated in Colombia because of a lack of proper registration and a need for specific and sensitive diagnostic methods (Higgins et al. 2012). Second, no cases of bovine TB were detected, which seems to reflect the progress of the Bovine Tuberculosis Eradication Program in Colombia, which has maintained the bovine TB prevalence below 1% (De Waard 2010). However, because the CFT is a screening test, negative results cannot be considered absolute proof of the absence of bovine TB in a given cow or herd. Thus, CFT only provides an indication if an animal develops an immune response to the antigen for *M. bovis* (De Kantor and Ritacco 2006).

Epidemiological surveillance of livestock production might support the government's planning initiatives and policies, programs, and procedures to protect animal and human health (Henrioud 2011; Donado-Godoy et al. 2015). The Ministry of Agriculture and Rural Development in Colombia assesses some economic benefits for livestock farmers who implement GFPs. Thus, GFP acceptance has been difficult to promote among small-holder farmers in the Sumapaz region because of the initial economic investment, the lack of knowledge about cost-benefits, the need for follow-up registration, and a lack of understanding about practices. Once livestock farmers accepted the advantages, it was possible to optimize the production and commercialization of dairy products and promote ecological practices (Seufert et al. 2012). Problems regarding the commercialization of contaminated dairy products and raw milk or infected cattle still persist in the Sumapaz region. Thus, we expect that the training and implementation of GFPs would contribute to the development of an agriculture sustainability model for livestock farmers in the Sumapaz region.

In conclusion, screening of food-borne and zoonotic pathogens is pivotal to control accurately potential outbreaks in the Sumapaz region. Although TB cases were not identified, brucellosis remains as a frequent zoonotic disease (4.2% seroprevalence). Occurrence of multidrug-resistant *E. coli* strains is relatively high and could represent a source of infection for farmers and consumers of dairy products. Further studies to identify the molecular basis of antibiotic resistance are required.

AMX, amoxicillin; CFTs, caudal-fold tuberculin tests; CIP, ciprofloxacin; DOX, doxycycline; EFX, enrofloxacin; ELISA, enzyme-linked immunosorbent assay; ERY, erythromycin; FOF, fosfomycin; GEN, gentamicin; GFPs, good farming practices; ICA, Colombian Agriculture and Livestock Institute; NCCLS, National Committee for Clinical Laboratory Standards; NOR, norfloxacin; OXY, oxytetracycline; PPD, purified protein derivative; SXT, trimethoprim-sulfamethoxazole; TB, tuberculosis

ACKNOWLEDGEMENTS

The authors would like to acknowledge the SERVIGAN-Sábana LTDA Laboratory for sample testing. We also would like to acknowledge all of the livestock farmers who generously participated in this study and Natalia Pardo, Juan F. Morales, and Juan C. López for technical support. We thank Adriana Paola Bendek and Michal Beker for proofreading this manuscript. This project was funded by the Committee of Livestock Farmers from the Sumapaz region (COMIGAN-Sumapaz), Colombian Association for Science Advance (Spanish abbreviation: ACAC), and Universidad de Cundinamarca. Overall project is knowledge transfer in Good Farming Practices to livestock farmers from the Sumapaz region, Fusagasugá, Cundinamarca.

REFERENCES

1. Abbasi, T. and Abbasi, S.A., 2016. Reducing the global environmental impact of livestock production: the minilivestock option. Journal of Cleaner Production, 112, 1754–1766.
2. Addo, K.K., Mensah, G.I., Aning, K.G., Nartey, N., Nipah, G.K., Bonsu, C., Akyeh, M.L. and Smits, H.L., 2011. Microbiological quality and antibiotic residues in informally marketed

raw cow milk within the coastal savannah zone of Ghana. International Journal of Environmental Research and Public Health, 16, 227–232.

3. Amagliani, G., Petruzzelli, A., Carloni, E., Tonucci, F., Foglini, M., Micci, E., Ricci, M., Di Lullo, S., Rotundo, L. and Brandi, G., 2016. Presence of *Escherichia coli* O157, *Salmonella* spp., and *Listeria monocytogenes* in raw ovine milk destined for cheese production and evaluation of the equivalence between the analytical methods applied. Foodborne Pathogens and Disease, 13, 626–632.
4. Azevedo, C., Pacheco, D., Soares, L., Romão, R., Moitoso, M., Maldonado, J., Guix, R. and Simões, J., 2016. Prevalence of contagious and environmental mastitis-causing bacteria in bulk tank milk and its relationships with milking practices of dairy cattle herds in São Miguel Island (Azores). Tropical animal health and production, 48, 451–459.
5. Aznar, M.N., Samartino, L.E., Humblet, M.F. and Saegerman, C., 2014. Bovine brucellosis in Argentina and bordering countries: update. Transboundary and Emerging Diseases, 61, 121–133.
6. Chang, Q., Wang, W., Regev-Yochay, G., Lipsitch, M. and Hanage, W.P., 2015. Antibiotics in agriculture and the risk to human health: how worried should we be?. Evolutionary Applications, 8, 240–247.
7. Crowder, D.W. and Reganold, J.P., 2015. Financial competitiveness of organic agriculture on a global scale. Proceedings of the National Academy of Sciences of the United States of America, 112, 7611–7616.
8. De Kantor, I.N. and Ritacco, V., 2006. An update on bovine tuberculosis programmes in Latin American and Caribbean countries. Veterinary Microbiology, 112, 111–118.
9. De Waard, J.H., 2010. ¿Ordeñando micobacterias del ganado? Impacto económico y en salud de Tuberculosis bovina y Paratuberculosis en Colombia. Revista MVZ Córdoba, 15, 2037–2040.
10. Donado-Godoy, P., Castellanos, R., León, M., Arevalo, A., Clavijo, V., Bernal, J., León, D., Tafur, M.A., Byrne, B.A., Smith, W.A. and Perez-Gutierrez, E., 2015. The establishment of the Colombian Integrated Program for Antimicrobial Resistance Surveillance (COIPARS): A pilot project on poultry farms, slaughterhouses and retail market. Zoonoses Public Health, 62, 58–69.
11. González, M., Ríos, R. and Mattar, S., 2007. Prevalencia de bacterias asociadas a la infertilidad infecciosa en bovinos de Montería, Colombia. Revista MVZ Córdoba, 12, 1028–1035.
12. Grace, D., 2015. Food safety in low and middle income countries. International Journal of Environmental Research and Public Health, 12, 10490–10507.
13. Haran, K.P., Godden, S.M., Boxrud, D., Jawahir, S., Bender, J.B. and Sreevatsan, S., 2012. Prevalence and characterization of *Staphylococcus aureus*, including methicillin-resistant *Staphylococcus aureus*, isolated from bulk tank milk from Minnesota dairy farms. Journal of Clinical Microbiology, 50, 688–695.
14. Henrioud, A.N., 2011. Towards sustainable parasite control practices in livestock production with emphasis in Latin America. Veterinary Parasitology, 180, 2–11.
15. Higgins, J., Stuber, T., Quance, C., Edwards, W.H., Tiller, R.V., Linfield, T., Rhyan, J., Berte, A. and Harris, B., 2012. Molecular epidemiology of *Brucella abortus* isolates from cattle, elk, and bison in the United States, 1998 to 2011. Applied and Environmental Microbiology, 78, 3674–3684.
16. ICA, 2004. Colombian Agriculture and Livestock Institute. Law 1513, Sanitary assessment for prevention, control and erradication of bovine tuberculosis in Colombia. Issued on July 15th, 2004. Available in: https://www.ptp.com.co/documentos/ Resolucion%201513.pdf
17. ICA, 2010. Colombian Agriculture and Livestock Institute. Bovine brucellosis: Prevention, diagnosis and control. Bogota, Colombia. Fourth Edition. 13 p. In: http://www.ica.gov.co/Areas/Pecuaria/Servicios/Enfermedades-Animales/Brucelosis-Bovina-(1)/ Brucelosis-Bovina4.aspx
18. Iweriebor, B., Iwu, C., Obi, L., Nwodo, U. and Okoh, A., 2015. Multiple antibiotic resistances among Shiga toxin producing *Escherichia coli* O157 in feces of dairy cattle farms in Eastern Cape of South Africa. BMC Microbiology, 15, 213.

19. Kateete, D.P., Kabugo, U., Baluku, H., Nyakarahuka, L., Kyobe, S., Okee, M., Najjuka, C.F. and Joloba, M.L., 2013. Prevalence and Antimicrobial Susceptibility Patterns of Bacteria from Milkmen and Cows with Clinical Mastitis in and around Kampala, Uganda, PLoS One, 8, e63413.
20. Landers, T.F., Cohen, B., Wittum, T.E. and Larson, E.L., 2012. A review of antibiotic use in food animals: perspective, policy, and potential. Public Health Reports, 127, 4–22.
21. Leibler, J.H., Dalton, K., Pekosz, A., Gray, G.C. and Silbergeld, E.K., 2016. Epizootics in industrial livestock production: preventable gaps in biosecurity and biocontainment. Zoonoses Public Health. doi:10.1111/zph.12292
22. Máttar, S., Calderón, A., Sotelo, D., Sierra, M. and Tordecilla, G., 2009. Detección de antibióticos en leches: un problema de salud pública. Revista de Salud Pública, 11, 579–590.
23. Mhone, T.A., Matope, G. and Saidi, P.T., 2011. Aerobic bacterial, coliform, *Escherichia coli* and *Staphylococcus aureus* counts of raw and processed milk from selected smallholder dairy farms of Zimbabwe. International Journal of Food Microbiology, 151, 223–228.
24. Morales, D.F. and Combariza D.A., 2004. Seroprevalencia de brucelosis en trabajadores de mataderos de municipios del Tolima (Colombia). Revista Ciencias de la Salud, 2, 15–23.
25. Ministerio de Salud y Proteccion Social, 2013. Law 1382 Maximum limits for the use of veterinary medicines in food from animal origin. Issued on 2nd May, 2013. 15 p. In: https://www.minsalud.gov.co/Normatividad_Nuevo/Resoluci%C3%B3n%201382%20de%202013.pdf
26. Rasheed, M.U., Thajuddin, N., Ahamed, P., Teklemariam, Z. and Jamil, K., 2014. Antimicrobial drug resistance in strains of *Escherichia coli* isolated from food sources. Revista do Instituto de Medicina Tropical de São Paulo, 56, 341–346.
27. Rivera, D., Rueda, O., Calderón, C., Marino, O., Gall, D. and Nielsen, K., 2003. Evaluación comparativa del método inmunoenzimático indirecto en leche para la detección de bovinos infectados con *Brucella abortus*, en hatos del departamento de Cundinamarca, Colombia. Scientific and Technical Review of the Office International des Epizooties (Paris) 22, 1065–1075.
28. Roess, A. A., Winch, P. J., Akhter, A., Afroz, D., Ali, N. A., Shah, R., Begum, N., Seraji, H. R., El Arifeen, S., Darmstadt, G. L., Baqui, A. H. and the Bangladesh Projahnmo Study Group 2015. Household animal and human medicine use and animal husbandry practices in rural Bangladesh: Risk factors for emerging zoonotic disease and antibiotic resistance. Zoonoses Public Health, 62, 569–578.
29. Seufert, V., Ramankutty, N. and Foley, J.A., 2012. Comparing the yields of organic and conventional agriculture, Nature, 485, 229–232.
30. Tique, V., González, M. and Mattar, S., 2009. Seroprevalencia de *Brucella abortus* en bovinos del departamento de Cordoba. Revista U.D.C.A Actualidad & Divulgación Científica 12, 51–59.
31. Tique, V., Daza, E., Álvarez, J. and Mattar, S., 2010. Seroprevalencia de *Brucella abortus* y ocurrencia de *Brucella melitensis* en caprinos y en ovinos de Cesar y Sucre. Revista U.D.C.A Actualidad & Divulgación Científica, 13, 133–139.
32. Van Boeckel, T.P., Brower, C., Gilbert, M., Grenfell, B.T., Levin, S.A., Robinson, T.P., Teillant, A. and Laxminarayan, R., 2015. Global trends in antimicrobial use in food animals. Proceedings of the National Academy of Sciences, 112, 5649–5654.

CHAPTER 9

Evaluation of Surveillance for Documentation of Freedom from Bovine Tuberculosis

Francisco Fernando Calvo-Artavia[1,*], Lis Alban[2] and Liza R. Nielsen[1]

[1]*Department of Large Animal Sciences, Faculty of Health and Medical Sciences, University of Copenhagen, 1870 Frederiksberg C, Denmark*

[2]*Risk Analysis Group, Department for Food Safety & Veterinary Issues, Danish Agriculture & Food Council, Axeltorv 3, Copenhagen, 1609 Denmark*

ABSTRACT

The objective was to study how surveillance for bovine tuberculosis (bTB) could be made more resource-effective in a bTB free country. A stochastic scenario tree model was developed to: (1) evaluate the sensitivity (CSe) of four surveillance system components (SSC) (*i.e.*, meat inspection of slaughtered domestic cattle, farmed deer and pigs, and tuberculin testing of adult export cattle) given that bTB would enter one of these components, (2) estimate the probability of freedom (PFree) from bTB over time, and (3) evaluate how future alternative programmes based on visual meat inspection would affect the confidence in freedom from bTB at the very low animal-level design prevalence 0.0002% and a low probabilities of introduction (1%). All, except the export cattle component reached a PFree above 96% within five years. The PFree was slightly reduced if surveillance was changed to visual inspection, e.g., PFree was reduced from 96.5% to 94.3% in the cattle component, and from 98.5% to 97.7% in the pig component after 24 years. In conclusion, visual meat inspection of pigs and cattle will only reduce the confidence in freedom from bTB slightly. However, with negligible probability of introduction (0.1%) the PFree could be maintained well above 99% in the cattle, pigs and deer components, which highlights the importance of rigid testing and quarantine procedures in trade of livestock.

Keywords: bovine tuberculosis; livestock; meat inspection; risk-based; freedom

1. INTRODUCTION

Documenting freedom from disease is an essential requirement for international trade of live animals and animal products. In accordance with the Agreement on the Application of Sanitary and Phytosanitary Measures (the SPS Agreement) of the World Trade Organization (WTO), science-based evidence is needed to determine the appropriate country sanitary level for international trade [1]. Meat inspection is one way of collecting such evidence. However, meat inspection is up for discussion in several parts of the world. Currently, the European Food Safety Authority (EFSA) is evaluating which public health hazards (biological and chemical) to be covered by meat inspection for several livestock species. Moreover, EFSA suggests using indicators (also called epidemiological criteria) for these hazards in food and animals, when meat inspection procedures do not adequately address these risks. An example of such an indicator could be the housing system [2].

When meat inspection was founded more than 100 years ago, bovine tuberculosis (bTB) was endemic in most countries. Since then, many countries have managed to eradicate this infection. One example is Denmark, which has been declared free from bTB since 1980 by the European Union (EU) (Decision 2004/320/EC). The requirements for a country to be given the bTB free status according to the OIE and EU are that bTB has been found in less than 0.2% and 0.1% of the herds, respectively [3,4], hence allowing the presence of a limited number of bTB cases in the country. A single case of bTB was diagnosed in Denmark in 1988, when one cow with clinical signs was identified at meat inspection. The case was subsequently confirmed by laboratory examination, but no further animals were found bTB positive. The infection has not been observed in cattle since then despite extensive surveillance activities. Furthermore, bTB has never been found in Danish free-ranging wildlife, making the risk of introduction to livestock herds through contact with wildlife negligible [5]. In 1994, the last out of 16 cases of bTB was found in farmed deer in Denmark.

The Danish surveillance programme for bTB consists of four independent components (*i.e.*, slaughtered domestic cattle, pigs and deer as well as tuberculin testing of adult export cattle), each with their own sampling and diagnostic processes. In slaughter cattle, pigs and deer surveillance relies on *post mortem* examination by palpation and/or incision of lymph nodes. For cattle these cover the retropharyngeal, mandibular, parotid, tracheobronchial, mediastinal and mammary lymph nodes, as well as palpation of lungs. In addition, most adult cattle for export are tuberculin tested because of export requirements [5]. For pigs, incisions into the mandibular lymph nodes as well as palpations of the intestinal lymph nodes are required. The current EU meat inspection Regulation (EC) No 854/2004 [6] requires that every single bovine carcass is examined for bTB as described above. This is time-consuming, costly and not necessarily very effective for the detection of infected animals in countries with a very low prevalence of disease.

The overall objective was therefore to evaluate the surveillance system for a country that has been declared free from bTB for many years to elucidate how and where surveillance could be made more effective, using Denmark as a study case. The specific aims of this study were:

1) to evaluate the sensitivity of each of the surveillance system components given that bTB would enter one of these components, but not necessarily spread between these due to limited contact and risk of transmission between the four components;

2) to estimate the probability of freedom from bTB in Denmark over time, and

3) to evaluate how future alternative programmes based on visual meat inspection and risk-based meat inspection would affect the probabilities of freedom from bTB compared to the current meat inspection system, at the very low animal-level design prevalence 0.0002% and a low probability of introduction of 1% and 0.1%.

The results are relevant for livestock industries and veterinary authorities in countries aiming at improving and modernising the meat inspection procedures, while lowering the resources spent on surveillance programme for bTB. Furthermore, the results can support international organisations such as EFSA and OIE in the decision-making processes concerning modernisation of meat inspection, and hereby ensure free trade on safe conditions.

2. RESULTS AND DISCUSSION

2.1. Results

The mean surveillance system component (SSC) sensitivity (CSe) of the individual SSCs in the current Danish surveillance programme for bTB and the alternative scenarios, together with their 95% credibility interval, are shown in Table 1. Under the current surveillance system for bTB, the pig SSC had the highest CSe (68%), mainly due to the large average annual number of units processed (n = 20,022,790), compared to the domestic cattle (32%), deer (43%) and export cattle (7%) SSCs. The CSe were generally higher under the current meat inspection system compared to the alternative surveillance system scenarios.

Figure 1 illustrates the temporal change in the posterior mean probability of freedom from bTB in Denmark, for the period between 1995 and 2012, under the current surveillance system. A design prevalence of 0.0002% was assumed for pigs and domestic cattle. For deer and export cattle the design prevalence was one infected animal out of the number of units processed during a one-year period. The probability of introduction of bTB to the SSCs was set to 0.1% and 1%, respectively, each year in the two illustrated scenarios. With a negligible probability of introduction of 0.1% the pig SCC provided the highest posterior mean probability of freedom at 99.8% (CI: 99.8%–99.9%), compared to 99.7% (CI: 99.6%–99.7%), 99.8% (CI: 99.7%–99.8%) and 97.1% (CI: 96.1%–98.1%) for the domestic cattle, deer and export cattle SSCs, respectively, by year 2012. The posterior mean probability of freedom for the domestic cattle SSC, reached a value above 99% after the end of the sixth year of surveillance (*i.e.*, 2001) and the maximum value was reached after the fourteenth year of surveillance (*i.e.*, year 2009), compared to the second year (*i.e.*, 1997) and fourth year (*i.e.*, 1999) for the pig SSC, and the fifth year (*i.e.*, 2000) and thirteenth year (*i.e.*, 2008) for the deer SSC, while for the export cattle SSC a value above 99% was never reached (Figure 1). In other words, basing the surveillance only on export cattle would not provide a good confidence in freedom from bTB compared to the other SSCs, and therefore the benefit of this surveillance activity would be inferior, if it was going to be used as the only surveillance activity. Basing the surveillance on slaughter pigs and/or cattle was found to increase the confidence in freedom from bTB markedly over time, given that the infection would enter one of these components and spread to the level of the design prevalence.

With a low probability of introduction of 1%, the pig SCC again provided the highest posterior mean probability of freedom at 98.5% (CI: 98.0%–98.8%), compared to 96.9% (CI: 96.4%–97.2%), 97.6% (CI: 96.7%–98%) and 87.4% (CI: 85.3%–89.5%) for the domestic cattle, deer and export cattle SSCs, respectively, by year 2012. The posterior mean probability of freedom never reached values above 98.5%.

Table 1. Results of simulations with 10,000 iterations for the sensitivities (CSe), with design prevalence and units processed for each of four surveillance system components (SSC) (*i.e.*, all slaughtered animals subjected to invasive meat inspection as well as cattle tested prior to export) in the current Danish surveillance programme for bovine tuberculosis and for four alternative meat inspection surveillance programmes involving more visual meat inspection.

Surveillance programme/SSC	Design prevalence	Units processed	CSe (95% CI)
Current surveillance system (Invasive in all slaughtered animals)			
Domestic cattle	0.000002	498,364[a]	0.32 (0.28–0.36)
Pig	0.000002	20,022,790	0.68 (0.52–0.80)
Deer[b]	0.000125	8,000	0.43 (0.31–0.53)
Export cattle[b]	0.00038	2,634	0.07 (0.05–0.09)
Alternative scenario 1 (100% visual inspection cattle)			
Domestic cattle	0.000002	498,364	0.18 (0.15–0.20)
Alternative scenario 2 (100% visual inspection pigs)			
Pig	0.000002	20,022,790	0.43 (0.30–0.54)

[a] Average number of cattle slaughtered between January 2004 to June 2012 in Denmark;
[b] Alternative surveillance programmes were not investigated for these SSCs.

Figure 1 also shows the results of the temporal discounting evaluating the posterior mean probability of freedom from bTB in Denmark for the two alternative surveillance programmes assuming a change in the meat inspection procedure starting in the year 2013. Out of the alternative scenarios at the end of the second study period (*i.e.*, 2013–2037), Scenario 2 where all pigs would be visually inspected had higher posterior mean probability of freedom at 99.8% (CI: 99.7%–99.8%), compared to 99.4% (CI: 99.3%–99.5%) for Scenario 1 where 100% of the domestic cattle would be visually inspected.

These results show that even with lower CSe for the alternative surveillance programme scenarios (Table 1) for the domestic cattle and pig SSCs, the posterior mean probability of freedom would be maintained well above 99% (Figure 1, top graph) if a negligible probability of introduction can be obtained. However, more realistically the probability of introduction is 1% and as illustrated in the bottom graph in Figure 1 this leads to posterior mean probabilities of freedom below 99% for the SSCs under the current surveillance programme and the alternative surveillance programmes. Moving to 100% visual inspection in cattle resulted in a posterior mean probability of freedom of 94.3% (CI: 93.5%–95.1%) by year 2037, and 100% visual inspection in pigs reached 97.6% (CI: 96.6%–98.1%).

The results of the scenario analyses showed that if the proportion of granulomas detected in the alternative surveillance system, based on visual inspection only, was set to 40% or 60% rather than the 50% used in the original model, no noteworthy changes were observed in the results.

If in the pig SCC the population of adults (2% of all slaughtered pigs) was inspected using the current meat inspection procedures and the rest of the pig population was only visually inspected, no noteworthy changes were observed in the results. Nor did it change the results much if the submission of detected lesions for bacteriological culture was set to 100% instead of 10%, as in the current scenario.

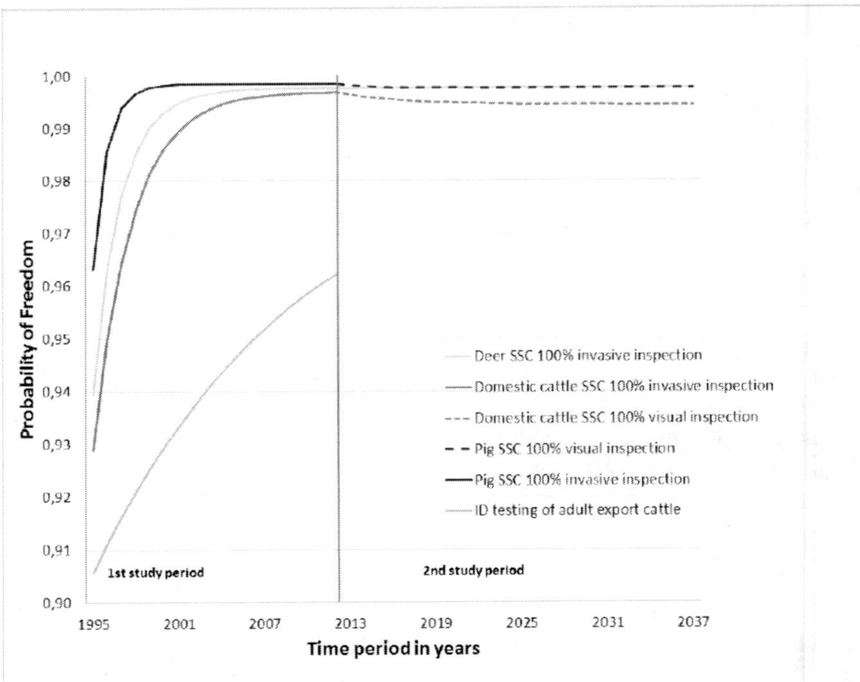

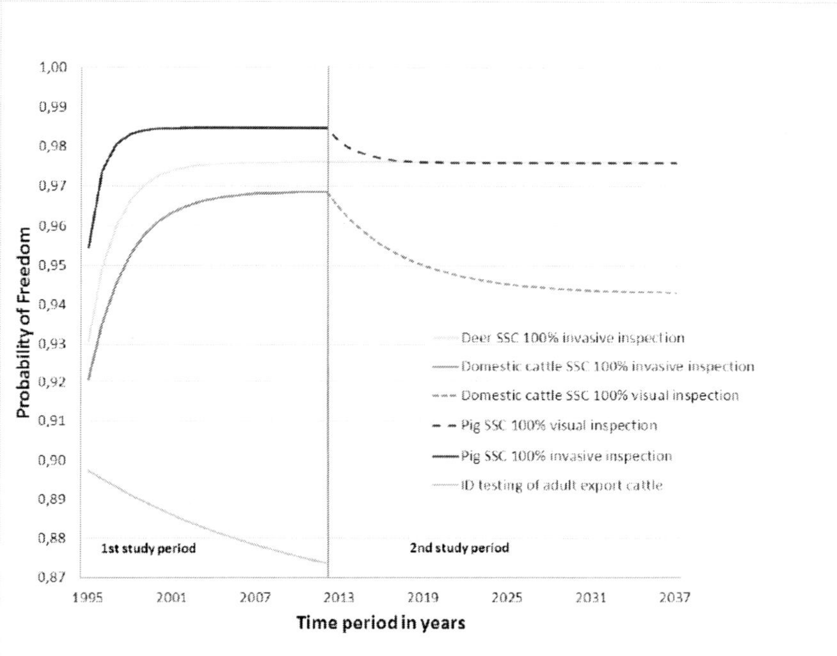

Figure 1. Temporal discounting, estimating the probability of freedom from bovine tuberculosis in Denmark, for the first study period (1995–2012) under the current surveillance system for the four surveillance system components and for the second study period (2013–2037) under the alternative surveillance programmes: Scenario 1 where all the domestic cattle would be subjected to visual meat inspection and Scenario 2 where all pigs would be subjected to visual meat inspection. Export cattle were tested using intradermal tuberculin test (ID). The top graph shows the results from the scenario with 0.1% and the bottom graph the scenario with 1% probability of introduction.

If the design prevalence was changed to 0.01% corresponding to the EU requirement of a herd-level design prevalence of 0.1% multiplied with an assumed animal-level prevalence within infected herds of 10%, the posterior mean probability of freedom would reach a maximum level of 99.9% at the end of the first year of surveillance and would be maintained that high throughout the two study periods.

If start probabilities of freedom in 1995 were set to 0.5 (corresponding to an uninformed prior) instead of 0.9, the posterior mean probability of freedom from bTB would reach values above 99% after the end of the fifth, tenth and thirteenth year of surveillance for the pig, deer and domestic cattle SSC. If changed to 0.7 instead of 0.9, the posterior mean probability of freedom from bTB would reach values above 99% after the end of the fourth, eighth and tenth year of surveillance for the pig, deer and domestic cattle SSC.

The most influential input parameters identified in the tornado plots for the current surveillance system and posterior probability of freedom outputs were the proportion of granulomas submitted for laboratory investigation in the pig SSC and the probability of granulomas present in young animals from the domestic cattle and pig SSC. In the case of the alternative surveillance system CSes and posterior probabilities of freedom outputs, the most influential input parameters were the proportion of granulomas present in the adult animals of the domestic cattle and pig SSC, and the proportion of granulomas submitted from the domestic cattle SSC.

2.2. Discussion

The present study has shown that a country officially recognized as being free from bTB for many years can retain a high confidence in being free, even if meat inspection procedures are changed to visually based procedures, but only if the probability of introduction is kept negligible. This underpins the importance of effective import testing and quarantine procedures when import of cattle into bTB free countries cannot be avoided. The prediction of freedom from bTB in Denmark was made using the SSCs independently, because of the lack of contact and hence, low risk of transmission of bTB between cattle, pigs and deer representing the different surveillance components. We believe that the introduction of bTB into one of the components is unlikely to lead to a spread to the other components; therefore the results are presented for each component separately. The benefit of not palpating and making incisions is that the inspector can have more time to focus on other food safety hazards (*e.g.,* faecal contamination), which might constitute a non-negligible food safety issue.

The presented results indicate that the current surveillance programme - requiring palpation and/or incisions of relevant lymph nodes of cattle, pigs and deer at meat inspection - has the highest CSe of detecting at least one slaughter animal with bTB at the design prevalence used in this study (*i.e.*, 32% for domestic cattle and 68% for pigs). This was compared to the alternative surveillance programmes based on more visual meat inspection, which all had CSe <50% (Table 1), if bTB was present at a design prevalence corresponding to 40 infected pigs in the pig SSC, and one infected animal in each of the cattle and deer SSCs. This national animal-level design prevalence is much lower than the OIE and EU requirements for a country to obtain the bTB free status. If we had used the design prevalences suggested by the OIE or EU then we would have expected to have around 300 herds with at least one positive animal before one bTB infected animal would be detected. This was considered an unacceptably high number of infected herds for a country that has been truly free for many years. Other studies have used a higher animal-level design prevalence than we did [7,8], but we find that it is appropriate to use such a low design prevalence for a slowly spreading disease compared to a fast spreading disease, as also stated by [9], in order to evaluate the ability of the system to detect the infection early, should it enter the country. For the deer SSC, the design prevalence used was higher than 0.0002% because we needed to be able to detect at least one infected deer. Therefore, the CSe appears to be higher for deer than for the domestic cattle component, but the two CSes are not directly comparable due to the different assumptions about underlying design prevalence. Still, the CSe in both SSCs indicate the probability of detecting one infected animal that goes to slaughter in each SSC. This is the lowest prevalence an infected country can have. Therefore, the used method provides very conservative estimates of the CSes and confidence in freedom. In many cases more than one animal would be infected in the SSCs, if bTB had been introduced.

Assuming a probability of 0.9 that Denmark was free of bTB at the beginning of 1995 and an annual probability of introduction of disease into the country of 0.1% and 1%, the posterior mean probability of freedom increased rapidly, except in export cattle under the 1% scenario. The sensitivity analyses showed that the posterior mean probability of freedom would be affected, if start probabilities of freedom in 1995 of 0.5 (corresponding to an uninformed prior) or 0.7 were assumed instead of 0.9. The posterior mean probability of freedom from bTB would reach values above 99% much later in time, which underpins the importance of striving to keep the bTB-free status continuously. The results indicated that it took around five years to obtain the highest obtainable confidence in freedom after bTB had been eradicated with the level of surveillance used in Denmark.

The probability of introduction of bTB into Denmark was set to 1% due to the fact that Denmark mainly imports a low number of animals and mostly from bTB free countries. Testing is performed before cattle are moved in from countries that are not officially free of bTB according to OIE criteria. The scenario with negligible probability of introduction (0.1%) was evaluated to investigate the effect of improving import biosecurity procedures such as import testing and quarantine procedures on the confidence of freedom over time. The number of import cattle was reduced from 1235 in 2009 to 170 in 2011, and a large proportion of the imported animals came from the bTB free countries

Sweden and Germany. Therefore, the assumption about the probability of introduction of 1% is believed to be a valid reasonable estimate for current Danish practices, but lowering this probability through improved testing and quarantine practices would be beneficial for the level of confidence in freedom.

For the second study period, assuming a change in the current meat inspection procedures to visual inspection with a negligible probability of introduction of 0.1%, the posterior mean probability of freedom from bTB decreased slightly for the domestic cattle alternative scenario (Scenario 1) by the end of 2013 from 99.7% to 99.6%. Furthermore, the pig alternative scenario (Scenario 2) had the highest posterior mean probability of freedom by the end of year 2037 at 99.8% (Figure 1). However, we also found that if the design prevalence was changed to 0.01% corresponding to the EU requirements, the posterior mean probability of freedom would be maintained above 99.8%, at least until 2037 in all SSCs, except for the adult export cattle in which it will be maintained above 98.4%. This provides further confidence in freedom for Denmark, even if visual inspection is introduced in both the pig and cattle components. In the EU, visual inspection is going to become standard practice in pigs from June 2014 onwards, according to the EU Commission Standing Committee on the food chain and animal health (SCFCAH), whereas visual inspection of cattle is less likely to be introduced in the near future.

Overall, the level of confidence in freedom from bTB remained close to the same regardless of how large a proportion of the cattle or pigs would be subjected to visual inspection, *i.e.*, if visual inspection was only used in the 98% of the pig SSC consisting of young slaughter pigs, or 80% rather than 100% of the adult cattle. The confidence in freedom remained high, mainly due to the high probability of freedom at the starting point (90%) and low probability of introduction, but also due to a high number of inspected animals in the surveillance components. It is important to notice that visual inspection performs much better in a population with a long history of surveillance and negligible probability of introduction, compared to a population higher probability of introduction and lower prior probability of freedom from bTB.

A posterior mean probability of freedom reaching levels above 96%, under the current and alternative surveillance programme scenarios, suggests that there is a low probability of undetected bTB cases in the country. However, in the Danish case the probability of freedom is calculated for a country where there have been no bTB cases in cattle since 1995 (more than 18 years), and therefore, the design prevalence was set much lower than the international standards. This provides a higher confidence in the level of freedom, even with a change in the meat inspection procedures, compared to countries that still have a few bTB cases in the country.

Bovine tuberculosis is a notifiable disease [10,11] due to its zoonotic nature and the economic implications for livestock and international trade [12], and it is considered an unwanted infection in most countries. Each year, suspect cases of tuberculosis are found primarily in cattle and pigs in Denmark. A publicly accessible database lists the suspects and show how these are dealt with [13]. In 2009, 15 suspicions in pigs were raised by abattoirs, which had detected granulomatous lesions in lymph nodes in a number of slaughtered pigs from nine herds. The epidemiological investigation revealed that most of the farms had used non-heat-treated bedding (peat), originating from countries not free from bovine tuberculosis, for the piglets. Because of the risk of bovine tuberculosis, the farms were placed under official veterinary supervision with movement restrictions. However, laboratory analyses revealed *Mycobacterium avium* in all cases and therefore the restrictions on the farms were lifted [5]. Subsequently, the Danish agricultural industry banned the use of non-heat-treated peat at pig farms, unless specific approval is obtained - as described in the Danish Product Standard [14].

The main routes for transmission of bTB to humans are the consumption of unpasteurized milk and milk products, and direct contact through aerosols or skin cuts [10,11]. In most developed countries, bTB in humans has been eradicated [15]. Pasteurisation of milk has resulted in a substantial decline in the incidence of human cases infected with *M. bovis*. In Denmark, between zero and two human cases of bTB were identified annually between 2007 and 2011. All cases consisted either of a reactivation of childhood infection in elderly people or immigrants infected in their home countries [16]. This suggests that human infection with *M. bovis* has become an occupational hazard in developed countries. This points to a benefit in using human cases of tuberculosis in the countryside for the tracing-back of infected premises in a country believed to be free from bTB.

3. METHODS

The stochastic scenario tree methodology described by Martin *et al.* (2007) [17] was used to estimate the sensitivity of each SSC and the country-level probability of freedom from bTB provided by each SSC. Historical data is used in this methodology to determine the confidence in freedom from an infection, therefore the bTB free status in our study is not based on a single year of data. Others have used this approach [18,19,20]. This model is based on the assumption that all final results of the surveillance system are negative [17], therefore referring to true freedom from infection as in the case of Denmark. This is different from the status defined by the OIE and EU, where bTB might be found in 0.2% and 0.1% of the herds, respectively, allowing the occurrence of bTB cases in the country [3,4].

The four components of the Danish surveillance programme for bTB were assessed separately to evaluate each their sensitivities, because of limited contact and hence low risk of transmission of bTB between the different surveillance components. The sensitivity of a SSC specifies the probability that at least one bTB test positive animal will be detected given that bTB has been introduced into the SSC and spread to the level of a predefined design prevalence (see below). The four SSCs of the current Danish surveillance programme for bTB are slaughtered domestic cattle, farmed deer and pigs as well as testing of adult cattle for export. Hence, there is no overlap between these components. All slaughtered domestic cattle, farmed deer and pigs are examined at meat inspection for clinical signs or lesions indicative of bTB, and suspect findings during meat inspection are tested by bacterial culture for confirmation. Adult cattle for export are tested with an intradermal tuberculin test (ID). If there is a suspect case of bTB, the herd is put under veterinary official supervision and all or a large proportion of the animals in the herd are tested with ID [5].

Due to very few imports of livestock into Denmark and negligible risk of infection from wildlife (*i.e.*, the two most obvious herd-level risk factors of introduction of bTB into livestock herds), we did not include herd-level risk factors in the model. The sensitivity of the component (CSe) was defined as the probability of detecting at least one bTB infected animal at a design prevalence of 0.0002% corresponding to a herd-level design prevalence of 0.2% multiplied with an animal-level design prevalence of 0.1% when inspecting or testing all units processed in the domestic cattle and pig SSCs. These two prevalence levels were stated in the OIE guidelines for a country to qualify as free from bTB [3]. However, for the deer and export cattle this design prevalence would result in less than one infected animal in those populations, therefore the design prevalence for these two SSCs was estimated as the 1/N, where N is the total number of deer and tested export cattle in the SCC in 2011, respectively. The diagnostic test sensitivity and number of units processed were also entered in the stochastic scenario tree model. The structure and contents of the scenario tree model is described in the sections below. The values for the key parameters entered in the stochastic model were estimated using data from on-going surveillance for bTB and information from the literature (Table 2).

3.1. Scenario Tree

Four scenario trees were constructed, one for each of the SSCs in the Danish bTB surveillance programme. The trees consisted of nodes dividing the population into risk groups. Within each of the risk groups the units had the same probability of detection, given the presence of disease (Figure 2). The probabilities of a positive outcome were estimated for each limb of the scenario tree by multiplication of the population proportions in the risk groups and the probabilities of infection and detection in each branch of the tree. The probabilities of detection resulting in a positive outcome (*i.e.*, an infected unit) were then combined to obtain the CSe. The scenario for the deer component differs from the cattle and pig components at the detection node section, because detailed information on lesions present, detected and submitted was not readily available, as it was for the other two components. Therefore, only one combined node for detection was used at this step in the deer SSC.

3.2. Population and Risk Category Node

Register data were obtained from the Danish Cattle Database (DCD) for the period from January 2004 to June 2012 of all recorded cattle slaughtered in Denmark. The average number of cattle slaughtered per year during this period was calculated (n = 498,364). Data describing the approximated average population of deer (n = 8000) in 2011 and pigs (n = 20,022,790) slaughtered from 2006 to 2011 were obtained from the Danish Agriculture & Food Council. The average number of cattle exported (n = 24,655) per year from 2009 to 2011 was obtained from the Danish Knowledge Centre for Agriculture. However, only 2634 of the adult cattle exported for breeding purposes were tested using the ID test (Annual Report on Zoonoses in Denmark 2011).

The first node in the tree was a risk category node at animal level (*i.e.*, age), which divided all slaughtered cattle into two branches with proportions of young ($Prop_Y$) and adult animals ($Prop_A$), respectively. In general, a young animal was less than or equal to one year old at slaughter and an adult animal was above one year old. However, in the pig SSC, sows and boars were considered as adult animals and finisher pigs were categorized as young animals (Table 2).

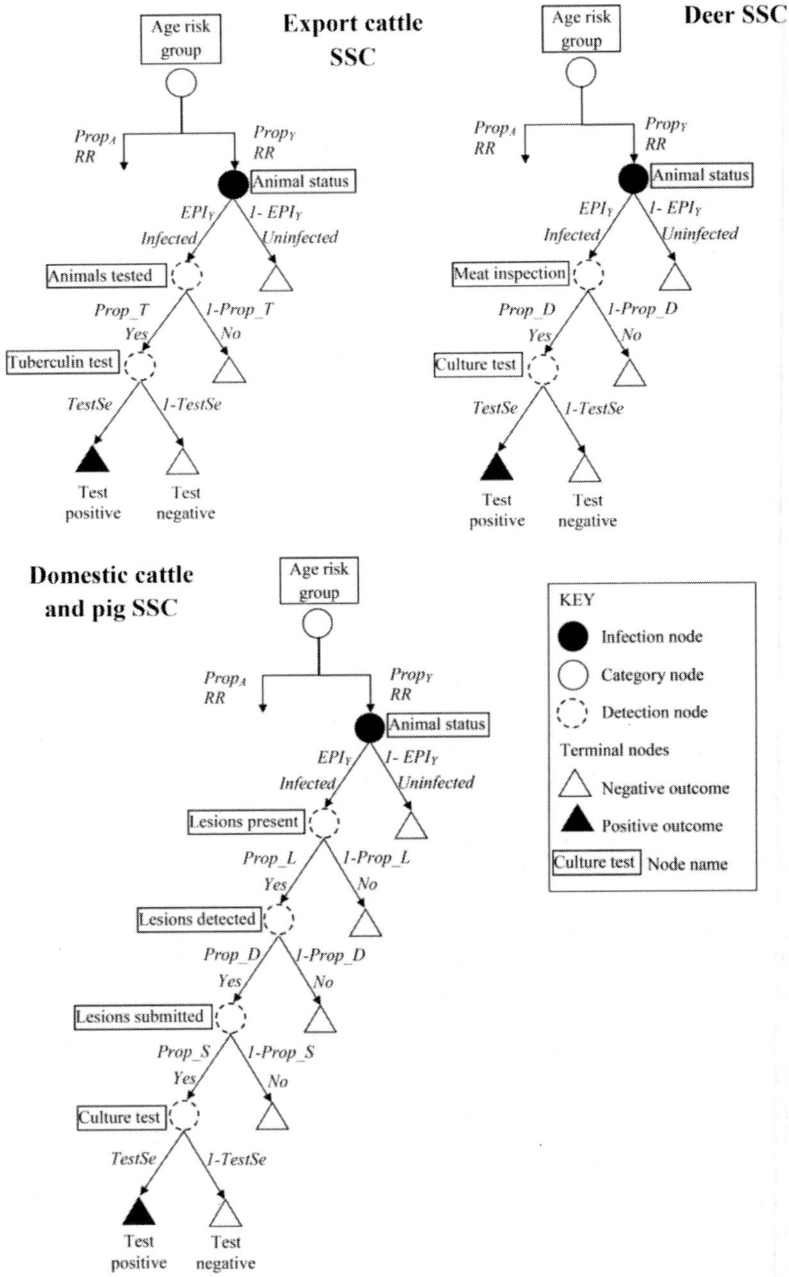

Figure 2. Conceptual scenario tree developed to evaluate the performance of the four surveillance system components (SSCs) for detection of bovine tuberculosis (bTB) at meat inspection in Denmark.

$Prop_A$ and *$Prop_Y$* are proportions of animals in the adult and young risk groups, respectively, and *RR* is the respective relative risk. *EPI_Y* is the effective probability of infection in the young animal group. *Prop_T*, *Prop_D*, *Prop_L* and *Prop_S* are the proportions of animals for each detection node according to the current invasive meat inspection procedures for bTB and its corresponding SSC.

Table 2. Input variables used to estimate the probability of freedom from bovine tuberculosis in Denmark represented either as fixed values or distributions, and sources of information.

Input variable		Input value and distribution	Source
Design prevalence (P_A) for domestic cattle and pigs		0.000002	Adjusted from [3]
Design prevalence (P_A) for deer		0.000125	Calculated: 1 animal out of 8000
Design prevalence (P_A) for exported cattle		0.00038	Calculated: 1 animal out of 2634
Proportion of young and adult animals *per* SSC *(PropUG)*			
Domestic cattle	Young	0.3	Danish Cattle Database (DCD)
	Adult	0.7	
Pig	Young	0.98	Danish Agriculture & Food Council
	Adult	0.02	
Deer	Young	0.9	Danish Agriculture & Food Council
	Adult	0.1	
Export cattle	Young	0.77	Danish Knowledge Centre of Agriculture, Cattle
	Adult	0.23	
Relative risk of infection between the young and adult groups *(RR)*		RiskPert (1.5, 2, 3)	[7]
Proportion of lesions present ($Prop_L$)			
	Adult animals	RiskBeta (84, 58)	[21]
	Young animals	50% of adults	
Proportion of lesions detected ($Prop_D$)			
	Current programme	RiskPert (0.9, 0.95, 0.99)	[7]
	Alternative programme	50% of the current	
Proportion of lesions submitted ($Prop_S$)			
	Pig SSC	RiskPert (0.005, 0.1, 0.15)	Expert opinion
	Other three SSC	RiskPert (0.7, 0.8, 0.9)	[7]
Meat inspection Se ($Prop_D$)		RiskPert (0.29, 0.6, 0.86)	[9,21-24]
Bacteriological culture test Se (*TestSe*)		RiskPert (0.92, 0.95, 0.98)	[7]
Intradermal tuberculin test Se (*TestSe*)		RiskPert (0.53, 0.7, 0.96)	[25]
Probability of introduction (*PIntro*)		0.001 and 0.01	Evaluation of import practices

3.3. Population and Risk Category Node

The next node in the scenario tree was the infection node, where the design prevalence (P_A) was used in the calculations to obtain the effective probabilities of infection (*EPI*) within each subpopulation defined by the previous category node. The relative risk (*RR*) of infection between the young and adult groups—irrespective of animal species—was based on [7] and modeled using a pert distribution. Then the relative risks for the adult animals compared to the young animals (Table 2), and the proportion of adult animals was used for calculation of the adjusted risk (AR) for the young animals (AR_Y), and multiplied by the *RR* to obtain the AR of the adult animals (AR_A) (Martin *et al.*, 2007). Finally, the ARs were multiplied by P_A to calculate the *EPI* for each risk group (EPI_Y and EPI_A) as shown in Equation 1 and Equation 2, respectively [17].

$$EPI_Y = \left(\frac{1}{RR \times Prop_A + (1 - Prop_A)}\right) \times P_A \tag{1}$$

$$EPI_A = \left(\frac{1}{RR \times Prop_A + (1 - Prop_A)} \times RR\right) \times P_A \tag{2}$$

Where EPI_Y is the effective probability of infection for the group of young animals, EPI_A is the effective probability of infection for the adult animals, *RR* is the relative risk for the adult animals compared to the young animals, $Prop_A$ is the proportion of all slaughtered cattle (or pigs or deer, respectively) in the group of adult animals and P_A is the design prevalence.

3.4. Detection Nodes

The next nodes correspond to the probability of detecting bTB in each group. For the domestic cattle and pig SSC, the meat inspection was divided into three nodes:

1) The proportion of lesions present (*i.e.*, granulomas) in the animals. This was set 50% lower for young animals than adult animals, and was modeled using a beta distribution based on an efficiency study of inspection procedures for detection of tuberculous lesions in cattle [21];

2) The proportion of granulomas detected at meat inspection, modeled using a pert distribution based on [7];

3) The proportion of granulomas submitted for confirmation by bacterial culture. Based on expert opinion, the proportions were set higher for domestic cattle than for pigs (Table 1), because around 90% of tuberculous lesions found in pigs are detected in the gastro-intestinal tract and according to the current meat inspection legislation, these lesions will not be submitted for confirmatory bacterial culture, as they are in the cattle abattoirs [26]. The last

detection node for domestic cattle and pigs SSCs was the probability of testing positive at culture, modeled using a pert distribution (Table 2) [7].

For the export cattle SSC the detection node was the probability of testing positive to the ID test, which was also modeled using a pert distribution, based on a field trial literature review by [25]. For farmed deer SSC, the detection node was the probability of tuberculous lesions detected at meat inspection. The sensitivity of meat inspection for the farmed deer SSC was modeled using a pert distribution based on literature [9,21,22,23,24].

3.5. Component and Overall System Sensitivity Calculations

To estimate the overall CSe of each SSC, the component unit sensitivity (CSeU) was first calculated by the sum of the estimated probabilities with a positive outcome at the end of each limb (Equation 3). These probabilities were the result of the product of all branch probabilities and proportions. For the tree of the domestic cattle SSC (Figure 2) this was:

$$CSeU = \sum_{k=1}^{2} PropUG_k \times EPI_k \times Prop_L \times Prop_D \times Prop_S \times TestSe \quad (3)$$

Where $PropUG_k$ is the proportion of animals on the k risk group (*i.e.*, adult or young animals), EPI_k is the effective probability of infection for k risk group, $Prop_L$ is the proportion of lesions present in the animals to be inspected, $Prop_D$ is the proportion of lesions detected at meat inspection, $Prop_S$ is the proportion of lesions submitted for bacteriological culture test and *TestSe* is the sensitivity of the bacteriological culture test.

Once the CSeU was calculated, then the probability that one or more positive units would be detected given that the SSC was infected at the design prevalence (CSe) was estimated for each of the four SSC (Equation 4).

$$CSe = 1 - (1 - CSeU)^n \quad (4)$$

Where *CSeU* is the component unit sensitivity and n is the number of units (*i.e.*, animals) processed in the SSC of interest.

3.6. Probability of Freedom

To calculate the posterior probability of freedom (*PostPFree*) (Equation 5), the prior probability of infection (*PriorPInf*) is needed. In this model, the *PriorPInf* was set to 0.1 for the first year of the study period (*i.e.*, 1995), because Denmark obtained the free status of bTB in 1980. After that a lone-standing case of bTB in a cow was detected and removed in 1988, and 16 cases were found in deer between 1988 and 1994. Since then no cases of bTB have been detected so the simulations were started from year 1995. Then the *PostPFree* from bTB was calculated for the first year as:

$$PostPFree_{Tj} = \frac{(1 - PriorPInf)}{(1 - PriorPInf) + (PriorPInf \times (1 - CSe))} \tag{5}$$

The first posterior probability of infection ($PostPInf_{T1}$) was calculated as 1 - *PostPFree*, which was then adjusted for the probability of introduction (*PIntro*) of disease to the SSC in question of 0.1% and 1%, respectively (Equation 6). The adjusted probability of infection ($PostPInf_{Adj}$) was used as the *PriorPInf* for the subsequent year and to calculate the adjusted probability of freedom ($PostPFree_{Adj}$) (Equation 7); this procedure was repeated for each year of the study period. This adjusted probability of freedom is the mean probability of freedom reported in the results section (Figure 1).

$$PostPInf_{Adj} = PostPInf_{T1} + Pintro - (PostPInf_{T1} \times PIntro) \tag{6}$$

$$PostPFree_{Adj} = 1 - PostPInf_{Adj} \tag{7}$$

3.7. Simulations

The models were set up in @Risk 6 (Palisade Corporation®, Middlesex, UK) and run with 10,000 iterations. The input parameters were described by use of distributions to account for the uncertainty in the parameter estimates (Table 2). Sensitivity analysis was performed by regression analysis for each scenario, identifying significant outputs, displayed as "tornado" type charts. Scenario analysis was conducted for two variables of specific interest, *i.e.*, the probability of introduction of bTB into Denmark, the proportion of granulomas detected, submission of detected lesions, design prevalence and prior probability of freedom estimate. Two scenarios were evaluated in the scenario analysis (1) when visual inspection was only used in the 98% of the pig SSC consisting of young slaughter pigs, and (2) where 80% rather than 100% of the adult cattle were visually inspected.

3.8. Modeling of Alternative Surveillance Systems Based on Visual Inspection

A new period was simulated (*i.e.*, 2013–2037), where alternative surveillance systems were assumed to be put in place. The assumption was that the meat inspection would change from the current invasive procedures (*i.e.*, incision and palpation of lymph nodes and palpation of lungs) to a visual inspection only in all or a proportion of units processed in the SCCs. To model the alternative surveillance systems the proportion of granulomas detected was reduced by 50% in units inspected visually. This reduction was only relevant for the domestic cattle and pig SSCs, due to the descriptions of the detection nodes in section 2.3. Two alternative surveillance systems were evaluated: Scenario 1 in which all domestic cattle were inspected visually and Scenario 2 where all pigs were inspected visually.

To estimate the probability of freedom for the alternative surveillance programme, the $PostPFree_{Adj}$ estimate of the last year (2012) of the previous study period was used to calculate the *PriorPInf* for the first year of the following study period, assuming that the alternative surveillance programme started in 2013.

4. CONCLUSIONS

In conclusion, for a country considered free from bTB, lowering the intensity of surveillance e.g., by introducing visual meat inspection of cattle and pigs, will not reduce the confidence in freedom from bTB markedly providing the probability of introduction of the infection can be kept low. However, with such a slow spreading infection with a long incubation time, new introduction might take a long time to be detected. Hence, both continued surveillance and rigid import testing and quarantine practices are recommended to acquire a negligible risk of introduction of bTB.

ACKNOWLEDGMENTS

The authors would like to thank the following people for providing data and good discussions about the methods and results: Camilla Brasch

Andersen, Danish Veterinary and Food Administration. Flemming Thune-Stephensen, Danish Agriculture and Food Council. Preben Willeberg, UC-Davis, California, USA/Former CVO from Denmark.

REFERENCES

1. Agreement on the Application of Sanitary and Phytosanitary Measures. World Trade Organization. Available online: http://www.wto.org (accessed on 5 November 2012).
2. EFSA. Technical specifications on harmonized epidemiological indicators for public health hazards to be covered by meat inspection of swine. *EFSA J.* **2011**, *9*, 2371. Available online: www.efsa.europa.eu/efsajournal (accessed on 28 December 2012).
3. Terrestrial Animal Health Code. World Organization for Animal Health. Available online: http://www.oie.int/index.php?id=169&L=0&htmfile=chapitre_1.11.6.htm (accessed on 28 December 2012).
4. EEC. Council Directive 64/432/EEC on animal health problems affecting intra-community trade in bovine animals and swine. *Off. J.* **1964**, *121*, 1977–2012. edu/DiseaseInfo/disease.php?name=bovine-tuberculosis (accessed on 28 December 2012).
5. Animal Health in Denmark 2011. Danish Veterinary and Food Administration. Available online: http://www.foedevarestyrelsen.dk/Publikationer/Alle% 20publikationer/ 2012 095.pdf (accessed on 28 December 2012).
6. Regulation (EC) No 854/2004 of the European Parliament and of the Council of 29 April 2004 laying down specific rules for the organizations of official controls on products of animal origin intended for human consumption. *Off. J.* **2004**, *L 226*, 83–127.
7. Sergeant, E.; Happold, J.; Hutchison, J.; Langstaff, I. *Evaluation of Australian Surveillance for Freedom from Bovine Tuberculosis*; Report Prepared for the Australian Biosecurity CRC for Emerging Infectious Disease; AusVet Animal Health Services: City, Country, 2010; pp. 1–42.
8. Welby, S.; Govaerts, M.; Vanholme, L.; Hooyberghs, J.; Mennens, K.; Maes, L.; van der Stede, Y. *Prev. Vet. Med.* **2012**, *106*, 152–161.
9. Wahlström, H.; Frössling, J.; Lewerin, S.S.; Ljung, A.; Cedersmyg, M.; Cameron, A. Demonstrating freedom from *Mycobacterium bovis* infection in Swedish farmed deer using non-survey data sources. *Prev. Vet. Med.* **2010**, *94*, 108–118.
10. Bovine tuberculosis. The Center for Food Security and Public Health. Available online: http://www.cfsph.iastate.edu/DiseaseInfo/disease.php?name=bovine-tuberculosis (accessed on 28 December 2012).
11. EFSA; ECDC. The European Union Summary Report on Trends and Sources of Zoonoses, Zoonotic Agents and Food-borne outbreaks in 2010. *EFSA J.* **2012**, *10*, 2597. Available online: www.efsa.europa.eu/efsajournal (accessed on 28 December 2012).
12. Michel, A.L.; Müller, B.; van Helden, P.D. *Mycobacterium bovis* at the animal-human interface: A problem, or not? *Vet. Microbiol.* **2010**, *140*, 371–381.
13. Fødevaretyrelsen. Available online: http://www.fodevarestyrelsen.dk/Dyr/ Dyresundhed_og_dyresygdomme/Dyresygdomme_og_zoonoser/Mistankedatabase.htm (accessed on 28 December 2012).
14. Videncenter for Svineproduktion. Available online: http://vsp.lf.dk/~/media/Files/DANISH/DANISH%20produktstandard/Produkt_Standard_UK.ashx (accessed on 28 December 2012).
15. Humblet, M.F.; Boschiroli, M.L.; Saegerman, C. Classification of worldwide bovine tuberculosis risk factors in cattle: A stratified approach. *Vet. Res.* **2009**, *5*, 40–50.
16. Annual Report on Zoonoses in Denmark 2011. National Food Institute, Technical University of Denmark. Available online: http://www.food.dtu.dk/upload/ Institutter/Food/Pdfer/annual_report 2011.pdf (accessed on 28 December 2012).

17. Martin, P.A.J.; Cameron, A.R.; Greiner, M. Demonstrating freedom from disease using multiple complex data sources: 1: A new methodology based on scenario trees. *Prev. Vet. Med.* **2007**, *79*, 71–97.
18. Alban, L.; Boes, J.; Kreiner, H.; Petersen, J.V.; Willeberg, P. Towards a risk-based surveillance for *Trichinella* spp. in Danish pig production. *Prev. Vet. Med.* **2008**, *87*, 340–357.
19. Frössling, J.; Ågren, E.C.C.; Eliasson-Selling, L.; Lewerin, S.S. Probability of freedom from disease after the first detection and eradication of PRRS in Sweden: Scenario-tree modeling of the surveillance system. *Prev. Vet. Med.* **2009**, *91*, 137–145.
20. Boklund, A.; Dahl, J.; Alban, L. Assessment of confidence in freedom from Aujeszky's disease and classical swine fever in Danish pigs based on serological sampling-effect of reducing the number of samples. *Prev. Vet. Med.* **2013**, *110*, 214–222.
21. Corner, L.A.; Melville, L.; McCubbin, K.; Small, K.J.; McCormick, B.S.; Wood, P.R.; Rothel, J.S. Efficiency of inspection procedures for the detection of tuberculous lesions in cattle. *Aust. Vet. J.* **1990**, *67*, 389–392.
22. De Kantor, I.N.; Nader, A.; Bernardelli, A.; Giron, D.O.; Man, E. Tuberculous infection in cattle not detected by slaughterhouse inspection. *J. Vet. Med.* **1987**, *B 34*, 202–205.
23. Asseged, B.; Woldesenbet, Z.; Yimer, E.; Lemma, E. Evaluation of abattoir inspection for the diagnosis of *Mycobacterium bovis* infection in cattle at Addis Ababa abattoir. *Trop. Anim. Health Prod.* **2004**, *36*, 537–546.
24. Norby, B.; Bartlett, P.C.; Fitzgerald, S.D.; Granger, L.M.; Bruning-Fann, C.S.; Whipple, D.L.; Payeur, J.B. The sensitivity of gross necropsy, caudal fold and comparative cervical tests for the diagnosis of bovine tuberculosis. *J. Vet. Diagn. Invest.* **2004**, *16*, 126–131.
25. De la Rua-Domenech, R.; Goodchild, A.T.; Vordermeir, H.M.; Herwinson, R.G.; Christiansen, K.H.; Clifton-Hadley, R.S. Ante mortem diagnosis of tuberculosis in cattle: A review of the tuberculin tests, gamma-interferon assay and other ancillary diagnostic techniques. *Res. Vet. Sci.* **2006**, *81*, 190–210.
26. Meat inspection circular (in Danish). Available online: https://www.retsinformation.dk/Forms/ R0710.aspx?id=139770&exp=1#Bil2 (accessed on 28 December 2012).

CHAPTER 10

Assessment of the Major Animal Health Problems and its conomic mpact on Beef Cattle Export Industry at Adama Quarantine Station, Central Ethiopia

Adege HB, Masebo NT*

Wolaita Sodo University, School of Veterinary Medicine, Wolaita Sodo, Ethiopia.

ABSTRACT

The current study was conducted with major objective of assessing the major animal health factors that cause rejection in international live animal trade at Adama export quarantine station and to estimate economic loss due to the rejection factors/disease. A cross-sectional study was conducted from October 2015 to April 2016 at two private feedlots found in Adama and Adami Tullu Eastern Ethiopia. From the total of 2697 bulls examined for the presence of physically and clinical abnormalities, 3.74% (101/2697) physical and clinical abnormalities were detected. From a total of 2596 bulls that were examined for the presence of FMD (Foot and mouth disease) and CBPP (Contagious Bovine Pleuropneumonia); 12.9% (334/2596) and 9% (233/2596) were found to be positive respectively. The results of the present study showed significant variation in the prevalence of FMD (Foot and mouth disease) and CBPP with respect to origin of bulls ($P < 0.05$). The study also revealed the overall loss of hard currency due to failure to export seroposative live animals and rejected due to physical and clinical abnormalities from the two exporter farms which were included in our study was 580,826 US$. The economic loss that is incurred from FMD, CBPP and physical and clinical abnormalities were 290,413 US$, 202,593.5 US$ and 87,819.5 US$ respectively. The results of this study indicate that FMD and CBPP are still important disease in the feedlots. Therefore prevention and control strategies are needed in the feedlots; and biosecurity of the feedlots must be improved so that to reduce the exposure of bulls to new infections.

Keywords: Economic Loss; CBPP; FMD; Physical and Clinical Abnormalities.

1. INTRODUCTION

Ethiopia is a largely rural country with an agrarian economy. Livestock are of economic and social importance both at the household and national levels, and have in the past provided significant export earnings [1]. The diverse agro-climatic conditions of Ethiopia make it very suitable for the production of different kinds of livestock. Most of the livestock are produced by pastoralists, agro-pastoralists, and smallholder mixed crop–livestock farmers and sold to private entrepreneurs operating in a marketing chain involving collection, fattening and transportation up to terminal markets [2].

Livestock produce about 30% of the agricultural gross domestic product (AGDP) in the developing world, and about 40% of the global GDP. Livestock play multiple roles in the livelihoods of people in developing communities, especially the poor [3]. In the majority of the rural areas of Ethiopia livestock production plays important role in the provision of draft power, food, cash, transportation, fuel, and, especially in pastoral areas, social prestige. In the highlands, oxen provided draft power in crop production [4].

Livestock development in Ethiopia is constrained, amongst other important factors, by a widely distributed disease. Out of the 15 OIE Lists-A diseases known for their rapid spread and serious socio-economic or public health consequences, and which affect international livestock trade, 8 of them are endemic in the country (FMD, CBPP, PPR, RVF, LSD, sheep pox, African Horse Sickness and Newcastle Disease) [5]. For cattle FMD, contagious bovine pleuropneumonia (CBPP) and brucellosis are the three most important diseases in Ethiopia according to livestock master plan [6]. The presence of these diseases makes it difficult for the country to access international livestock markets. Among these diseases FMD and CBPP are the major hindrance to international trade [7]. Importing countries frequently ban imports of meat and livestock from the horn as a result of outbreaks of livestock diseases [5]. Therefore, the objective of the present study was to assess the major health and physical factors that causes rejection in international live animal trade at Adama export quarantine station and to estimate their economic loss.

2. MATERIALS AND METHODS

2.1 Study Area

The study was conducted at private export quarantine farms of beef cattle found in Adama and surroundings. Adama city, the previous capital of Oromia region is located in the eastern shewa zone of Oromia, at 8.55°N 39.27°E at elevation of 1712 meters, 99 Km southeast of Addis Ababa. It has annual rainfall and temperature ranging from 400mm-800mm and 13.9°C-27.7°C; respectively. The city sits between the base of an escarpment to the west and the Great Rift Valley to the east [8].

2.2 Study Animals

The study population was apparently healthy bulls that were ready to be exported, with average age range 1.5-4 years; and according to the information obtained from the feedlots and quarantine station the origin of the bulls were of Borana area, Yabello town market and Bale area. A total of 2697 beef cattle were examined to assess the major factor that cause detention and rejection in international live animal trade in the study area.

2.3 Study Design

A cross sectional study was conducted from October 2015 to April 2016 to assess the major health and physical factors that causes rejection in international live animal trade at Adama export quarantine station and to estimate their economic loss. Vaccination status, origins, body condition score, body weight of the animal in the feedlot was recorded. The body condition score was categorized into three groups such as poor, medium and good according to Nicholson and Butterworth [9].

2.4 Physical and Clinical Examination of Animals

Veterinary clinical investigation methods were applied to assess the physical factors that lead to the rejection of animals from international live animal trade.

2.5 Blood Sample Collection and Processing

Blood collection for serological examination was carried out in each export quarantine farms (feedlots). About 7-10 ml of blood was collected from the jugular vein of the animal into sterile plane vacationer's tube and venject needle. Then each animal was tagged and the respective blood samples were labeled accordingly. This blood was let down to clot for about 2-3 hours at room temperature then serum was stored at 4°C till serum extraction, usually within 24 hours. Then serum was extracted and dispensed into cryovials in national veterinary institution (NVI). Serum was stored at -20°C; each serum sample was subjected to laboratory test through the OIE recommendation diagnostic tool [10]. Antibodies against the 3ABC non-structural polyproteins of FMD virus in cattle sera were determined using an indirect enzyme-linked immunosorbent assay (ELISA) according to the procedure obtained by De Diego et al., [11]. A competitive ELISA to detect antibodies against small colony forming Mycoplasma mycoides mycoides (MmmSC) in cattle sera was used according to the procedure obtained by Le Goff and Thiaucourt [12].

2.6 Economical Loss Analysis Due to Rejected Animal

The price of the live animal varies based the market demands. During the study period the data obtained from Adama animal quarantine centre shows that the minimum weight expected from the animal to be export was 320 kg and the unit price/kg costs was 1.85 US$. Thus to calculate the amount cost (money) of a single bull, the exact weight in kg was taken and multiplied with the price per Kg [13].

2.7 Data Entry and Statistical Analysis

Data were classified, filtered, coded using Ms Excel, and was transferred to Statistical Package for Social Sciences software version 20 (SPSS software V.20). Thereafter, it was analyzed according to the different variables. Pearson chi-square was used to evaluate the statistical significance of variables. In all analyses, confidence interval (CI) was 95% and significance was set at $P \leq 0.05$.

3. RESULTS

3.1 Major Physical and Clinical Observed Abnormalities

From the total of 2697 bulls examined for the presence of physical and clinical abnormalities, 3.74% (101/2697) physical and clinical abnormalities were detected. The distribution of physical and clinical abnormalities based on their different origins, body condition score and weight range of animals were summarized in Table 1.

Table 1. The prevalence of physical and clinical observed abnormalities from adami tulu and adama feedlots.

Types of physical abnormalities							
Origin of bulls	Long horn	Skin disease	Deformed leg and hoof	Blindness	Swelling and hernia	Bulky abdomen	Total rejected bulls (%)
Bale (581)	6 (1.03%)	10 (1.72%)	4 (0.68%)	1 (0.17%)	1 (0.17%)	2 (0.34%)	3.94%
Yabelo (600)	2 (0.33 %)	3 (0.5%)	5 (0.83%)	1 (0.16%)	1 (0.16%)	2 (0.33%)	2.31%
Borena (1516)	13 (0.85%)	15 (0.98%)	14 (0.92%)	3 (0.19%)	13 (0.85%)	5 (0.33%)	4.12%
Total (2697)	21 (0.778%)	28 (1.038%)	23 (0.775%)	5 (0.185)%	15 (0.556%)	9 (0.333%)	101 (3.74%)

3.2 Estimation of economic losses due to physical abnormalities

Direct economic loss was calculated based on the rejection rate of bulls from international market due to physical and clinical abnormalities. The total financial loss incurred due to physically and clinically observed abnormalities was 87,819.5 US$ as shown in Table 2.

3.3 Seroprevalence of Foot and Mouth Disease (FMD)

From a total of 2596 bulls that were examined for the presence of FMD antibodies by 3ABC ELISA techniques 12.9 % (334/2596) were found to be positive. The prevalence of foot and mouth disease (FMD) based on origins; body condition score and weight range of bulls were summarized in Table 3.

Table 2. Economic Losses of Beef Cattle (n=101) Rejected Due to Physical and Clinical Abnormalities from Adami Tulu and Adama Feedlots.

Name of the feedlot	Total animal examined	Total animal rejected	Average live weight	Rejected bulls (%)	Unit price/ kg US$	Total loss in US$
Adami Tulu	1881	81	470	4.30%	1.85	70,429.50
Adama	816	20	470	2.45%	1.85	17,390
Total	2697	101		3.74%		87,819.50

Table 3. Prevalence of foot and mouth disease (fmd) in relation to origins, body condition score, body weight range and site of feedlots.

	No of animals examined	No of positive	Prevalence (%)	P-value
Origin				
Bale	557	26	4.7	
Borena	1453	228	15.7	0.00**
Yabello	586	80	13.7	
Body condition score				
Medium	1797	279	15.5	0.00**
Good	799	55	6.9	
Body weight (kg)				
320-420	882	173	19.6	
421-520	1541	145	9.4	0.00**
521-620	173	16	9.2	
Farm				
Adami Tulu	1725	233	13.5	0.17
Adama	871	101	11.6	
Total	2596	334	12.9	

** Statistical significant difference at $p < 0.05$

3.4 Estimation of economic losses due to FMD

In the present study the economic losses due to FMD was calculated based on the rejection rate of animals from international export was 290,413 US$ as summarized in Table 4.

Table 4. Economic losses of beef cattle (n = 334) rejected due to fmd in adama and adami tulu feedlots.

Name of the feedlot	Total animals examined	Average live weight	No of rejected animal	Prevalence	Unit price/kg US$	Total loss in US$
Adama	871	470	101	11.60%	1.85	87,819.5
Adami Tulu	1725	470	233	13.50%	1.85	202,593.5
Total						290,413.0

3.5 Seroprevalence of contagious bovine pleuropneumonia (CBPP)

From a total of 2596 bulls that were examined for the presence of CBPP 9% (233/2596) were found to be positive.

Table 5. Prevalence of contagious bovine pluropnumonia (cbpp) in relation to origins, body condition score, body weight range and site of feedlots.

	No of animal examined	No of positive	Prevalence (%)	P-value
Origin				
Bale	557	22	3.9	
Borena	1453	166	11.4	0.000**
Yabello	586	45	7.7	
Body condition score				
Medium	1797	205	11.4	0.000**
Good	799	28	3.5	
Body weight (kg)				
320-420	882	77	8.7	
421-520	1541	146	9.5	0.026**
521-620	173	10	9.2	
Farm				
Adami Tulu	1725	176	10.2	0.020**
Adama	871	57	6.5	
Total	2596	233	9%	

** Statistical significant difference at $p < 0.05$.

The prevalence of contagious bovine pleuropneumonia (CBPP) based on origins; body condition score and weight range of bulls were summarized in Table 5.

3.6 Estimation of economic losses due to Contagious Bovine Pleuropneumonia (CBPP)

The total financial loss incurred due to CBPP was 202,593.5 US$ as summarized in Table 6.

Table 6. Economic losses of beef cattle (n = 233) rejected due to cbpp in adama and anami tulu feedlots.

Rejection factors	Total animals examined	Average live weight	No of rejected animal	Preva-lence	Unit price/ kg US$	Total loss in US$
Adama	871	470	57	10.20%	1.85	49,561.5
Adami Tulu	1725	470	176	6.50%	1.85	153,032
Total						202,593.5

4. DISCUSSION

In this study a total of 2697 bulls intended for export which were found in two feedlots located in central Ethiopia were assessed. Based on the examination; the major physical abnormalities that cause live animal export detention and rejection were long horn, skin diseases, deformities of leg and hoof, blindness, swelling and hernia and bulky abdomen. Poor farm management and lack of screening of the bulls for these abnormalities during purchasing or collection has a great role for this result. Generally the two private farms persuaded to receive this tremendous loss with above mentioned problems, with live animal rejection thus rejected animals served for local markets with lower price. Bulls those pass the clinical and physical examinations were assessed further for the presence of disease like foot and mouth disease (FMD) and contagious bovine pleuropneumonia (CBPP). Transboundary diseases, especially CBPP, FMD, and RVF hinder smooth trade flows of live animals and livestock products to the markets at domestic, regional and international level. The increased occurrence of these diseases suggests that there are certain weaknesses that are not being addressed [14].

The overall high seroprevalence of FMD (12.9 %) reported in this study is shows that FMD is still existing problem in feedlots business. Similar studies done previously in export quarantine shows almost similar finding on prevalence of FMD [15]; reported overall prevalence of FMD (14.5 %) in Adama –Modjo feedlots in bulls originated from Borena area, Southern Ethiopia and [16] also reported an overall prevalence of (12.9%) at Adama-Modjo Livestock Export Industry, Eastern Ethiopia. According to study performed by [17] in livestock export quarantine stations in central Ethiopia; majority of the feedlots practice all in/all out management system and almost none of them practice cleaning and disinfection between herds and flocks regularly. They also indicated that hygienic practices of feedlot workers and visitors are none existed. This could be one of the major factors that FMD and other infectious disease like CBPP still exist in the feedlots. The present finding of FMD was higher than the findings of [18] who reported, (5.53 %) on exported bulls of Borana and Jimma origin of Ethiopia; also study performed by [19] indicated that an overall prevalence of 5.6% in Afar Regional State while [20] reported (5.59%) in Dessie Zuria veterinary clinic and Kombolcha area of Ethiopia which is lower result from this study finding; this could be attributed to season of the year, type of animals selected for the study or age of the animal, ecological and management factors [21]. On the other hand, the find of this study was low when compared to report from Borena and Gujji zone (24.57%) and Bale zone (21.49%) of Ethiopia [22, 23].

In this study (9%) seroprevalence of CBPP recorded shows that also it is important diseases in the feedlots. Study done by [24] in prioritized transboundary diseases in borena bulls market chain of Ethiopia indicates that the bulls are usually introduced into feedlots without quarantine and also he indicated that vaccination of bulls started after two weeks of registration by Adama veterinary controlling authority. The length of time from introduction of first batch of bulls to vaccination is sufficient to transmit the disease to healthy bulls. This could be the reason that transboundary diseases like FMD and CBPP still existing in the feedlots which serve as quarantine since this trend is not changed [24]. The finding of this study is in agreement with reports of [17] in older animals at different export quarantine of Ethiopia [21]; at Adama-Modjo Livestock Export Industry, Eastern Ethiopia and [25] in Dessanech District of South Omo Zone, South Eastern Ethiopia. The prevalence of CBPP obtained in this study was lower as compared to [26] who reported from Southern Zone of Tigray Regions, Northern Ethiopia. The lower seroprevalence in the current study might be due to the fact that, bulls intended for finishing were purchased based on their good body condition and health, the probability of purchasing chronically ill or recovered animals with poor body condition is therefore very low. This might also account for lower prevalence of diseases in the feedlots and bulls purchased after physical examination and brought to the farms and vaccinated for endemic diseases including CBPP. However the prevalence of this study was higher than report of [27] in bulls originated from Borena Pastoral Area of Southern Ethiopia.

The prevalence discrepancy recorded above with two disease show the fact between the current and the previous studies differ each other might be due to the variation exist in the epidemiology of the disease as well as cattle management system of the different study areas. In this study all animals were apparently healthy bulls purchased after physical examination and brought to the farms and vaccinated for endemic disease including FMD and CBPP. During this study over 55.9% of the animals were from Borana origin. Also this variation in result among researcher might be attribute to season of the year, furthermore, animal in field condition may have a chance of movement, contact with wild animals; production system and composition of animal species have got their own role in the epidemiology of the disease. As a result the Borana area is not free from FMD and CBPP. Even bulls were purchased from market where continuous commingling of animals from different localities. This shows that the bulls for export are not free from the FMD and CBPP due to the origins of the animals besides other reasons[28,29].

ACKNOWLEDGMENTS

The authors are very grateful to the owners of the feedlots to let the study to take place and providing any information that is required; and also our thanks also goes to Adama quarantine station office for facilitating and providing any assistance needed. All contributions and supports are gratefully acknowledged.

CONCLUSION

The result of this study indicates that FMD and CBPP is an important disease problem in live animal export industry of Ethiopia and are well established infection in bulls of the feedlots. A strong veterinary service structure has to be established to strengthen prevention and control of transboundary disease, and there should be surveillance network at national level, co-ordinate data collection, reporting any signs of the disease and implement appropriate control and preventive measures throughout the country. Further prevention and control strategies are needed in the feedlots; and biosecurity of the feedlots must be improved so that to reduce the exposure of bulls to new infections.

REFERENCES

1. Gebre Mariam S, Amare S, Baker D, Solomon A, Davies R (2013) Study of the Ethiopian live cattle and beef value chain. ILRI Discussion Paper 23, Nairobi, Kenya.
2. Hailemariam T, Getachew L, Dawit A, Asfaw N (2008) Live animal and meat export value chains for selected areas in Ethiopia: Constraints and opportunities for enhancing meat exports. ILRI Discussion Paper 12. ILRI Nairobi, Kenya.
3. Frans S, Aldo S, Siboniso M (2010) The Role of Livestock in Developing Communities: Enhancing Multifunctionality. The Technical Centre for Agricultural and Rural Cooperation CTA, Wageningen, Netherlands.
4. Ethiopian Economic Association (EEA). A research report on land tenure and agricultural development in Ethiopia, Addis Ababa, Ethiopia, October 2002.
5. Wondwosen Asfaw (2003) Influence of animal diseases and sanitary regulations on livestock export trade and cases of export restrictions. Challenges and Opportunities of Livestock Marketing in Ethiopia: Proceedings of the 10th annual conference of the Ethiopian Society of Animal Production (ESAP) held in Addis Ababa, Ethiopia.
6. Livestock master plan (LMP) (2013) Livestock health priorities in the Ethiopia livestock master plan, Ethiopia.
7. Zewdie S (2004) Current status of veterinary services in Ethiopia. In proceedings of a national workshop on managing animal health constraint to export marketing of meat and livestock. 27-28.
8. NMSA: National Metrological SeasonalAgenc (2006) Agro-meteorology bulletin, Addis Ababa, Ethiopia.
9. Nicholson M, Butterworth T (1986) A guide to body condition score in zebu cattle. international livestock center for Africa, Addis Ababa, Ethiopia, 29.
10. Office International Des Épizooties (2002) Manual of Standards for Diagnostic Tests and Vaccines.
11. De Diego M, Brocchi E, Mackay D, De Simone F (1997) The use of the non-structural polyprotein 3ABC of FMD virus as a diagnostic antigen in ELISA to differentiate infected from vaccinated cattle. Arch Virol. 142(10): 2021–2033.
12. Le Goff C, Thiaucourt F (1998) A competitive ELISA for specific diagnosis of contagious bovine pleuropneumonia (CBPP). Vet Microbiol. 60(2-4): 179–191.
13. Adama District Agricultural Office (2008) Document on the Live Animal Export Industry Market Bulletin. Adama, Ethiopia.
14. Eighth Session of the Committee on Food Security and Sustainable; Development and Regional Implementation Meeting for the Twentieth Session of the Commission on Sustainable Development (E/ECA/CFSSD), Addis Ababa, Ethiopia, 2012.
15. Gezahegn A, Girma Z, Berhanu A (2014) Sero-prevelance of foot and mouth disease (FMD) and associated economic impact on central Ethiopia cattle feed lots. J vet med animal health. 6(5): 154-158.
16. Tadesse BT (2014) Prevalence of the Major Infectious Animal Diseases Affecting Livestock Trade Industry in Ethiopia. J of Bio Agri and Healthcare. 4(17): 76-82.
17. Gezahegn A, Samson L (2014) Biosecurity practice in central Ethiopian cattle feedlots. Its implication for live cattle export. Int j livestock product. 5(11): 181-187.
18. Bedru H (2006) Seroprevalence Study of Foot and Mouth diseases in Export Bulls of Borana and Jimma origin, Ethiopia. Faculty of Veterinary Medicine, Addis Ababa University, Debrezeit, Ethiopia.
19. Jembere S (2008) Participatory epidemiology and seroprevalence of Foot and Mouth Disease in Afar pastoral region, Ethiopia, Faculty of Veterinary Medicine, Addis Ababa University, Debreziet, Ethiopia.
20. Abraha G, Ahmed I (2013) Seroprevalence of foot and mouth disease in cattle in Dessie Zuria and Kombolcha area, Ethiopia. Vet res. 6(1): 1-4.

21. Habtamu M, Desta B, Tesfaye R, Ashenafi F, Fufa A (2011) Study on the prevalence of Foot and Mouth Disease in Borana and Guji Zones, Southern Ethiopia. Vet World. 4(7): 293-296.
22. Masinga D (2008) Seroprevalence Study of Foot and Mouth Disease in Cattle In Bale Zone, Ethiopia. Faculty of Veterinary Medicine Addis Ababa University, Debre Zeit, Ethiopia. 17-19.
23. Gezahegn A (2012) Risk assessments of prioritized transboundary diseases in borena bulls market chain and its implication for international trade. Addis Ababa University, College of Veterinary Medicine and Agriculture, Bishoftu, Ethiopia.
24. Kassaye D, Molla W (2012) Sero-prevalence of contagious bovine pleuropneumonia at export quarantine centers in and Around Adama, Ethiopia. Trop Anim Health Prod. 45(1): 275- 9.
25. Molla B, Delil F (2015) Mapping of major diseases and devising prevention and control regimen to common diseases in cattle and shoats in Dassenech district of South Omo Zone, South- Western Ethiopia. Trop Anim Health Prod. 47(1): 45-5.
26. Teshale T, Temesgen T, Tsigabu N, Birhanu H, Solomon W, et al., (2015) Epidemiological Status of Contagious Bovine Pleuro Pneumonia in Southern Zone of Tigray Regions, Northern Ethiopia. Anim Vet Sci. 3(1): 32-36.
27. Gezahegn A, Samson L, Berhanu H (2015) Sero-prevalence of Contagious Bovine Pleuropneumonia (CBPP) in bulls originated from Borena Pastoral Area of Southern Ethiopia. Trop Animm Health Prod. 47(5): 983-7.
28. Megersa B, Beyena B, Abunna F, Regessa A, Amenu K, et al., (2008) Risk factors for foot and mouth disease seroprevalence in indegionaous cattle in Southern Ethiopia: the effect of production system. Trop Anim Health Prod. 41(6): 891-8.
29. Aftos F (2007) Foot and mouth Disease (FMD), The center for Food security and public Health, Iowa State University, College Veterinary Medicine. 1-3.

CHAPTER 11

Towards an Ethological Animal Model of Depression? A Study on Horses

Carole Fureix[1*], Patrick Jego[1], Séverine Henry[1], Léa Lansade[2], Martine Hausberger[1]

1 Universite´ de Rennes 1, UMR CNRS 6552 Ethologie Animale et Humaine, Rennes, France

2 Universite´ de Tours, Institut Franc¸ais du Cheval et de l'Equitation, UMR 6175 INRA-CNRS Physiologie de la Reproduction et des Comportements, Nouzilly, France Editor: Georges Chapouthier, Universite´ Pierre et Marie Curie, France

ABSTRACT

Background

Recent reviews question current animal models of depression and emphasise the need for ethological models of mood disorders based on animals living under natural conditions. Domestic horses encounter chronic stress, including potential stress at work, which can induce behavioural disorders (*e.g.* "apathy"). Our pioneering study evaluated the potential of domestic horses in their usual environment to become an ethological model of depression by testing this models' face validity (*i.e.* behavioural similarity with descriptions of human depressive states).

Methodology/Principal Findings

We observed the spontaneous behaviour of 59 working horses in their home environment, focusing on immobility bouts of apparent unresponsiveness when horses displayed an atypical posture (termed *withdrawn* hereafter), evaluated their responsiveness to their environment and their anxiety levels, and analysed cortisol levels. Twenty-four percent of the horses presented the withdrawn posture, also characterized by gaze, head and ears fixity, a profile that suggests a spontaneous expression of "behavioural despair". When compared with control "non-withdrawn" horses from the same stable, withdrawn horses appeared more indifferent to environmental stimuli in their home environment but reacted more emotionally in more challenging situations. They exhibited lower plasma cortisol levels. Withdrawn horses all belonged to the same breed and females were over-represented.

Conclusions/Significance

Horse might be a useful potential candidate for an animal model of depression. Face validity of this model appeared good, and potential genetic input and high prevalence of these disorders in females add to the convergence. At a time when current animal models of depression are questioned and the need for novel models is expressed, this study suggests that novel models and biomarkers could emerge from ethological approaches in home environments.

1. INTRODUCTION

Depression is a major mood disorder with a high prevalence of morbidity (especially in women) and a heritable component. Severe forms of depression occur in 2–5% of the US population [1], [2]. Early ethological models of depression were based on observations of socially-deprived young monkeys and chicks that, after a protestation phase, gave way to "behavioural despair" by becoming hypoactive, silent and unresponsive to their environment *e.g.* [3]. These models have been replaced by more rapid and repeatable tests with rodents, such as the Porsolt and the tail suspension tests, where animals are considered to be depressed when they stop "resisting" *e.g.* [2], [4].

However, nowadays most reviews converge in questioning current animal models of depression [2], [5], [6]. After decades of using well-controlled animal models, the outcome however is a discrepancy between the positive influence of some drugs on these models and the lack of their efficiency for humans [7]. Nestler *et al.* [2] emphasized the need for better animal models and for biomarkers of depression and they regretted that none were available. They concluded that at that time no *bona fide* animal model of biomarker of depression existed, and that the researches involving animal models of mood disorders desperately needed a "focus on ethologically informed models of mood disorders with animals living under more natural conditions", *i.e.* models focused on observing animals' spontaneous behaviour in their home environments. A first step towards validating an ethological model is to insure its "face validity", that is a comparability of symptoms, such as loss of interest in the environment, behavioural despair, apathy and possible anxiety-related behaviour [4]. Depression and anxiety may be associated and correspond to a continuum [8].

Ethological animal models may help to address the area of peripheral markers of depression, as for example increased or decreased cortisol levels in community/hospital groups of patients have been debated ([8]–[10], see also [11] in animals), and they should be useful to test aetiology and genetic components of this disorder. Animal emotionality is considered a reliable model of human anxiety and has a heritable component [12]. Despite recent attempts to develop novel models (*e.g.* guinea pigs' isolation-induced calling paradigm or gerbils' foot tapping response, [6]) no satisfactory animal model has emerged. Ethological approaches, focused on animals' spontaneous behaviour in their home environments, might prove to be useful especially if the environmental conditions offered to animals share features with environmental conditions known to induce depression in human. For instance, stress at work, and especially interpersonal stressors, may lead to a variety of negative and durable effects, such as depression [13]. Domestic horses may encounter social and spatial restriction (as several domestic or lab species) but also share with humans the characteristic of working on a daily basis and have then "interpersonal" interactions (with other working horses, working with a "boss" who is the human who manages or rides it) [14]. Horses' studies, mimicking in animals conditions in which human depression is known to occur, might then open a "window" onto the complex aetiology of human depression.

According to recent studies (though not designed in order to compare collected data with human depression characteristics), "apathy", "withdrawal" and/or "inhibition" can occur in sows, pigs, sheep or dogs submitted to social/spatial restrictions or to repeated aversive stimuli inducing chronic stress [15]–[17]. Recently, two studies respectively suggested or reported the occurrence of "apathetic", "unresponsive", "switched off" horses in a restricted domestic environment [18] or under harsh conditions in developing countries [19]. Previous observations performed by the authors in horses from riding school also revealed that part of this population appears to display an atypical posture (standing motionless with eyes open, stretched neck and similar height between neck and back), when horses seem to display an apparent unresponsiveness, *i.e.* seem to have "withdrawn". In the present study, we hypothesized that this symptomatology in working horses, termed *withdrawn* hereafter, could suggest human depression characteristics. We tested this hypothesis by observing horses' spontaneous behaviour in their home environment, evaluating their reactivity to their environment and by taking physiological samples (plasma cortisol levels assessment). This pioneering study aims to evaluate the potential of domestic horses in their usual environment to become a potential ethological model by testing this model's face validity, before performing further more invasive studies. We focused on two elements: spontaneous expression of "behavioural despair" and unresponsiveness to a variety of environmental stimuli (tactile/visual, human/non-human).

2. MATERIALS AND METHODS

All experiments comply with current French laws (Centre National de la Recherche Scientifique) related to animal experimentation and were in accordance to the European Communities Council Directives of 24 November 1986 (86/609/EEC). Mostly behavioural observations were performed. Total duration of the blood sampling procedure did not exceed 1 min, and non-painful sampling was confirmed by the absence of any retreat behaviour of the horses during the procedure. Animal husbandry and care were under management of the riding schools staff, as this experiment involved only horses "from the field" (no laboratory animals).

2.1 Subjects

Fifty-nine horses from three riding schools were observed between January and June 2007 (**TABLE 1**). Activities and housing conditions in the schools were similar. In all cases, the horses were kept singly in 3 m * 3 m individual straw-bedded boxes. Each box was cleaned once a day (in the morning). Animals were fed commercial pellets three times a day and hay was provided *ad libitum* once a day. Each box was equipped with an automatic drinker. Horses worked in riding lessons for 4–12 hours a week, with at least 1 free day each week (closing day). Riding lessons involved children and teenagers and were mainly related to indoor (instruction) and outdoor activities, including a few competition activities. Both geldings (n=44) and mares (n=15) were tested. Sixty-eight percent of the horses were French Saddlebreds, equally distributed among the centres. Other horses belonged to a variety of breeds or were unregistered animals. They were between 5 and 20 years old (median =12, 1st quartile =9, 3rdquartile =14).

Table 1. Study population: horses' breed, sex and age for each riding school.

	School 1	School 2	School 3
Breed			
French Saddlebred	6	25	9
Connemara	1	–	–
French Trotter	1	1	1
Thoroughbred	1	–	–
French pony	1	–	–
Hanoverian	–	1	–
Anglo-Arab	–	1	–
Criolo	–	–	1
Iberian	–	–	1
Unregistered	2	1	6
Sex			
Mare	4	6	5
Gelding	8	23	13
Age			
mean ± error standard (range)	12.6±3.1 (8–17)	11.9±2.5 (6–16)	11.2±4.9 (5–20)

2.2 Behavioural Observations in the Box

Horses were observed in their box using a focal sampling method: all behaviours of the focal animal were recorded continuously during 5-min sessions. Observations were made during three periods: 9–11 a.m., 2–5 p.m. and half an hour before meals (*i.e.* between 6.30–7.30 a.m., 11.30–12.00

a.m. or 5.30–6.00 p.m. according to riding school schedule). All horses were observed 6 times (2 sessions per period =30 min total/horse). All observations were made by a single observer. All behaviours were recorded and special attention was given to withdrawn bouts, *i.e.* bouts of immobility when horses displayed a unusual posture, characterized by horses standing with eyes open, stretched neck (open jaw-neck angle) and similar height between neck and back (**FIG.1**). During observations, this posture was easily distinguished from observation of the environment, when the horse's neck is held higher [20], and from resting, when "in the standing posture, the horse is supported usually by only 3 legs with the slope of the neck lower and rounder than when attentive and alert. The muscles relax, the ears rotate laterally, the eyelids and lips get droopy and the eyes close. In the extreme, the crest of the neck may drop 20° or more below horizontal, with the dorsal surface of the head sometimes reaching vertical" ([21], see also [20] for objective inter-behavioural comparisons of postures).

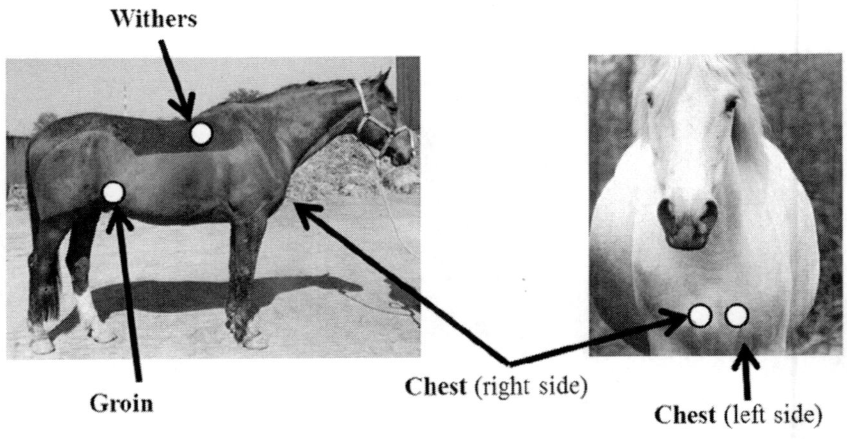

Figure 1. The withdrawn posture of "withdrawn" horses.

Pictures of a horse a) in a withdrawn posture, b) standing non-resting and c) resting. Withdrawn posture is characterized by a similar height between the horse's neck and back (the nape – withers – back angle approximately 180°) and a stretched neck (obtuse jaw-neck angle). This posture is distinguished from postures associated with observation of the environment (for which the neck is higher), and resting, when eyes are at least partly closed and the horse's neck is rounder [20], [21]. Note that the restricted size of the box (3 m * 3 m) prevented the authors from taking a picture of the whole horse displaying the withdrawn posture, as we chose to use the same lens in order to limit shape distortion between pictures.

"Apathetic", "unresponsive", "switched off" horses" are described as having "dull eyes looking nowhere" [18], [19]and therefore special attention was paid to eye and ear fixity (and position) when horses displayed the withdrawn posture. Ear position is an interesting indicator of a horse's internal state [21]–[23]and welfare (Fureix *et al.* in prep) and we have defined positions by referring to studies of other species [24]: axial ear (perpendicular to the head – rump axis), forward ear (tip of the ear towards the front at an angle of more than 30° from the perpendicular) or backward ear (tip of the ear towards the back at more than 30° from the perpendicular). Discomfort, pain and stress appear to be associated with ears in a backward position [21]–[23], Fureix *et al.* in prep]. Head movements, gaze durations and ears movements during these bouts were therefore compared to those of four other non-working horses (two geldings and two stallions, 13–20 years old) living under natural conditions (*e.g.* not socially deprived, not confined, not regularly exercised) and observed for 30 min each while standing non resting.

2.3 Testing Responsiveness to Environment

1) To tactile stimulations. Reactions to tactile stimuli are common in horses as an adaptive response to the presence of insects such as flies. Horses react by a muscular twitching **[25]**. Here we estimated tactile reactivity using von Frey filaments (Stoelting, IL, USA), a procedure first used by Redua et al. **[26]**on horses and adapted by Lansade et al. **[27]**. A filament consisted of a hard plastic body extended by a nylon thread and is calibrated to exert a specific force on the skin (namely 0.008, 0.02, 1 and 300 g/matter in our study). They were applied perpendicularly on an animal's skin until the nylon thread started to bend. Muscular twitching (reaction used by horses to drive away flies **[25]**), was recorded.

Sensory reactivity was tested by the same unfamiliar experimenter (blind to the result of behavioural observations) in each individual box outside horses' working times. The horse was lightly restrained and von Frey filaments were applied at the basis of the horse's withers, groin and chest. Both sides of the horse were tested. Von Frey filament (0.008, 0.02, 1 and 300 g/matter) types, areas (withers, groin and chest, **FIG. 2**) and sides were tested in a random order for each horse. The tests involved two sessions (separated by 2 hours) during which two von Frey filaments were tested at 10 minutes interval. For instance, following a pre-established random order, a given horse was tested first at 08:30 p.m. using the 0.02 and 300 g/matter von Frey filaments, then re-tested at 10:30 p.m. using the other von Frey filaments (namely 0.008 and 1 g/matter in this example). The response was coded in a binary form (trembling/no trembling), in accordance with Lansade *et al.* [27].

Figure 2. Test areas for tactile sensitivity assessment: withers, chest and groin.

2) To human approach. Horses may be surprised by the sudden appearance of a human at their box door and their reactions can then vary from friendly approach to indifference or even aggression [28]. A previous study performed in the same 59 horses revealed inter-individual variations in the horses' responses to a sudden human approach: *e.g.* 12% of them reacted friendly at least once, while 51% of them displayed at least one aggressive reaction [29], strongly suggesting that, even if horses from riding school might be habituated to people suddenly appearing at the door of the box, a high proportion of them still reacts (either negatively or positively) to the situation. We hypothesised that "depressed" horses would be indifferent to such a stimulation. The experimenter, walking slowly along the corridor, appeared suddenly at the closed door of the box while the horse was feeding (hay, straw), head down. She recorded the horse's first reaction, following Hausberger and Muller's [28] scoring method. Thus, five scores ranged from very "friendly" to very aggressive behaviour: the horse looks at the experimenter with upright ears and approaches: A; the horse looks at the experimenter with upright ears and remains where it is: B; the horse shows no evidence of directed attention towards the experimenter (no change in behaviour, no gaze towards the person): C; the horse looks at the experimenter with ears laid back and remains where it is: D; the horse looks at the experimenter with ears laid back and approaches with a threatening posture (neck lowered, head extended or even exposed incisors): E. Each horse was tested five times, at different times of the day, yielding five scores for each horse for each test. For instance, a given horse could score "C" in the first, second and third sudden approach tests, "B" in the fourth test and "C" again in the last test. This would yield five scores: CCCBC for that horse.

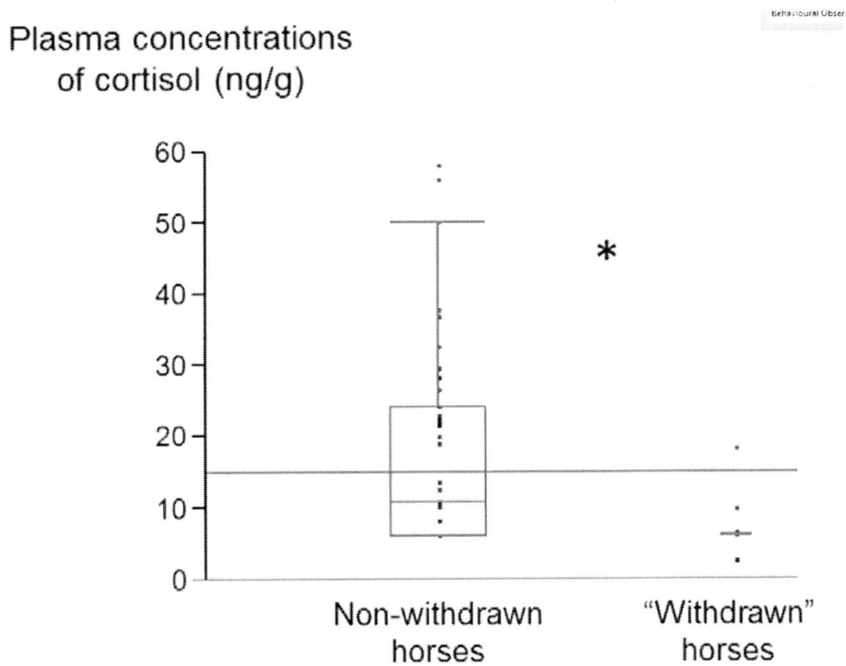

Figure 3. Behavioural characteristics of "withdrawn" bouts of "withdrawn" horses compared to horses standing non-resting observed under natural conditions.

(a) rates of ear position changes (number of ear position changes per number of withdrawn/standing non-resting bouts); (b) rates of head position changes (number of head position changes per number of withdrawn/standing non-resting bouts, and (c) gaze durations of withdrawn/standing non-resting bouts (s) for "withdrawn" horses and horses standing non-resting observed under natural conditions. Data are given as boxplot diagrams showing medians (lines in the boxes), 25 and 75% quartiles (boxes) and minimum and maximum ranges (whiskers). Unusual features of "withdrawn" horses are the absence of ear and head movements and a fixed gaze.

2.4 Are "Depressed" Horses also more Anxious?

Reactions towards a novel object while being released in a large environment (arena) are considered to reflect the horses' level of emotionality [30] and more precisely nervousness [31]or anxiety[32]–[34]. Our procedure and measures were the same as Wolff et al.'s [30](see also [31], [32], [34]. A novel object, *i.e.* a cage formed by metallic rails and long red fluorescent ribbons (100 * 80 * 80 cm), was placed in a familiar arena where the horse was released for 5 min. As in the above cited studies, the horse's behaviour was recorded by a motionless observer using instantaneous scan sampling [35] every 10s. The behavioural patterns sampled were: (a) standing; (b) exploration (characteristic slow walk of a quiet horse in a calm situation), (c) sustained walk, (d) trot, (e) passage, (f) gallop, (g) vigilance and (h) tail posture (detailed in *e.g.* [30]). Rare or brief behavioural patterns such as snorts, pawing, defecation, rolling, whinnying were recorded *ad libitum*. To rank the reactivity of horses, we used an emotionality index based both on behavioural patterns and their frequencies of occurrence; this index proved useful in previous studies. Values were attributed to the behavioural patterns according to their degree of specificity and corresponding level of arousal. These values were exploration =1, sustained walk =2, trot or gallop =3, vigilance =4 and passage, snorting or tail raised =6 ("5" was used in another test assessing reaction to social separation and is whinnying, not scored here). These values were multiplied by the number of times the corresponding pattern was observed. Note that these values only give a ranking indication and do not represent data on ratio scale (a horse with an index twice as high as that of another horse was not necessarily twice as reactive).

2.5 Physiological Data: Cortisol Measures

Horses' cortisol levels were assessed by blood sampling. In order to minimise the aversive effects of blood sampling *e.g.* [36], each horse was lightly restrained by one unfamiliar experimenter, gently petted and systematically given a food reward at the end of the sampling (total duration of this procedure <1 min). Four horses became highly agitated when their neck was rubbed with the alcoholic cotton and were not sampled. Seven ml of blood were collected in heparinised polypropylene tubes (BD Vacutainer®). Samples were kept in crushed ice until centrifugation (with a maximal delay between sampling and centrifugation of 15 min) and then aliquots of plasma were immediately placed on dry ice and stored at −20°C for further processing. Blood samples were collected between 18:00 and 19:00 p.m. Two samples were taken for each horse: after a day's work and after a day's rest. Plasma cortisol levels were measured using Immunotech kits for cortisol determination (Beckmann and Coulter), usually used for human cortisol plasma determinations and modified so that it could be used for equine plasma (Fureix et al. in prep).

2.6 Data and Statistical Analysis

Analyses were conducted using Statistica© 7.1 software (accepted p level at 0.05). Some differences appeared in horses' responses between riding schools, but that did not affect the present question and the results, and those data will be published in another paper.

As mentioned above, special attention was given to bouts of immobility when a horse displayed an atypical posture (**FIG.1**) when they seemed to have "withdrawn". In order to investigate further the features of this spontaneously expressed "withdrawn" posture, gaze durations and ears movements during these bouts were compared to those of four other non-working horses living under natural conditions observed while standing non-resting. The differences in gaze durations and ears movements were however not statistically investigated here due to low number of horses living under natural conditions.

Unresponsiveness to a variety of environmental stimuli, anxiety and cortisol measures were investigated in withdrawn horses by comparing them with non-withdrawn control animals from the same stables. Further analysis including the presence/absence of withdrawn bouts instead of frequencies of occurrence yielded more clear-cut results. Thus, our analysis involved horses displaying the withdrawn posture at least once (called "withdrawn", or W horses hereafter) and non- withdrawn off horses (never displaying an immobility bout: non-W). Horses' response to each von Frey stimulation was coded in a binary form (trembling =1/not trembling =0), that yielded a total tactile reactivity score adding up all a horse's responses (4 filaments * 3 areas * 2 sides). Further analysis compared horses with "high" tactile reactivity scores (with scores ≥12; 12 being the median score for our sample, *i.e.* the 50% more reactive horses) to horses with "low" tactile reactivity scores (the other 50%). Indifference towards a human was related to the number of C scores in the sudden approach test (see below). Further analysis of the novel object test data compared "highly reactive" horses (*i.e.* with emotionality index ≥38; 38 being the 3rd quartile for our sample, *i.e.* the 25% more anxious horses here) to "less reactive" horses (the other 75%).

As data were not normally distributed, we used non-parametric statistical tests [37]. Chi- square tests compared "withdrawn" to non- withdrawn horses in relation to their breed, sex, riding school and reactions to stimulations (*i.e.* highly reactive or not to tactile stimuli and to the novel object). We used one sample chi-square tests when chi-squared tests of association were impossible because more than 20% of the expected frequencies were below 5 [37]. Mann Whitney tests compared behavioural occurrences, frequencies and cortisol levels between "withdrawn" and "non- withdrawn" horses. Spearman correlation tests correlated frequencies of withdrawn posture and behavioural occurrences (*e.g.* number of "C" responses in the human sudden approach test) and cortisol concentrations. Descriptive statistics are median values (Med), followed by 1st (Q1) and 3rd (Q3) quartiles, range (minimum – maximum).

3. RESULTS

3.1 Behavioural Characteristics of "Withdrawn" Horses

In all, 24% of the 59 horses from riding school were observed displaying at least once the withdrawn posture (standing with eyes open, stretched neck, similar height between neck and back) up to 4 times each in 30 minutes. All of these horses were of the same breed (French Saddlebred, SF), which was overrepresented as no horse of another breed presented this syndrome (35% of the SF; 0% of the other breeds, $\chi^2{}_1=8.72$, $p<0.01$). Females were also overrepresented: while one third of the females presented this syndrome (33%, $\chi^2{}_1=1.67$, $p>0.05$), only 20% of the geldings did ($\chi^2{}_1=15.36$, $p<0.001$). Finally, more horses from one riding school presented this syndrome than did horses from the other two schools (school $_1$:0%, school $_2$:45%, school $_3$:6%, $\chi^2{}_2=14.15$, $p<0.001$).

The "withdrawn" posture was also characterized by:

1) The absence of ear and head movements during a bout (that could last from 17 to 97s, Med =27.88, *Q*1=22.75, Q3=34). This is an unusual trait as horses under natural conditions moved their ears up to 9 times (Med =1.23, *Q*1=0.30, Q2=3.07, **FIG. 3A**) and their head up to 11 times (Med =1.51, *Q*1=0.72, Q2=2.16, **FIG. 3B**) during a standing non-resting bout. The ears of all the horses except one (its ears were in an axial position in two of the four withdrawn postures bouts observed for this horse) were directed backwards during a withdrawn posture.

2) A fixed gaze with no eye movements during the whole bout that lasted from 17 to 97 seconds without any visible eye movements (Med =27.88, *Q*1=22.75, Q3=34)) (median gaze duration while standing non-resting: 15.32, *Q*1=11.95, *Q*3=19.44 under natural conditions, **FIG. 3C**).

3.2 Responsiveness to Environmental Stimuli

1) Tactile responsiveness.

"Withdrawn" horses showed a lowered responsiveness to tactile stimuli. Thus, only half of the "withdrawn" horses (7/14 horses) had a high total reactivity score (***i.e.*** a score ≥12, $\chi^2{}_1$=0.00, p>0.05), whereas most of the other (***i.e.*** that never displayed the withdrawn posture) horses from the same stables had high scores (32/45, $\chi^2{}_1$=8.02, p<0.01) (**FIG. 4A**). This was confirmed when concentrating on particular body areas (high tactile reactivity: **withers**: W horses: 9/14, $\chi^2{}_1$=1.14, p>0.05, non-W horses: 36/45, $\chi^2{}_1$=16.2, p<0.001; **groin**: W horses: 9/14, $\chi^2{}_1$=1.14, p>0.05, non-W horses: 41/45, $\chi^2{}_1$=30.42, p<0.001). Few horses of either category reacted to chest stimulations (W horses: 1/14, $\chi^2{}_1$=10.29, p<0.01, non-W horses: 5/45, $\chi^2{}_1$=27.22, p<0.001).

(**A**) Proportion of horses with high reactivity scores (*i.e.* ≥12; as 12 is the median score, thus including the 50% more reactive horses) for reactions to von Frey filaments applications, and (**B**) "C" scores (indifference, *i.e.* no change in activity) in the sudden human approach test. Data for (**B**) are given as boxplot diagrams showing medians (lines in the boxes), 25 and 75% quartiles (boxes) and minimum and maximum ranges (whiskers). Only half of the "withdrawn" horses had a high total reactivity score, whereas most of other horses (*i.e.* that never displayed the withdrawn posture) did. "Withdrawn" horses also displayed more often indifference to a sudden human approach than non- withdrawn off horses. Chi-square and Mann Whitney tests, ** p<0.01.

2) Reaction to sudden approach of a human.

"Withdrawn" horses clearly differed from non-withdrawn horses from the same stables as they generally reacted mostly with indifference (*i.e.* no change in activity) (Mann Whitney test, n_W =14, $n_{non\text{-}W}$ =45, Med_W =1, Q1=0.75, Q3=3; $Med_{non\text{-}W}$ =0, Q1=0, Q3=1; U =169.5, p<0.01, **FIG. 4B**). Frequency of no response to the sudden human approach did not differ according to which stable horses were kept (Kruskall-Wallis test, $H_{(2, N=59)}$ =4.67, p>0.05). Interestingly, duration of their withdrawn posture was correlated with the frequency of no response to the approach of a human (Spearman correlation test, N =59, r_S =0.36, p =0.01).

3.3 Emotional level/anxiety?

"Withdrawn" horses tented to show higher reactions when confronted with a novel object: 36% (more than one third) of them were in the first reactive quartile of the population (5/14, χ^2_1=1.14, p>0.05), whereas only 24% of the non-withdrawn horses from the same stables reacted as strongly (11/45, χ^2_1=11.76, p<0.001).

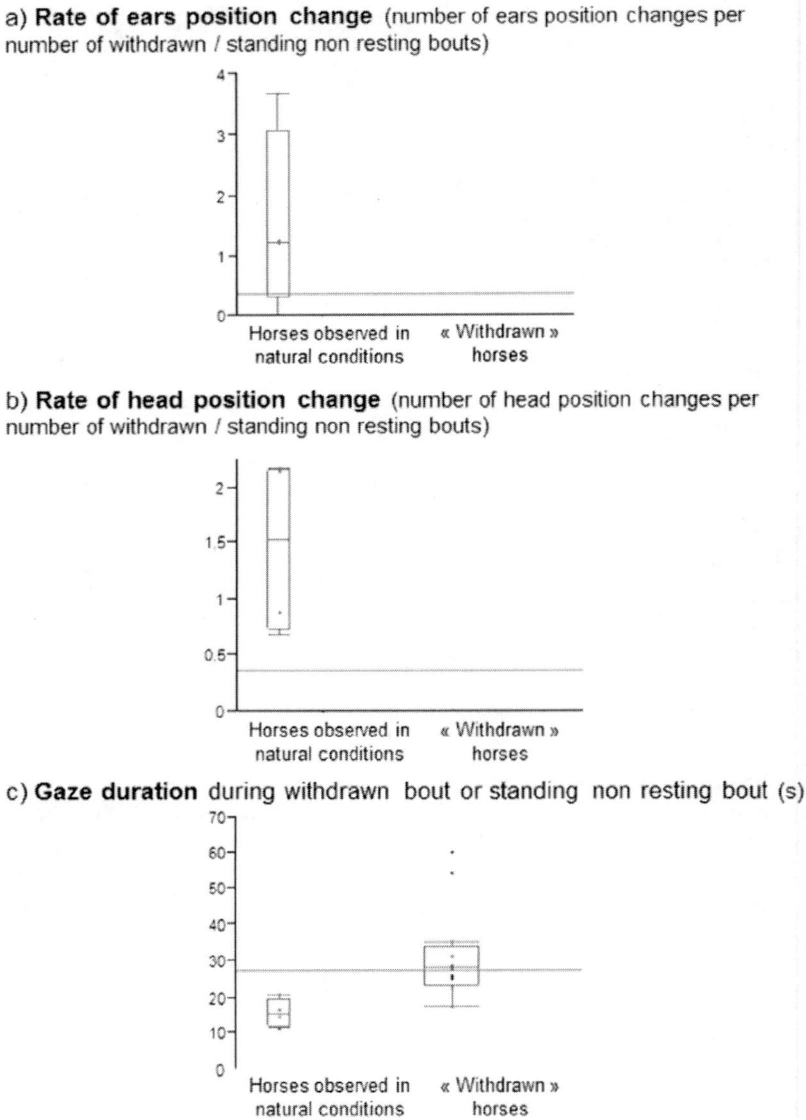

Figure 4. **Low responsiveness to environmental stimuli by "withdrawn" horses.**

3.4 Physiological Data: Cortisol Assays

Plasma concentrations of cortisol (pC) varied from 2.5 to 57.9 ng/ml after a day's work and from 3.0 to 35.7 ng/ml after a day's rest. Interestingly, "withdrawn" horses had lower pC concentrations after a day's work (Mann Whitney test, $n_W = 12$, $n_{non\text{-}W} = 43$, $Med_W = 6$, Q1=6, *Q*3=6.21, $Med_{non\text{-}W} = 10.65$, *Q*1=6, *Q*3=24.09, U =144.00, $p<0.05$) (**FIG. 5**). Moreover, the frequency of withdrawn posture was negatively correlated with pC concentrations after a day's work (Spearman correlation, N =55, $r_s = -0.30$, $p<0.05$).

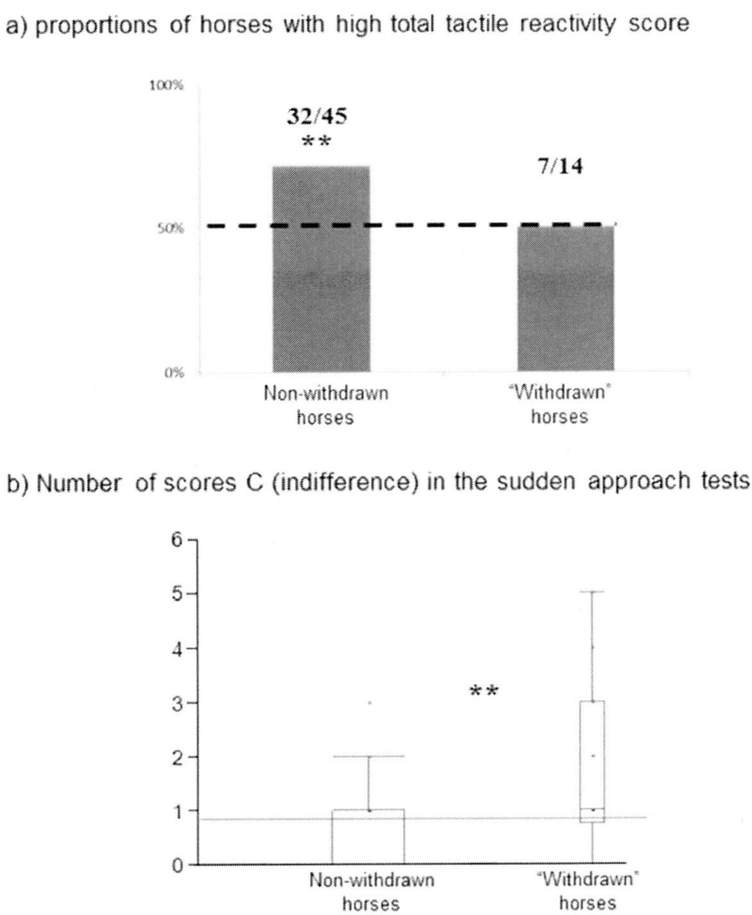

Figure 5. Mean plasma cortisol concentrations after a day's work for "withdrawn" horses and "non- withdrawn" horses.

Data are given as boxplot diagrams showing medians (lines in the boxes), 25 and 75% quartiles (boxes) and minimum and maximum ranges (whiskers). "Withdrawn" horses had lower cortisol concentrations than non- withdrawn horses (*i.e.* that never performed the withdrawn posture). Mann Whitney test, * $p<0.05$.

4. DISCUSSION

Our observations of horses in their usual domestic environment have led us to spot individuals displaying a particular behavioural and postural profile that presents strong similarities with a "depressive syndrome". These animals, displaying an atypical posture (stretched neck) and characterized by their unusual gaze, head and ears fixity, were also more indifferent than the others to environmental (tactile and visual) stimuli in their home environment. However, they reacted more emotionally in other more challenging situations. Finally, these "depressed" horses exhibit lower plasma cortisol levels. All these characteristics present strong similarities with some aspect of the depressive states of humans and other animal models *e.g.* [4], [6], [7]. Moreover, breed effect (*i.e.* suggested genetic input) and a higher prevalence in females add to the convergence.

Withdrawn horses, like other domestic animals living under unfavourable conditions, showed "apathy" (sows: [15]; horses: [18], [19], "withdrawal" sheep: [16], socially isolated chicks: [38] and lower reactivity to environmental stimulations (pigs: [39], [40]). Pigs submitted to unpredictable electric shocks first become agitated and then inactive [41].

The flatness and fixity of the "withdrawn" horses' posture when more weight appeared to be put on the fore body and forelegs may reveal chronic pain [42]but also an apparent overloading of the fore body, which evoke the slumped posture of depressive monkeys *e.g.* [3], [43]and some depressive patients [44]. The backward ears position suggests discomfort/pain or stress [21]–[23] and altered welfare (Fureix et al. in prep). Decreased eye contact is also observed in depressed patients [5], whereas "withdrawn" horses just gaze "nowhere" and do not attempt to establish eye contact with humans or other horses.

As for human depression, horse "depression" corresponds to a multifaceted syndrome: apathy and loss of interest, lower reactivity but higher anxiety. These horses surprisingly displayed higher emotional responses when facing a challenging situation (novel object in a familiar environment), suggesting, as in depressive humans, a higher level of anxiety *e.g.* [32]. Such emotional reactions have been shown to be influenced by genetic (breed, sire) and environmental (type of work, management) factors in horses *e.g.*

[34], with high interindividual differences in all horse populations tested *e.g.*[30], [33], [45]–[48].

Lower cortisol levels in "withdrawn" horses are in accordance with the results of some studies of depression in communities, which is a more usual environment for humans than hospital settings [8], [10]. Reduced cortisol levels are also observed in chronic fatigue syndrome and posttraumatic stress disorders *e.g.* [8], [49], [50] (see also [11] in animals). The lower cortisol levels observed here are also partly in agreement with those observed in other horses submitted to stress-inducing management conditions, presenting lowered cortisol responses in a corticotrophin releasing hormone challenge test [50]. Data for depressive patients are controversial (*e.g.* in [49]) and for some authors, cortisol levels are not a reliable biomarker [6]. However, the lower levels observed here may reflect a depression of the hypothalamo-pituitary-adrenocortical axis, that is, a profound disturbance of the physiological system *e.g.* [11], [51], [52].

Our results suggest that horses' "depressive" states may reflect genetic inputs – as one breed was over represented in our sample – and environmental factors certainly have an effect, as descriptions of "apathetic" horses all correspond to domestic situations when horses experience social, spatial and/or feeding restrictions and potentially stress-inducing work [14], [18], [19], [32]. The prevalence of females displaying this syndrome is another intriguing convergence. These findings suggest that horses as humans and other species may be particularly sensitive to environmental conditions *e.g.* [49] that could induce them to develop "depressive syndromes".

These results open a promising line of investigation of what impaired welfare states could look like in horses. Indeed, though considerable attention has been put on "abnormal" behaviours, *i.e.* stereotypic behaviours (see for instance [53] for a review), little interest has been put on this species on chronic states where horses "switch off", becoming unresponsive and apathetic. Moreover, it has been recently suggested that negative experiences linked to training may add to the effects of management style (*e.g.* social, spatial restrictions) and lead to behavioural despair in horses [18]. Our results suggest that estimations of gaze or body fixity and of body posture might indicate depressive-like state. Adding to the attempts of finding suitable animal welfare indicators, this study would make formally testing the previous hypothesis more feasible.

Beyond direct implications of this work for welfare assessment in horses, horses might be a potential useful candidate to become an animal model for depression, as face validity appears to be high (behavioural similarities) at this stage and construct validity (shared aetiology) may prove fulfilled in future studies when genetic determinism and shared environmental stressors (work constraints, social restriction...) with humans are taken into consideration. Further studies should involve work on anhedonia (loss of pleasure, a core symptom of human depression), cognitive biases (*i.e.* tendency to make negative judgements about events and to interpret ambiguous stimuli unfavourably, typical of depressed humans *e.g.* [54], heritability, epidemiology and drugs effects. If confirmed by these studies, horses may well be a particularly useful ethological animal model of human depression. At a time when current animal models of depression are strongly questioned [2], [5], [6]and the need for novel models expressed [2], [7], this study suggests that novel models and biomarkers may well emerge from ethological approaches in the home environment. Estimations of gaze or body fixity and of body posture may reveal to be interesting new biomarkers.

ACKNOWLEDGMENTS

The authors are grateful to the managers of the riding schools for allowing us to work with their horses and all the staff for their help and cooperation. The authors are grateful to Dr A. Cloarec for correcting the English of this manuscript.

REFERENCES

1. Nestler E, Barrot M, Di Leone R, Eisch A, Gold SJ, et al. (2002) Neurobiology of Depression. Neuron 34: 13–25.E. NestlerM. BarrotR. Di LeoneA. EischSJ Gold2002Neurobiology of Depression.Neuron341325
2. Nestler E, Gould E, Manji H (2002) Preclinical models: status of basic research in depression. Biol Psychiatry 52: 503–528.E. NestlerE. GouldH. Manji2002Preclinical models: status of basic research in depression.Biol Psychiatry52503528
3. Harlow H, Suomi S (1974) Induced Depression in Monkeys. Behav Biol 12: 273–296.H. HarlowS. Suomi1974Induced Depression in Monkeys.Behav Biol12273296
4. Deussing JM (2006) Animal models of depression. Drug discovery today: disease models 3: 375–383.JM Deussing2006Animal models of depression.Drug discovery today: disease models3375383
5. O'Neill M, Moore N (2003) Animal models of depression: Are there any? Hum Psychopharmacol Clin Exp 18: 239–254.M. O'NeillN. Moore2003Animal models of depression: Are there any?Hum Psychopharmacol Clin Exp18239254
6. Matthews K, Christmas D, Swan J, Sorrell E (2005) Animal Models of Depression and Antidepressant Activity. Neurosci Biobehav Rev 29: 503–513.K. MatthewsD. ChristmasJ. SwanE. Sorrell2005Animal Models of Depression and Antidepressant Activity.Neurosci Biobehav Rev29503513
7. McArthur R, Borsini F (2006) Animal models of depression in drug discovery: A historical perspective. Pharmacol Biochem Behav 84: 436–452.R. McArthurF. Borsini2006Animal models of depression in drug discovery: A historical perspective.Pharmacol Biochem Behav84436452
8. Strickland PL, Deakin JFW, Percival C, Dixon J, Gater RA, et al. (2002) Bio-social origins of depression in the community - Interactions between social adversity, cortisol and

serotonin neurotransmission. Br J Psychiatry 180: 168–173.PL StricklandJFW DeakinC. PercivalJ. DixonRA Gater2002Bio-social origins of depression in the community - Interactions between social adversity, cortisol and serotonin neurotransmission.Br J Psychiatry180168173

9. Maes M (1994) Cytokines in major depression. Biological Psychiatry 36: 498–499.M. Maes1994Cytokines in major depression.Biological Psychiatry36498499
10. Moore B (2002) Cortisol, stress and depression. Br J Psychiatry 181: 348–348.B. Moore2002Cortisol, stress and depression.Br J Psychiatry181348348
11. Romero LM, Dickens MJ, Cyr NE (2009) The reactive scope model - A new model integrating homeostasis, allostasis, and stress. Horm Behav 55: 375–389.LM RomeroMJ DickensNE Cyr2009The reactive scope model - A new model integrating homeostasis, allostasis, and stress.Horm Behav55375389
12. Flint J, Corley R (1996) Do animal models have a place in the genetic analysis of quantitative human behavioural traits? J Mod Med 74: 515–521.J. FlintR. Corley1996Do animal models have a place in the genetic analysis of quantitative human behavioural traits?J Mod Med74515521
13. O'Brien A, Terry DJ, Jimmieson NL (2008) Negative affectivity and responses to work stressors: an experimental study. Anxiety stress coping 21: 55–83.A. O'BrienDJ TerryNL Jimmieson2008Negative affectivity and responses to work stressors: an experimental study.Anxiety stress coping215583
14. Hausberger M, Gautier E, Biquand V, Lunel C, Jego P (2009) Could work be a source of behavioural disorders? A study in horses. PloS One 4: e7625.M. HausbergerE. GautierV. BiquandC. LunelP. Jego2009Could work be a source of behavioural disorders? A study in horses.PloS One4e7625
15. Broom DM (1987) Applications of neurobiological studies to farm animal welfare. In: Wiepkema PR, van Adrichem PWM, editors. DM Broom1987Applications of neurobiological studies to farm animal welfare.PR WiepkemaPWM van Adrichemeditors. Biology of stress in farm animals: an integrative approach.The Hague: Martinus Nijhoff Publishers. The Hague: Martinus Nijhoff Publishers.
16. Fordham DP, Algahtani S, Durotoye LA, Rodway RG (1991) Changes in plasma-cortisol and beta-endorphin concentrations and behavior in sheep subjected to a change of environment. Anim Production 52: 287–296.DP FordhamS. AlgahtaniLA DurotoyeRG Rodway1991Changes in plasma-cortisol and beta-endorphin concentrations and behavior in sheep subjected to a change of environment.Anim Production52287296
17. Beerda B, Schilder MBH, Van Hooff J, De Vries HW, Mol JA (1999) Chronic stress in dogs subjected to social and spatial restriction. I. Behavioral responses. Physiol Behav 66: 233–242.B. BeerdaMBH SchilderJ. Van HooffHW De VriesJA Mol1999Chronic stress in dogs subjected to social and spatial restriction. I. Behavioral responses.Physiol Behav 66233242
18. Hall C, Goodwin D, Heleski C, Randle H, Waran N (2008) Is there evidence of learned helplessness in horses? J Appl Anim Welfare Sci 11: 249–266.C. HallD. GoodwinC. HeleskiH. RandleN. Waran2008Is there evidence of learned helplessness in horses?J Appl Anim Welfare Sci11249266
19. Burn CC, Dennison TL, Whay HR (2010) Relationships between behaviour and health in working horses, donkeys, and mules in developing countries. Appl Anim Behav Sci 126: 109–118.CC BurnTL DennisonHR Whay2010Relationships between behaviour and health in working horses, donkeys, and mules in developing countries.Appl Anim Behav Sci126109118
20. Fureix C, Hausberger M, Seneque E, Morisset S, Baylac M, et al. (2011) Geometric morphometrics for ethologists: improving the comparative study of behavioural postures. Natwiss 98: 583–592.C. FureixM. HausbergerE. SenequeS. MorissetM. Baylac2011Geometric morphometrics for ethologists: improving the comparative study of behavioural postures.Natwiss98583592
21. Waring G (2003) Horse Behavior, second edition. Norwich, New York: Noyes Publications/William Andrew Publishing. 442 p. G. Waring2003Horse Behavior, second edition. Norwich, New York: Noyes Publications/William Andrew Publishing.442 p

22. Von Borstel UU, Duncan IJH, Shoveller AK, Merkies K, Keeling LJ, et al. (2009) Impact of riding in a coercively obtained Rollkur posture on welfare and fear of performance horses. Appl Anim Behav Sci 116: 228–236.UU Von BorstelIJH DuncanAK ShovellerK. MerkiesLJ Keeling2009Impact of riding in a coercively obtained Rollkur posture on welfare and fear of performance horses.Appl Anim Behav Sci116228236
23. Pritchett LC, Ulibarri C, Roberts MC, Schneider RK, Sellon DC (2003) Identification of potential physiological and behavioral indicators of postoperative pain in horses after exploratory celiotomy for colic. Appl Anim Behav Sci 80: 31–43.LC PritchettC. UlibarriMC RobertsRK SchneiderDC Sellon2003Identification of potential physiological and behavioral indicators of postoperative pain in horses after exploratory celiotomy for colic.Appl Anim Behav Sci803143
24. Reefmann N, Kaszas FB, Wechsler B, Gygax L (2009) Ear and tail postures as indicators of emotional valence in sheep. Appl Anim Behav Sci 118: 199–207.N. ReefmannFB KaszasB. WechslerL. Gygax2009Ear and tail postures as indicators of emotional valence in sheep.Appl Anim Behav Sci118199207
25. Saslow CA (2002) Understanding the perceptual world of horses. Appl Anim Behav Sci 78: 209–224.CA Saslow2002Understanding the perceptual world of horses.Appl Anim Behav Sci78209224
26. Redua MA, Valadao CAA, Duque JC, Balestrero LT (2002) The pre-emptive effect of epidural ketamine on wound sensitivity in horses tested by using von Frey filaments. Vet Anaesth Analg 29: 200–206.MA ReduaCAA ValadaoJC DuqueLT Balestrero2002The pre-emptive effect of epidural ketamine on wound sensitivity in horses tested by using von Frey filaments.Vet Anaesth Analg29200206
27. Lansade L, Pichard G, Leconte M (2008) Sensory sensitivities: Components of a horse's temperament dimension. Appl Anim Behav Sci 114: 534–553.L. LansadeG. PichardM. Leconte2008Sensory sensitivities: Components of a horse's temperament dimension. Appl Anim Behav Sci114534553
28. Hausberger M, Muller C (2002) A brief note on some possible factors involved in the reactions of horses to humans. Appl Anim Behav Sci 76: 339–344.M. HausbergerC. Muller2002A brief note on some possible factors involved in the reactions of horses to humans.Appl Anim Behav Sci76339344
29. Fureix C, Jego P, Sankey C, Hausberger M (2009) How horses (Equus caballus) see the world: humans as significant objects. Anim Cogn 12: 643–654.C. FureixP. JegoC. SankeyM. Hausberger2009How horses (Equus caballus) see the world: humans as significant objects.Anim Cogn12643654
30. Wolff A, Hausberger M, LeScolan N (1997) Experimental tests to assess emotionality in horses. Behav Processes 40: 209–221.A. WolffM. HausbergerN. LeScolan1997 Experimental tests to assess emotionality in horses.Behav Processes402 09221
31. LeScolan N, Hausberger M, Wolff A (1997) Stability over situations in temperamental traits of horses as revealed by experimental and scoring approaches. Behav Processes 41: 257–266.N. LeScolanM. HausbergerA. Wolff1997Stability over situations in temperamental traits of horses as revealed by experimental and scoring approaches. Behav Processes41257266
32. Hausberger M, Muller C, Lunel C (2011) Does Work Affect Personality? A Study in Horses. Plos One 6: e14659.M. HausbergerC. MullerC. Lunel2011Does Work Affect Personality? A Study in Horses.Plos One6e14659
33. Visser EK, van Reenen CG, van der Werf JTN, Schilder MBH, Knaap JH, et al. (2002) Heart rate and heart rate variability during a novel object test and a handling test in young horses. Physiol Behav 76: 289–296.EK VisserCG van ReenenJTN van der WerfMBH SchilderJH Knaap2002Heart rate and heart rate variability during a novel object test and a handling test in young horses.Physiol Behav76289296
34. Hausberger M, Bruderer U, Le Scolan N, Pierre JS (2004) Interplay between environmental and genetic factors in temperament/personality traits in horses (Equus caballus). J Comp Psychol 118: 434–446.M. HausbergerU. BrudererN. Le ScolanJS Pierre2004Interplay between environmental and genetic factors in temperament/personality traits in horses (Equus caballus).J Comp Psychol118434446

35. Altmann J (1974) Observational study of behaviour: sampling methods. Behav 49: 227–267.J. Altmann1974Observational study of behaviour: sampling methods. Behav 49227267
36. Palme R, Rettenbacher S, Touma C, El-Bahr SM, Mostl E (2005) Stress hormones in mammals and birds - Comparative aspects regarding metabolism, excretion, and noninvasive measurement in fecal samples. Trends Comp Endocrinol Neurobiol 1040: 162–171.R. PalmeS. RettenbacherC. ToumaSM El-BahrE. Mostl2005Stress hormones in mammals and birds - Comparative aspects regarding metabolism, excretion, and noninvasive measurement in fecal samples.Trends Comp Endocrinol Neurobiol 1040162171
37. Siegel S, Castellan J (1988) Nonparametric statistics for the behavioral sciences, second edition. New York: McGraw-Hill. S. SiegelJ. Castellan1988Nonparametric statistics for the behavioral sciences, second edition.New York: McGraw-Hill
38. Sufka KJ, Feltenstein MW, Warnick JE, Acevedo EO, Webb HE, et al. (2006) Modeling the anxiety–depression continuum hypothesis in domestic fowl chicks. Behav Pharmacol 17: 681–689.KJ SufkaMW FeltensteinJE WarnickEO AcevedoHE Webb2006Modeling the anxiety–depression continuum hypothesis in domestic fowl chicks.Behav Pharmacol 17681689
39. Hutson GD, Dickenson LG, Wilkinson JL, Luxford BG (1993) The response of sows to novel visual, olfactory, auditory and tactile stimuli. Appl Anim Behav Sci 35: 255–266.GD HutsonLG DickensonJL WilkinsonBG Luxford1993The response of sows to novel visual, olfactory, auditory and tactile stimuli.Appl Anim Behav Sci35255266
40. Hutson GD, Ambrose TJ, Barnett JL, Tilbrook AJ (2000) Development of a behavioural test of sensory responsiveness in the growing pig. Appl Anim Behav Sci 66: 187–202.GD HutsonTJ AmbroseJL BarnettAJ Tilbrook2000Development of a behavioural test of sensory responsiveness in the growing pig.Appl Anim Behav Sci66187202
41. Jensen KH, Pedersen LJ, Nielsen EK, Heller KE, Ladewig J, et al. (1996) Intermittent stress in pigs: Effects on behavior, pituitary-adrenocortical axis, growth, and gastric ulceration. Physiol Behav 59: 741–748.KH JensenLJ PedersenEK NielsenKE HellerJ. Ladewig1996 Intermittent stress in pigs: Effects on behavior, pituitary-adrenocortical axis, growth, and gastric ulceration.Physiol Behav59741748
42. Cauvin E (1997) Assessment of back pain in horses. Equine pract 19: 522–533.E. Cauvin1997Assessment of back pain in horses.Equine pract19522533
43. Harlow H, Plubell P, Baysinger C (1973) Induction of Psychological Death in Rhesus Monkeys. J Autism Child Schizophr 3: 299–307.H. HarlowP. PlubellC. Baysinger 1973Induction of Psychological Death in Rhesus Monkeys.J Autism Child Schizophr3299307
44. Wallbott HG (1998) Bodily expression of emotion. Eur J Soc Psychol 28: 879–896.HG Wallbott1998Bodily expression of emotion.Eur J Soc Psychol28879896
45. Visser EK, van Reenen CG, Hopster H, Schilder MBH, Knaap JH, et al. (2001) Quantifying aspects of young horses' temperament: consistency of behavioural variables. Appl Anim Behav Sci 74: 241–258.EK VisserCG van ReenenH. HopsterMBH SchilderJH Knaap 2001Quantifying aspects of young horses' temperament: consistency of behavioural variables.Appl Anim Behav Sci74241258
46. Christensen J, Keeling L, Nielsen B (2005) Responses of horses to novel visual, olfactory and auditory stimuli. Appl Anim Behav Sci 93: 53–65.J. ChristensenL. KeelingB. Nielsen2005Responses of horses to novel visual, olfactory and auditory stimuli.Appl Anim Behav Sci935365
47. Lansade L, Bouissou M-F, Erhard HW (2008) Fearfulness in horses: A temperament trait stable across time and situations. Appl Anim Behav Sci 115: 182–200.L. LansadeM-F BouissouHW Erhard2008Fearfulness in horses: A temperament trait stable across time and situations.Appl Anim Behav Sci115182200
48. Visser EK, Ellis AD, Van Reenen CG (2008) The effect of two different housing conditions on the welfare of young horses stabled for the first time. Appl Anim Behav Sci 114: 521–533.EK VisserAD EllisCG Van Reenen2008The effect of two different housing conditions on the welfare of young horses stabled for the first time.Appl Anim Behav Sci114521533

49. Miller GE, Chen E, Zhou ES (2007) If it goes up, must it come down? Chronic stress and the hypothalamic-pituitary-adrenocortical axis in humans. Psychol Bull 133: 25–45.GE MillerE. ChenES Zhou2007If it goes up, must it come down? Chronic stress and the hypothalamic-pituitary-adrenocortical axis in humans.Psychol Bull1332545
50. Michaud K, Matheson K, Kelly O, Anisman H (2008) Impact of stressors in a natural context on release of cortisol in healthy adult humans: A meta-analysis. Stress 11: 177–197.K. MichaudK. MathesonO. KellyH. Anisman2008Impact of stressors in a natural context on release of cortisol in healthy adult humans: A meta-analysis.Stress11177197
51. McEwen BS (2003) Mood disorders and allostatic load. Biol Psychiatry 54: 200–207.BS McEwen2003Mood disorders and allostatic load.Biol Psychiatry54200207
52. Mormede P, Andanson S, Auperin B, Beerda B, Guemene D, et al. (2007) Exploration of the hypothalamic-pituitary-adrenal function as a tool to evaluate animal welfare. Physiol Behav 92: 317–339.P. MormedeS. AndansonB. AuperinB. BeerdaD. Guemene2007 Exploration of the hypothalamic-pituitary-adrenal function as a tool to evaluate animal welfare.Physiol Behav92317339
53. Mills DS (2005) Repetitive movement problems in the horse. In: McDonnell DSMaSM, editor. pp. 212–227.DS Mills2005Repetitive movement problems in the horse.DSMaSM McDonnell212227editor.The Domestic Horse, The Origins, Development and Management of its Behaviour.Cambridge: Cambridge University Press.pp. pp.
54. Harding E, Paul E, Mendl M (2004) Cognitive bias and affective state. Nature 427: 312–312.E. HardingE. PaulM. Mendl2004Cognitive bias and affective state.Nature427312312

CHAPTER 12

Detection of infectious bronchitis virus 793B, avian metapneumovirus, *Mycoplasma gallisepticum* and *Mycoplasma synoviae* in poultry in Ethiopia

S. Hutton[1], J. Bettridge[1], R. Christley[1,2], T. Habte[3], K. Ganapathy[1]

[1]*Institute of Infection and Global Health, School of Veterinary Science, University of Liverpool, Leahurst Campus, Neston, Cheshire CH64 7TE, UK*

[2]*NIHR Health Protection Research Unit in Emerging and Zoonotic Infections, Liverpool, UK*

[3]*Ethiopian Institute of Agricultural Research, Debre Zeit, Ethiopia*

ABSTRACT

A survey was conducted into respiratory infectious diseases of poultry on a chicken breeder farm run by the Ethiopian Institute of Agricultural Research (EIAR), located in Debre Zeit, Ethiopia. Oropharyngeal swabs were collected from 117 randomly selected birds, and blood was taken from a subset of 73 of these birds. A combination of serological and molecular methods was used for detection of pathogens. For the first time in Ethiopia, we report the detection of variant infectious bronchitis virus

(793B genotype), avian metapneumovirus subtype B and *Mycoplasma synoviae* in poultry. *Mycoplasma gallisepticum* was also found to be present; however, infectious laryngotracheitis virus was not detected by PCR. Newcastle disease virus (NDV) was not detected by PCR, but variable levels of anti-NDV HI antibody titres shows possible exposure to virulent strains or poor vaccine take, or both. For the burgeoning-intensive industry in Ethiopia, this study highlights several circulating infectious respiratory pathogens that can impact on poultry welfare and productivity.

Keywords: IBV, aMPV, Mg, Ms, Ethiopia, Serology, PCR

1. INTRODUCTION

Intensification of agriculture in Africa is accelerating, thanks to novel technologies, financial initiatives, changing social infrastructure and private sector engagement (Pretty et al. 2011), and policies promoting increases in consumption and productivity for a range of livestock species (Sumberg and Thompson 2013). Ethiopian entrepreneurs are setting up large, intensively managed flocks of exotic breeds, particularly in areas close to Addis Ababa (Mammo 2012). Government-owned poultry multiplication centres throughout the country (Pagani and Wossene 2008), non-governmental organisations and private individuals also distribute intensively reared chickens to smallholders (Pagani and Wossene 2008; Anon 2014). As a result, more urban and suburban households now keep flocks of 50 to 1000 birds under semi-intensive management (FAO 2008). The close links between intensive and smallholder farms facilitate disease spread, exacerbated by low biosecurity, and poor access to veterinary inputs and expertise among smaller producers (Sambo et al. 2014).

Respiratory infections have major impacts on poultry welfare and productivity, yet are largely unstudied in most African countries (Owoade et al. 2006), where attention has focussed on the notifiable avian influenza (AI) and Newcastle disease (ND) viruses. Other common respiratory pathogens worldwide include infectious bronchitis virus (IBV), infectious laryngotracheitis virus (ILTV) and avian metapneumovirus (aMPV) (Cavanagh and Gelb Jr. 2008; Gough and Jones 2008; Guy and Garcia 2008). These cause direct losses also interact with bacteria such as *Mycoplasma gallisepticum* (Mg) and *Mycoplasma synoviae* (Ms), which exacerbate respiratory diseases. Serology shows ND is widespread throughout Ethiopia (Tadesse et al. 2005; Zeleke et al. 2005; Regasa et al. 2007; Mazengia et al. 2010; Chaka et al. 2012), and virulent and lentogenic strains have been isolated (Chaka et al. 2013; Fentie et al. 2014). Vaccines, produced locally, are used in commercial farms, multiplication centres (Dessie and Jobre 2004) and more recently in field trials for village birds (Nega et al. 2012). Low pathogenicity AI was last reported in 2006 and high pathogenicity AI has never been identified in Ethiopia. ILT and IBV (serotype not stated) were last reported in 2000 and 2001, respectively, but no information is available on aMPV (O.I.E. 2015), which is under-reported in many parts of the world due to a lack of diagnostic facilities (Gough and Jones 2008). To our knowledge, the scientific literature contains no reports of these last three in Ethiopia; therefore, circulating genotypes, important for identification of appropriate vaccines, are entirely unknown. Mg has been identified in sick birds from commercial farms and in village chickens from live markets both serologically (Alamargot et al. 1985) and through culture and PCR (Bekele 2015). Ms appears to be previously unreported in Ethiopia.

The poultry farm at the Ethiopian Institute of Agricultural Research (EIAR), Debre Zeit, represents an important centre for research, farmer training in poultry production and distribution of birds among smallholders. As such, it has vital links with farmers, especially intensive and semi-intensive producers. In 2012, respiratory disease was a major cause of morbidity and mortality; therefore, this study screened the EIAR flocks for IBV, aMPV, Newcastle disease virus (NDV), ILTV, Mg and Ms.

2. MATERIALS AND METHODS

2.1 Flock and samples

Sampling was conducted over a 2-week period in May 2012. At the time of study, the farm consisted of 11 sheds, nine of which were populated. Three sheds which housed grandparent birds were situated separately to the main body of the poultry production premises. The main premises consisted of a hatching house; incubator and brooder sheds (both disused); one large new breeder shed; two multi-age multi-breed sheds containing parent birds and multiplier flocks and two sheds containing indigenous chicken ecotypes. These last two were not sampled due to local concern over biosecurity, as any losses incurred in this flock would interrupt an experimental breeding programme. All birds on the farm were vaccinated against NDV, but unfortunately records of vaccination schedules for the birds sampled were not available.

Groups were defined as birds of the same age and breed. These were mostly kept in the same shed, with the exception of some groups, where birds were split between the parent and multiplier sheds.

Table 1 Breeder flocks examined for serological evidence of exposure to NDV, IBV or aMPV by ELISA and the presence of IBV, aMPV, Mg and Ms by PCR

Table 1 Breeder flocks examined for serological evidence of exposure to NDV, IBV or aMPV by ELISA and the presence of IBV, aMPV, Mg and Ms by PCR

House	Group	Breed	Age (months)	Number of birds examined for NDV, IBV and aMPV ELISA antibodies	NDV vaccination index	PCR detection in pooled OP swabs collected from *n* of birds				
						n	IBV	aMPV	Mg	Ms
Grandparent 1	1[a]	Dominant CZ Rhode Island Red	22.5	10	133.2	10	–	–	–	+
Grandparent 2	2	Dominant CZ	22.5	10	278.1	15	+	–	–	–
Grandparent	3[b]	Hubbard JV	9	10	43.5	10	–	–	–	+
Breeder	4[c]	Hubbard JV	9	20	40.1	20	–	–	–	+
Parent (multi-age in single house)	5	Hybrid	7	3	12.3	5	–	–	+	+
	6	Kokok	23	3	15.1	7	–	–	–	+
	7	Hybrid	9	3	128.1	5	–	–	–	+
	8	Fayoumi	7	3	34.9	6	–	–	+	+
	9	Kokok	31	3	55.0	5	–	–	–	+
	10	Fayoumi	31	5	5.8	8	–	–	–	+
	11	Dominant CZ Rhode Island Red	28	3	64.5	6	–	–	–	+
Chick	12	Hybrid	0.5	Not done	Not done	20	–	–	–	+

[a] One dead bird was necropsied; no IBV, aMPV, Mg or Ms was detected by PCR

[b] One dead bird was necropsied; no IBV, aMPV or Mg was detected, but Ms was detected in the trachea and turbinate by PCR

[c] One dead bird was necropsied; no Mg or Ms was detected, but IBV 793B and aMPV subtype B were detected in the turbinate by PCR

There were seven sheds sampled in total, and the number of groups varied per shed. All groups were of mixed sex and, depending on the group size, were housed in up to eight pens within the same shed (Table 1.) Where groups were split between pens, at least one bird was selected from each pen. Birds were selected from pens by convenience sampling; avoiding birds by the pen doors and using the most competent keepers for catching to reduce selection of slow, weak birds and thus selection bias. Selected birds were first swabbed oropharyngeally and then blood sampled (1 ml via brachial vein). Blood samples were collected into a plain syringe and allowed to clot at room temperature before the serum was removed and stored at −20 °C. The majority of grandparent and breeder birds were both swabbed and blood sampled, whereas only around half of the parent birds were bled. These were kept in more, but smaller groups, thus three to five sera per pen were collected and two to four additional birds per pen were swabbed. Chicks were not blood sampled, due to possible adverse effects of the bleeding procedure and interference of maternally derived antibodies with serology results.

2.2 Serum antibody detection

ELISA kits (BioChek, Reeuwijk, The Netherlands) were used to detect antibodies for NDV, IBV and aMPV in 73 sera. Tests were carried out in a laboratory within the EIAR research facility in Debre Zeit. Tests were performed according to the manufacturer's instructions, but with the following modifications: all incubation times for NDV and IBV plates were reduced by 5 min, and a 10-min reduction was applied to aMPV plate incubation periods. This was to compensate for the higher-than-recommended room temperature of the laboratory (30 °C instead of between 17 and 24 °C). Samples were tested in triplicate, and individual antibody titres were calculated according to the manufacturer's instructions. For the NDV antibody results, as birds were vaccinated against this pathogen, the vaccination index [$VI = (\text{mean titre})^2/(\text{St Dev} \times 100)$] was calculated for each group according to van Leerdam and Bosman (2006), which relates the mean group titre to the coefficient of variation, and so provides a simple evaluation of the effectiveness of a vaccination.

2.3 Antigen detection

Oropharyngeal swabs were pooled; two to three swabs from birds within the same group were placed in a sterile bijou containing 2 ml of sterile water, vortexed for 3 s and 80 µl of supernatant was transferred onto an FTA card. In addition, three birds isolated in hospital pens within the grandparent and breeder sheds were sacrificed by cervical dislocation and post mortem examination was carried out. Tissues of the trachea ($n = 3$), lung ($n = 3$), oviduct ($n = 3$), kidney ($n = 3$) and turbinate ($n = 2$) were collected, ground until smooth with a pestle, sterile sand and 3 ml of sterile distilled water. After two freeze-thaw cycles and centrifugation at 2000*g* for 5 min, 80 µl of each supernatant was applied to an FTA card. FTA cards, stored at room temperature, were transported under license to the Poultry Respiratory Diseases Group, University of Liverpool, UK, for further processing.

Forty-one pooled oropharyngeal and 14 processed tissue samples on FTA cards were processed, where each circle of the FTA card containing a sample was removed using sterile scissors and forceps and placed in a bijou containing 800–1000 µl of TE buffer (10 mM Tris–HCl, 0.1 mM EDTA, pH 8.0), vortexed and incubated at room temperature for 10 min (Abdelwhab et al. 2011). The supernatant was then used to extract viral RNA for detection of NDV, IBV and aMPV. For IBV, NDV and aMPV detection, RNA extraction was performed on TE buffer using the QIAamp Viral RNA Mini Kit (Qiagen Ltd., Hilden, Germany), and for detection of ILTV and both Mg and Ms, DNA was extracted from TE buffer using the QIAamp DNA Mini Kit (Qiagen Ltd., Germany). All extractions were according to the manufacturer's instructions and stored at −20 °C until required.

The extracted RNA or DNA were tested for PCR using s for NDV (Aldous et al. 2003), ILTV (Diallo et al. 2010), IBV and aMPV (Cavanagh et al. 1999; Worthington et al. 2008; Al-Shekaili et al. 2015a; Ganapathy et al. 2015). A commercial PCR kit (Adiavet-myco-AV PCR, Adiagen, St Brieuc, France) was used to detect the presence of Mg or Ms DNAs. Positive IBV PCR reactions were sequenced, analysed and compared against reference strains as described before (Al-Shekaili et al. 2015a; Ganapathy et al. 2015).

3. RESULTS

aMPV Serum samples were considered to be positive for aMPV if titres ≥1656, as per the manufacturer's guidelines. Sixteen (21.9%) samples tested positive (Fig. 1a). All the OP swabs were negative for aMPV by RT-PCR. A turbinate from a necropsied bird from group 3 was positive for aMPV subtype B.

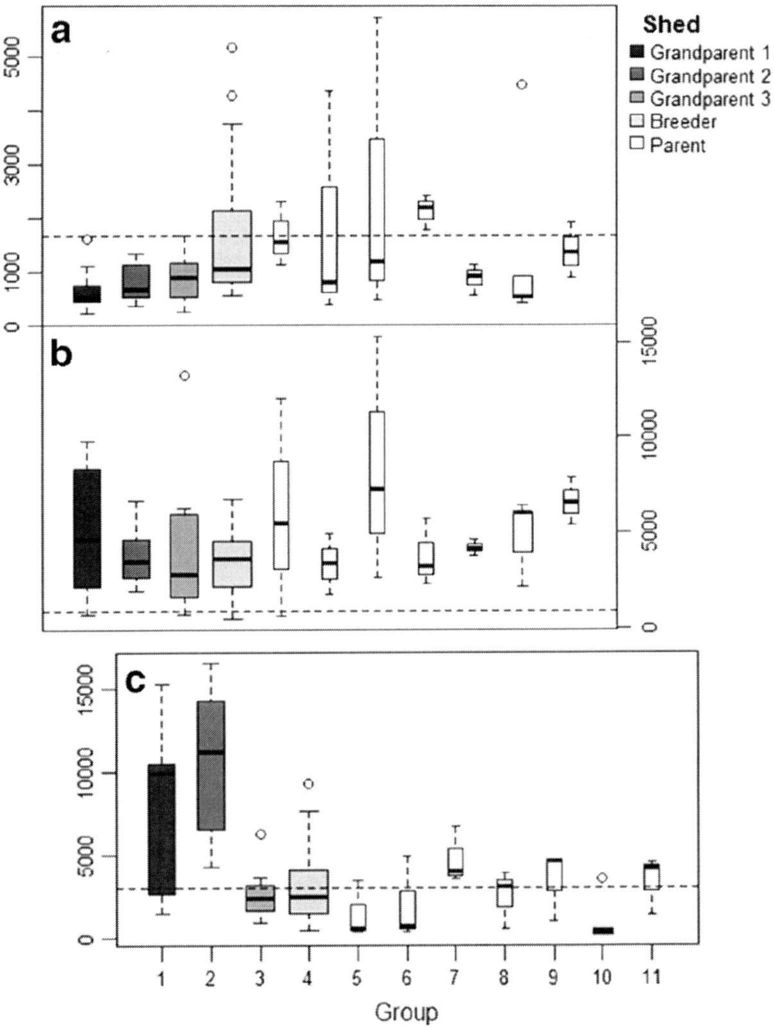

Figure. 1 Antibody titres for 73 birds against **A** avian metapneumovirus, **B** infectious bronchitis virus and **C** Newcastle disease virus. *Widths of the bars* represent the number of samples tested for each group. *Dotted horizontal lines* show positive cut-off values

IBV Using the manufacturer's cut-off of 834, 69 (94.5%) birds were serologically positive for IBV (Fig. 1b). By RT-PCR, only one (group 2) of the pooled OP swab was positive for IBV. Positive PCR reaction was also detected in the same turbinate sample from group 3, which was also positive for aMPV (above). Sequencing revealed these IBVs belong to 793B with 92–95% relatedness to the French isolate FR-94047-94 (GenBank accession number: AJ618987).

NDV ELISA results demonstrated considerable variation in the antibody titres, both within and between the groups (Fig. 1c). Three of the groups within the parent shed (groups 5, 6 and 10) had a VI below the lower limit of 20 suggested by the manufacturer for birds vaccinated with live vaccine (Table 1). One of the grandparent groups (group 2) was outside the upper limit of the normal range (250) with a VI of 278. However, this was below the cut-off value of 300, values above which might suggest field challenge. No NDV was identified in the OP or tissue samples by RT-PCR.

ILTV Serological testing for ILTV was not performed. No ILTV detected in any of the samples when examined by PCR.

Mycoplasmas No Mg or Ms serology was carried out. PCR examination of oropharyngeal swabs identified Mg in the two groups of birds in the parent shed (groups 5 and 8). Ms was identified in 11 of the 12 groups; only one group of grandparent stock (group 2) was negative. Ms was also identified in the trachea and turbinate samples of one of the sacrificed hospitalised birds from one of the grandparent sheds.

4. DISCUSSION

To our knowledge, this is the first publication to report IBV and aMPV in Ethiopia, and the first time that aMPV and Ms have been identified in the country. Sequencing data of IBV isolates identified 793B genotype with a 92–95% similarity to a French isolate FR-94047-94, which is closely related to virulent 4/91 (Cavanagh et al. 2005). IBV vaccine is not used on the farm; therefore, this is presumed to be a field infection. The strain 793B is one of the most common strains and has previously been isolated from West African countries (Ducatez et al. 2009) and Tunisia (Bourogâa et al. 2009). The aMPV subtype B is a very common strain in other parts of Africa and was also unlikely to have been vaccinated in origin, as aMPV vaccines are not in use on this farm, and to the best of our knowledge, not currently available in Ethiopia. Knowledge of the strains present can help inform vaccine selection in the event that vaccination programmes for respiratory pathogens are implemented in the future.

Both Mg and Ms were identified on this farm, with Ms being found in almost all sampled groups. Both these respiratory pathogens transmit readily in birds within the same airspace, with up to 100% of birds in a flock becoming infected (Kleven and Ferguson-Noel 2008). Human traffic is also an important contributor to spread between sheds (Mohammed et al. 1987). Other risk factors for *Mycoplasmas* include overcrowding, poor ventilation and hygiene, introduction of new birds and multiple age flocks. High standards of hygiene and biosecurity are essential for controlling the spread of these pathogens. In addition, as vertical transmission can also occur (Michiels et al. 2016), maintaining breeding stocks free of infection is desirable. However, in situations where multiple ages and types of poultry are in close proximity to each other and to backyard flocks, this may be difficult or impossible, and vaccinations or antimicrobials may be required to reduce transmission (Kleven and Ferguson-Noel 2008; Ley 2008). The EIAR farm is primarily a research station and trials different breeds imported from different parts of Ethiopia and abroad. Therefore, it has to continually receive and maintain several small groups of poultry of different ages, and even with quarantine precautions, it is unlikely that eradication will be feasible, as an all-in all-out system will never be practical with the current housing facilities.

The extreme variability and low VI of at least three groups of birds sampled would suggest that improvements could be made in the effectiveness of the NDV vaccine administration or problems with the vaccines themselves. Although part of this variation will be due to the mix of ages and types of birds kept, route and frequency of administration, it may be desirable for the farm to investigate this finding further. Whilst the presence of NDV was not detected (by PCR) on the farm, our findings would suggest that a proportion of birds may be vulnerable to virulent NDV challenges. The range and magnitude of the serological results also provided evidence to suggest repeat exposure of the birds to IBV and aMPV, supported by antigenic identification of both pathogens, suggesting circulation within the flocks. Again, increased biosecurity measures may be of assistance in reducing transmission, but may be insufficient alone, due to the multiple ages of flocks making complete disinfection of premises impossible. Other management practices, such as ensuring adequate ventilation and reducing overcrowding, and the use of vaccination may be necessary to reduce production losses (Cavanagh and Gelb Jr. 2008).

Further studies are needed to establish the presence of these pathogens in other flocks in the local area and further across the region. A recent review of IBV in Africa (Khataby et al. 2016) contained no papers from East Africa, although numerous strain variants, including some unique to particular countries, have been reported in North and West Africa and southern parts of the continent. Reports from other countries worldwide, such as Bangladesh, Mexico, Oman and Algeria, where the poultry sectors are transforming, all indicate that these respiratory pathogens are a common and widespread source of concern to the developing industries (Islam et al. 2014; Rivera-Benitez et al. 2014; Al-Shekaili et al. 2015b; Sid et al. 2015).

Ethiopia and other countries in East Africa are likely to face the same problems as intensification of poultry farming continues.

Future sampling should include backyard stock, as birds are frequently distributed from this, and other poultry centres, to smallholders, where they mix with the indigenous populations. Poultry traders are also known to sell chickens intended for retail to smallholder producers for breeding, and disease outbreaks in villages are often suspected to be caused by poultry introduced from markets to a household in the village (Chaka et al. 2011). Whilst trading live birds through markets is a known risk for introducing NDV, it is also highly likely to disseminate other pathogens, which are largely unstudied in village poultry. For the burgeoning-intensive industry, initiating control programmes involving higher biosecurity and vaccination for these important respiratory pathogens may be an essential component to boost welfare and productivity, and trialling such programmes at this research centre would be an ideal way to promote and disseminate such information among local producers through the existing training programmes available at the centre.

ACKNOWLEDGEMENTS

We would like to thank Anne Forrester and Marisol Collins for the help with sample processing and importation of samples from Ethiopia to the UK, and BioChek for the kind provision of the ELISA kits. SH was supported by grants from the Institute of Infection and Global Health (University of Liverpool) and the Wellcome Trust. We also thank the Biotechnology and Biological Sciences Research Council (BBSRC), the UK Department for International Development (DFID) and the Scottish Government for providing the funding for the 'Reducing the impact of infectious disease on poultry production in Ethiopia' project under the Combating Infectious Diseases of Livestock for International Development (CIDLID) programme (BB/H009396/1).

REFERENCES

1. Abdelwhab, E.M., Lüschow, D., Harder, T.C., Hafez, H.M., 2011. The use of FTA® filter papers for diagnosis of avian influenza virus. Journal of Virological Methods 174, 120–122.
2. Alamargot, J., Mengistu, A., Gebreab, F., 1985. [Avian pathology in Ethiopia. Examination of 198 necropsies performed 1983–1984 at the Debre-Zeit Faculty of Veterinary Medicine]. Revue d'elevage et de medecine veterinaire des pays tropicaux 38, 130–137.
3. Aldous, E.W., Mynn, J.K., Banks, J., Alexander, D.J., 2003. A molecular epidemiological study of avian paramyxovirus type 1 (Newcastle disease virus) isolates by phylogenetic

analysis of a partial nucleotide sequence of the fusion protein gene. Avian Pathology 32, 239–257.

4. Al-Shekaili, T., Baylis, M., Ganapathy, K., 2015. Molecular detection of infectious bronchitis and avian metapneumoviruses in Oman backyard poultry. Res Vet Sci 99, 46–52.
5. Al-Shekaili, T., Baylis, M., Ganapathy, K., 2015. Molecular detection of infectious bronchitis and avian metapneumoviruses in Oman backyard poultry. Research in Veterinary Science 99, 46–52.
6. Anon, 2014. Chick Distribution Model Helps Ethiopia's Small Poultry Farmers. The Poultry Site.
7. Bekele, L., 2015. Isolation, molecular identification and vaccine trial of Mycoplasma gallisepticum in Ethiopia. Department of Microbiology, Immunology and Veterinary Public Health. Addis Ababa University, Debre Zeit, Ethiopia.
8. Bourogâa, H., Miled, K., Gribâa, L., Behi, I.E., Ghram, A., 2009. Characterization of New Variants of Avian Infectious Bronchitis Virus in Tunisia. Avian Diseases 53, 426–433.
9. Cavanagh, D., Gelb Jr., J., 2008. Infectious Bronchitis. In: Saif, Y.M. (Ed.), Diseases of Poultry 12th Ed. Blackwell Publishing Ltd., Ames, Iowa, 117–135.
10. Cavanagh, D., Mawditt, K., Britton, P., Naylor, C.J., 1999. Longitudinal field studies of infectious bronchitis virus and avian pneumovirus in broilers using type-specific polymerase chain reactions. Avian Pathology 28, 593–605.
11. Cavanagh, D., Picault, J.P., Gough, R., Hess, M., Mawditt, K., Britton, P., 2005. Variation in the spike protein of the 793/B type of infectious bronchitis virus, in the field and during alternate passage in chickens and embryonated eggs. Avian Pathology: Journal of the W.V.P.A 34, 20–25.
12. Chaka, H., Goutard, F., Duboz, R., Bisschop, S.P.R., Thompson, P.N., 2011. Poultry market contact networks in Ethiopia: implications for disease spread and surveillance. 380–382.
13. Chaka, H., Goutard, F., Bisschop, S.P.R., Thompson, P.N., 2012. Seroprevalence of Newcastle disease and other infectious diseases in backyard chickens at markets in Eastern Shewa zone, Ethiopia. Poultry Science 91, 862–869.
14. Chaka, H., Goutard, F., Gil, P., Abolnik, C., Almeida, R.S.D., Bisschop, S., Thompson, P.N., 2013. Serological and molecular investigation of Newcastle disease in household chicken flocks and associated markets in Eastern Shewa zone, Ethiopia. Tropical Animal Health and Production 45, 705–714.
15. Dessie, T., Jobre, Y., 2004. A review of the importance and control of Newcastle disease in Ethiopia Ethiopian Veterinary Journal 8, 71–81.
16. Diallo, I.S., Taylor, J., Gibson, J., Hoad, J., De Jong, A., Hewitson, G., Corney, B.G., Rodwell, B.J., 2010. Diagnosis of a naturally occurring dual infection of layer chickens with fowlpox virus and gallid herpesvirus 1 (infectious laryngotracheitis virus). Avian Pathology 39, 25–30.
17. Ducatez, M.F., Martin, A.M., Owoade, A.A., Olatoye, I.O., Alkali, B.R., Maikano, I., Snoeck, C.J., Sausy, A., Cordioli, P., Muller, C.P., 2009. Characterization of a new genotype and serotype of infectious bronchitis virus in Western Africa. Journal of General Virology 90, 2679–2685.
18. FAO, 2008. Poultry Sector Country Review. Food and Agriculture Organization of the United Nations, Rome.
19. Fentie, T., Heidari, A., Aiello, R., Kassa, T., Capua, I., Cattoli, G., Sahle, M., 2014. Molecular characterization of Newcastle disease viruses isolated from rural chicken in northwest Ethiopia reveals the circulation of three distinct genotypes in the country. Tropical Animal Health and Production 46, 299–304.
20. Ganapathy, K., Ball, C., Forrester, A., 2015. Genotypes of infectious bronchitis viruses circulating in the Middle East between 2009 and 2014. Virus research 210, 198–204.
21. Gough, R.E., Jones, R.C., 2008. Avian Metapneumovirus. In: Saif, Y.M. (Ed.), Diseases of Poultry. Blackwell Publishing, Ltd, Ames, Iowa, 100–110.
22. Guy, J.S., Garcia, M., 2008. Laryngotracheitis. In: Saif, Y.M. (Ed.), Diseases of Poultry 12th Ed. Blackwell Publishing Ltd., Ames, Iowa, 137–152.

23. Islam, M., Hassan, J., Khan, M.S.R., 2014. Seroprevalence of Mycoplasma gallisepticum infection in backyard and commercial layer chickens in Bhola district, Bangladesh. J Adv Vet Anim Res 1, 11–15.
24. Khataby, K., Fellahi, S., Loutfi, C., Mustapha, E.M., 2016. Avian infectious bronchitis virus in Africa: a review. Veterinary Quarterly 36, 71–75.
25. Kleven, S.H., Ferguson-Noel, N., 2008. Mycoplasma synoviae infection. In: Saif, Y.M. (Ed.), Diseases of Poultry. Blackwell Publishing Ltd., Ames, Iowa, 845–856.
26. Ley, D.H., 2008. *Mycoplasma gallisepticum* Infection. In: Saif, Y.M. (Ed.), Diseases of Poultry, 12th Edition. Blackwell Publishing Ltd. Ames, 807–834.
27. Mammo, M., 2012. Chicken production scenarios and the headway options for improvement in Ethiopia. World's Poultry Science Journal 68.
28. Mazengia, H., Tilahun, S.B., Negash, T., 2010. Newcastle Disease and Infectious Bursal Diseases are Threat to Village Chicken Production in Two Districts of Amhara National Regional State, Northwest Ethiopia. IUP Journal of Life Sciences 4, 62–72.
29. Michiels, T., Welby, S., Vanrobaeys, M., Quinet, C., Rouffaer, L., Lens, L., Martel, A., Butaye, P., 2016. Prevalence of Mycoplasma gallisepticum and Mycoplasma synoviae in commercial poultry, racing pigeons and wild birds in Belgium. Avian Pathology: Journal of the W.V.P.A, 1–28.
30. Mohammed, H.O., Carpenter, T.E., Yamamoto, R., 1987. Evaluation of Factors Associated with Infection of Commercial Layers with Mycoplasma gallisepticum and M. synoviae. Avian Diseases 31, 470–476.
31. Nega, M., Moges, F., Mazengia, H., Zeleke, G., Tamir, S., 2012. Evaluation of I2 thermostable Newcastle disease vaccine on local chickens in selected districts of western Amhara. Online Journal of Animal and Feed Research (OJAFR) 2, 244–248.
32. O.I.E., 2015. World Animal Health Information Database (WAHID) Interface.
33. Owoade, A.A., Ducatez, M.F., Muller, C.P., 2006. Seroprevalence of Avian Influenza Virus, Infectious Bronchitis Virus, Reovirus, Avian Pneumovirus, Infectious Laryngotracheitis Virus, and Avian Leukosis Virus in Nigerian Poultry. Avian Diseases 50, 222–227.
34. Pagani, P., Wossene, A., 2008. Review of the new features of the Ethiopian poultry sector; biosecurity implications. In: Nations, F.a.A.O.o.t.U. (Ed.).
35. Pretty, J., Toulmin, C., Williams, S., 2011. Sustainable intensification in African agriculture. International Journal of Agricultural Sustainability 9, 5–24.
36. Regasa, C., Fufa, D., Berihanu, S., Hunduma, D., 2007. Seroprevalence of Newcastle disease in backyard chickens in mid Rift Valley of Oromia, Ethiopia. In: E., C., E., C., C., D., D., M., J.F., R., F., R. (Eds.), 12th International conference of the Association of Institutions of Tropical Veterinary Medicine, Montpellier, 121.
37. Rivera-Benitez, J.F., Martínez-Bautista, R., Ríos-Cambre, F., Ramírez-Mendoza, H., 2014. Molecular detection and isolation of avian metapneumovirus in Mexico. Avian Pathology 43, 217–223.
38. Sambo, E., Bettridge, J., Dessie, T., Amare, A., Habte, T., Wigley, P., Christley, R.M., 2014. Participatory evaluation of chicken health and production constraints in Ethiopia. Preventive Veterinary Medicine.
39. Sid, H., Benachour, K., Rautenschlein, S., 2015. Co-infection with Multiple Respiratory Pathogens Contributes to Increased Mortality Rates in Algerian Poultry Flocks. Avian Diseases 59, 440–446.
40. Sumberg, J., Thompson, J., 2013. Revolution Reconsidered: Evolving Perspectives on Livestock Production and Consumption. STEPS Centre, Brighton, UK.
41. Tadesse, S., Ashenafi, H., Aschalew, Z., 2005. Seroprevalence Study of Newcastle Disease in Local Chickens in Central Ethiopia. International Journal of Applied Research in Veterinary Medicine 3, 25–29.
42. van Leerdam, B., Bosman, H., 2006. Vaccination index: A new concept to evaluate ELISA vaccination results. Interpretation and Application of Results Manual. BioChek, 14–16.

43. Worthington, K.J., Currie, R.J.W., Jones, R.C., 2008. A reverse transcriptase-polymerase chain reaction survey of infectious bronchitis virus genotypes in Western Europe from 2002 to 2006. Avian Pathology 37, 247–257.
44. Zeleke, A., Sori, T., Gelaye, E., Ayelet, G., 2005. Newcastle Disease in Village Chickens in the Southern and Rift Valley Districts in Ethiopia. International Journal of Poultry Science 4, 507–510.

CHAPTER 13

Detection of inflammatory bowel disease by proton magnetic resonance spectroscopy (^{1}H MRS) using an animal model

Sonal Varma[1,2], Ranjana Bird[3], Michael Eskin[2], Brion Dolenko[1], Jayadev Raju[3] and Tedros Bezabeh*[1,2]

[1]*National Research Council Institute for Biodiagnostics, Winnipeg, Canada*

2Department of Human Nutritional Sciences, University of Manitoba, Winnipeg, Canada and [3]Department of Biology, University of Waterloo, Waterloo, Canada

ABSTRACT

Background

The aim of this study was to analyze the potential of proton magnetic resonance spectroscopy (1H MRS) in diagnosing early inflammatory bowel disease (IBD).

Methods

Thirty male Sprague Dawley rats were fed 2% carrageenan in their diet for either 1 or 2 weeks. ^{1}H MRS was performed ex-vivo on colonic mucosal samples (n = 123) and the spectra were analyzed by a multivariate method of analysis. The results of the multivariate analysis were correlated with histological analysis performed using H & E stain for the presence of inflammation in the samples from each group.

Results

Multivariate analysis classified the samples in their respective groups with an accuracy of 82%. Our region selection algorithm identified four regions in the spectra as being discriminatory. The metabolites assigned to these regions include creatine, phosphatidylcholine, the -CH_2HC= group in fatty acyl chain, and the glycerol backbone of lipids. The differences in concentration of these metabolites in each group offer insight into the biochemical changes occurring during IBD and confer diagnostic potential to ^1H MRS as a tool to study colonic inflammation in conjunction with biopsy.

Conclusion

1H MRS is a sensitive tool to detect early colonic inflammation in an animal model of IBD.

1. BACKGROUND

Inflammatory Bowel Disease (IBD) is a common condition in North America with rising incidence in previously less affected areas worldwide [1]. It is classified as ulcerative colitis and Crohn's disease. IBD shows a bimodal peak of appearance of the disease; typically, one in the population aged 15–25 years and another peak later in life around 60 years of age [2]. Recently, it has been shown that the incidence of IBD is increasing in the pediatric population [1]. It is known that ulcerative colitis is associated with an increased risk of colorectal cancer. The incidence of colorectal cancer is estimated to be 2% in people who have had ulcerative colitis for 10 years, approaches 8% after 20 years and 18% after 30 years of having the disease [3]. As such, people diagnosed with ulcerative colitis at a younger age are exposed to a significant risk of developing colorectal cancer. IBD has a multifactorial etiology that includes an interplay of genetic, environmental and immunological factors [2]. Diet is one of the major environmental factors that is being postulated to be associated with IBD and cancer.

Dietary fat may be one of the critical determinants of pathogenesis and remission in IBD [4]. Polyunsaturated fatty acids (PUFA), specifically omega-3 fatty acids, are currently being explored for therapeutic assistance for colon cancer. Studies analyzing this effect in IBD and its transition to colon cancer are however, very limited. Recently, Meister, et al. showed that *in vitro* treatment of colon biopsies from IBD patients with omega-3 fatty acid reduced inflammation [5]. In a crossover study by Aslan et al., IBD patients on omega-3 supplements showed clinical improvement, and in 72% of them it was possible to stop or reduce the medication. However, no histopathological or biochemical improvement was observed in the patients [6]. The studies so far are controversial and have pointed towards the fact that there are many critical limitations in conducting human studies on dietary influence in IBD. Since some of these patients may be on medical therapy, having a pure placebo-control group will be very difficult. Moreover, taking frequent biopsies in these patients may not be possible.

To overcome such design limitations in studies meant to evaluate the effect of diets in the window between IBD and colon cancer, the use of an animal model that closely resembles human ulcerative colitis becomes imperative. Many chemicals have been tried to induce experimental colitis in animals [7]. Degraded lambda-carrageenan has been shown to cause inflammation similar to human IBD in a rodent model [8]. In a preliminary study (unpublished observation), our lab has shown that carrageenan can cause colitis without massive ulceration and may be suitable for long term studies on IBD as also suggested by others [9]. Based on these observations, we have used Sprague-Dawley rats fed with 2% degraded carrageenan as our model for IBD.

Another factor in conducting studies on IBD is the availability of a sensitive tool to assess inflammation. Currently, histopathological analysis is the gold standard for evaluating inflammation. As biochemical and molecular changes are known to precede histological manifestation of a disease, assessing biochemical changes would provide an opportunity to diagnose IBD at an early stage, allowing timely intervention. Magnetic resonance spectroscopy (MRS) is a technique that is extensively used to characterize the metabolic profiles of tissues both in vivo and ex vivo [10, 11]. ^{1}H MRS is a robust and widely used technique that identifies proton signals from various biological samples [12, 13]. As proton-containing metabolites are very prevalent in tissues, ^{1}H MRS is well equipped to detect the presence and relative change in concentration of these metabolites. ^{1}H MR spectra show a complex array of metabolites present in the tissue, which can make the interpretation very difficult. To overcome this problem, a computer-based multivariate analysis method is being used. Multivariate analysis is particularly useful when the spectra from different experimental groups show only subtle differences in the intensity of metabolite peaks [14]. Together with ^{1}H MRS, it constitutes the critical technique used in the rapidly expanding field of 'metabonomics' [15]. Bezabeh et al. have previously shown that ^{1}H MRS has a potential to detect early inflammation, and also to differentiate between ulcerative colitis and Crohn's disease [10].

2. METHODS

2.1 Experimental design and sample collection

Thirty male Sprague-Dawley rats (Charles River, Canada) were divided into three groups of 10 animals each. The animals in the control group were fed a regular American Institute of Nutrition-76A (AIN-76A) diet for 2 weeks; animals in the 1 week treated group were fed a regular diet for 1 week followed by 2% carrageenan (Sigma, St. Louis, MO, USA) by weight for another 1 week; animals in the 2 weeks treated group received 2% carrageenan by weight with their diet for 2 weeks. The diets were made in-house and the ingredients were purchased from Harlan Teklad, Toronto, Ontario, Canada. All the animals were sacrificed at the end of the 2nd week by carbon dioxide asphyxiation. After sacrificing the animals, the colon was excised and the mucosal layer scraped off using a glass slide as described by Briere, et al. [16]. The mucosal specimens were stored in Phosphate Buffered Saline mixed with deuterium oxide (PBS/D_2O) in cryovials and kept frozen at -80°C until processed. All the animal work was conducted at the University of Waterloo. The animal care and handling procedures were done in accordance with the Animal Ethics guidelines at the University of Waterloo and the National Research Council of Canada, Winnipeg. The mucosal specimens were coded and then shipped on dry ice to the Institute for Biodiagnostics, National Research Council of Canada, Winnipeg for 1H MRS and histological assessment.

2.2 ^{1}H MRS experiments

Each specimen was thawed and cut into 5–7 mm long pieces, which yielded on average 18–20 pieces. ^{1}H MRS was conducted on alternate pieces, resulting in a total of 123 spectra. The samples were positioned in glass capillary tubes for ^{1}H MRS experiments as described by Kuesel, et al. [17]. Briefly, the colon sample was placed in a glass capillary tube filled with PBS/D_2O with one end plugged. This capillary was inserted into the NMR tube containing 300 μl of PBS/D_2O solution and 5 μl of chemical shift reference, 3-Trimethylsilyl-Propionic acid-D4- sodium salt (TSP). ^{1}H MRS was conducted on these samples using a Bruker Avance 360 MHz spectrometer at 25°C with presaturation of the water signal. The acquisition parameters included: 90° pulse at 9.30 μsec, number of scans = 256, spectral width = 4990.02 Hz, relaxation delay = 3 sec and time domain data points = 8 K. The samples were fixed in 10% neutral buffered formalin immediately after ^{1}H MRS experiments for histological assessment.

2.3 MRS data analysis

For multivariate analysis, each MRS spectrum was normalized by dividing the data points by the total spectral area and then aligned to the reference peak due to TSP. The region between 0.5 and 4.5 ppm was selected to eliminate the excess water signal at 4.7 ppm. First derivatives were taken, and rank ordering was done on the resulting data to eliminate baseline differences between the spectra. For control vs each of the treatment groups, random subsets of the data were used to train a genetic algorithm-based subregion selection method [18]. A histogram was constructed and the regions that stood out as being selected most often were used to develop the final classifier: linear discriminant analysis (LDA) with coefficients optimized using a bootstrapping method [19].

2.4 Histological analysis

The formalin fixed samples were embedded in paraffin blocks and cut into 5 μm sections. The sections were stained with Hematoxylin and Eosin (H&E) stains. The slides were analyzed under a Nikon inverted microscope at 40× resolution and pictures were taken using a Nikon CoolPix 5000 digital camera. Before making paraffin blocks, the samples were coded to eliminate measurement bias. The samples were decoded only after recording the findings of histological analysis.

2.5 Blood analysis

Blood samples were sent to the laboratory services division at the University of Guelph for complete and differential cell count. Carrageenan has been shown to induce systemic response [8], hence we speculated an increase in the white cell count in the blood of treated animals.

3. RESULTS

Multivariate analysis was performed to compare the spectra from the control group with both treatment groups. Multivariate analysis showed a significant overlap between the spectral data of the 1-week treated group and both the control and the 2-week treated groups. This indicates that after a period of 1 week, there is some inflammation in the colon, but not significant enough to set it apart from the normal colon. The analysis of the control and 2-week treated groups yielded results as shown in Table 1. All of the spectra (n = 123) were included in this analysis without correlating the data with histology.

Table 1 Multivariate analysis results of all MR spectra without histological confirmation (n = 123)

Control Vs. 2 weeks Carrageenan-treated group					
Desired Class	N	Classification		Accuracy	Overall Accuracy
		Correct	False		
Control	63	50	13	79.4%	77.2%
Treated	60	45	15	75.0%	

After conducting multivariate analysis on the complete dataset, histological assessment was performed. Figure 1 shows representative spectra from the control group and the 2 week-treated group along with the corresponding H & E stained images from the same specimens. In the histological assessment, it was observed that some samples had significant contamination by muscle tissue from the deeper layers of the colonic wall. These samples (n = 10) were eliminated from the analysis. The remaining 113 samples were evaluated for the degree of inflammation. The histological analysis of these samples revealed that some of the samples from the control group were actually inflamed and some samples from the 2-week treated group did not show any signs of inflammation. Also, it was noticed that there were differences in some of the spectral features from individual samples within the same group. This is possibly due to the fact that some of these animals did not develop any significant inflammation. In addition, the inflammation can also be patchy and may not be present uniformly throughout the colon [20]. However, we could not give a plausible explanation for colonic inflammation observed in the control group. Such spectra (n = 29) were removed from the full dataset (Table 1) and analyzed separately. Henceforth, these spectra will be referred to as 'aberrant spectra'.

After correlating with histology, the remaining 113 spectra were subjected to multivariate analysis. It was necessary to train the classifier on 'pure' spectra (n = 84) so as to distinctly identify the characteristics of inflamed and non-inflamed samples. Therefore, the aberrant spectra from both control and 2 weeks treated group (n = 29) were not used to develop the classifier for the multivariate analysis. Table 2 shows the results of the multivariate analysis of the main dataset (n = 84). The accuracy with which these samples were assigned to their respective groups increased from 77.2%, in the initial analysis, to 82%.

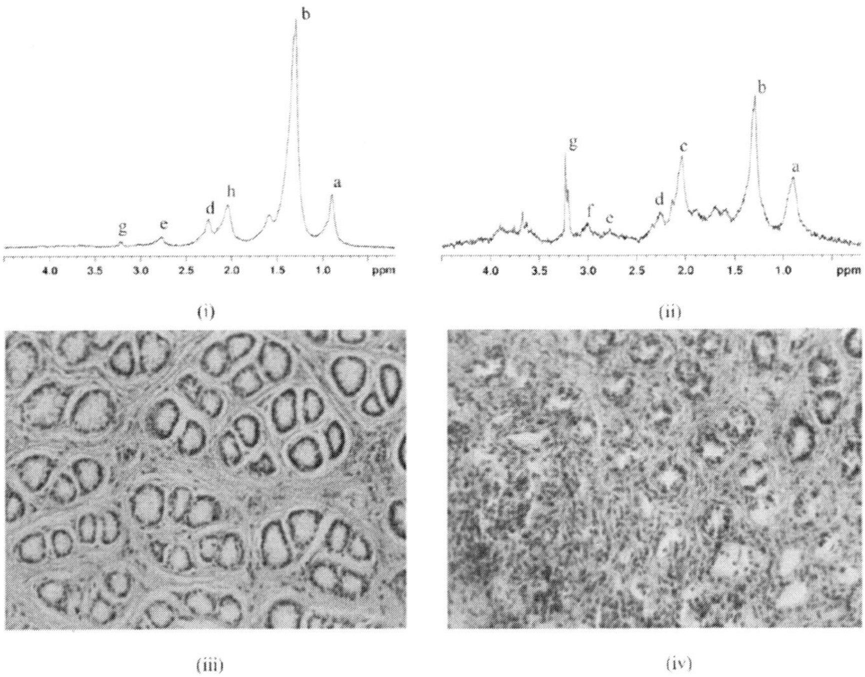

Figure 1 Representative ^{1}H MR spectra (360 MHz) and the histology sections from the control group and 2 week-treated group; (i) MR spectrum of specimen from control group (ii) MR spectrum of specimen from 2 week-treated group (iii) H&E stained section of the same control group specimen as above (iv) H&E stained section of the same specimen from 2 week-treated group as above. The metabolites represented by the labeled peaks in the MR spectra are as follows: (a) -CH3 group of fatty acids in lipids, (b) -CH_2 chain of fatty acids, (c) glutamic acid, d) -CH_2-COO group of fatty acids, (e) polyunsaturated fatty acids, (f) creatine, (g) total cholines (tCho), (h)-CH_2-CH= group of unsaturated fatty acids.

The aberrant dataset (n = 29) was separately submitted to the classifier developed using the 'pure' spectra. Out of the 16 'control' samples found to be inflamed on histological assessment, the multivariate analysis classified 10 as '2 weeks treated', i.e. inflamed. Similarly, out of the 13 carrageenan-treated samples that were found to be inflammation-free on histological assessment, the multivariate analysis classified 9 as 'control'.

Table 2 Results of multivariate analysis with histological confirmation (n = 84)

Control Vs. 2 weeks Carrageenan-treated group					
Desired Class	N	Classification		Accuracy	Overall Accuracy
		Correct	False		
Control	38	31	7	81.6%	82.1%
Treated	46	38	8	82.6%	

The reason why the four "control" samples were classified as "inflamed" by our analysis remains unclear. Further investigation needs to be undertaken on such samples to understand the exact reason for the misclassification.

One of the reasons for not attaining 100% accuracy in the MRS analysis may be the intragroup variation, particularly in the carrageenan-treated group. The degree of inflammation widely varied within the same group. Differences in intensities of metabolite peaks were observed among individual spectra from the same group of animals and in fact, from the same animal. The averaging of these variations in spectral intensities may have resulted in blunting of the effect of certain metabolites during multivariate analysis. This problem can be overcome by using a larger sample size to reduce the effect of individual variations.

Other factors that may raise concerns about the variability in the results include the use of CO_2 asphyxiation method to euthanize the animals, leading to acute cell injury and mechanical injury during sample preparation. We felt that since we were using the same method of euthanization for all groups, this variable was common to both groups. In addition, acute hypoxic events need some time to be translated as biochemical change in the colon. The initial response to CO_2 asphyxiation is respiratory acidosis and the brunt is on organs that are metabolically more active like brain and heart. There may also have been some tissue trauma and alteration in metabolic profile during stripping of colonic mucosa from the deeper layers. Every effort was made to minimize trauma during sample collection procedure. This procedure was performed by one person in an effort to keep the processing time and force reasonably uniform for all the samples.

Our region selection algorithm identified 4 regions of the spectra that were most discriminatory, i.e. were selected most often by the genetic algorithm. These included spectral resonances due to the $-CH_2HC=$ group in fatty acyl chain of triglycerides (2.79–2.83 ppm), creatine (3.00 ppm), phosphocholine (3.56–3.60 ppm) and glycerol backbone of lipids (4.03–4.05 ppm).

4. DISCUSSION AND CONCLUSION

Martin et al. have characterized the metabolic profile of rodent intestinal tissue including proximal and distal colon using high resolution ^{1}H MRS [21]. They observed significant levels of phosphatidylcholine, creatine, glycerol and triglycerides in the normal colon. Phosphatidylcholine is the predominant phospholipid present in the cell membranes [22] including colonic mucosal cells [23]. It is also an important constituent of the inflammatory mediator, platelet activating factor (PAF). PAF is crucial for critical steps in inflammatory cascade such as platelet activation and increase in vascular permeability. The active species of PAF have been shown to increase significantly in mucosal biopsies from ulcerative colitis patients [24]. PAF levels in stool are considered diagnostic and indicative of severity in patients with ulcerative colitis [25]. The increased levels of phosphatidylcholine seen in our study may be because of increased levels of PAF in colitis. Increased apoptosis and cell turnover in IBD may also contribute to increased phosphatidylcholine concentration in colonic mucosa.

Creatine is a metabolite present in normal tissues and serves as an energy reservoir. It is converted via a reversible reaction into phosphocreatine by adding a high energy phosphate bond in the presence of creatine kinase. Phosphocreatine is broken down to creatine while releasing energy as ATP whenever required. Significant concentrations of creatine have previously been detected in human colonic mucosa [26]. The BB isoenzyme of creatine kinase has been isolated from normal [27] and infracted colonic mucosa [28]. As colonic inflammation causes cellular damage and increased requirement for energy, the creatine concentrations may be elevated in these cells.

The region around 2.8 ppm denotes the peak from protons in polyunsaturated fatty acids (PUFA) [29]. PUFAs are critical for maintaining the membrane structure and function of various cells in the body [30]. Linoleic acid is an important constituent of the lipid bilayer of cell membranes and a precursor for another PUFA, arachidonic acid. Arachidonic acid, in turn, is the precursor of potent inflammatory mediators such as LT-B4 and PG-G2.

These inflammatory markers are increased in many disorders in the human body including IBD [31]. Nishida et al., observed an increase in arachidonic acid concentration in the mucosa of patients with ulcerative colitis [32]. Nieto et al. also observed an increase in linoleic acid concentration in the colonic mucosa of rats 2 weeks after inducing experimental colitis [33]. Increase in glycerol resonance in our samples may be reflective of an increase in the total lipids in the colonic mucosa during inflammation.

These changes in colonic mucosal lipid profile are similar in human ulcerative colitis as well as experimental colitis in animals.

It should be noted that the above metabolites are seen in normal as well as diseased states in colonic mucosa. The difference in their intensities is what is more significant and indicative of a disease process as opposed to only the presence or absence of these metabolites. The change in the overall profile of the spectrum, and not changes in the intensity of only one or two specific peaks, is a better indicator of the disease process.

To assess the systemic effects of colonic inflammation, we analyzed blood samples from animals in the control (n = 5) and 2 week-treated group (n = 6). As IBD is a chronic inflammation, we compared the lymphocyte counts in these animals. In 10 out of these 11 animals, the histological assessment of inflammation agreed with their respective lymphocyte counts. In one animal, the samples were classified as inflamed by histology whereas the lymphocyte counts were not raised. Since the samples from this animal were kept in the refrigerator for longer period of time than any other sample (over 2 days), some changes might have occurred in the tissue specimens interfering with the results of the histological analysis.

The results of this study need to be further reinforced by correlating the findings of 1H MRS with systemic markers of inflammation such as leukotriene B4 levels, serum C-reactive protein, albumin and thrombocytosis. In addition, testing the reproducibility of these results in other models of colonic inflammation such as IL-10 knock-out mice and dextran sulphate-induced colitis model will help validate this new role of 1H MRS. Moreover, there may be critical differences in the degree and chronicity of inflammation in different parts of the colon. Based on the results of this study, further work needs to be undertaken to examine the differences in inflammatory profile between distal, middle and proximal colon.

In conclusion, this study demonstrates that ^{1}H MRS combined with multivariate analysis is a highly sensitive technique to distinguish between inflamed and non-inflamed colonic tissue. In addition, it also suggests that ^{1}H MRS can serve as a tool to study the sequential progression of IBD and explore the effect of various dietary and therapeutic agents on this process. This model will be helpful to approach a multitude of questions like whether and how dietary practices can affect the development and progression of IBD. It will also enable us to overcome some of the limitations faced in human studies.

ACKNOWLEDGEMENTS

We thank the Natural Sciences and Engineering Research Council of Canada (NSERC) for the strategic project grant to R. P Bird, M. Eskin and T. Bezabeh. We would also like to thank Saro Bascaramurty for her invaluable assistance with histology.

REFERENCES

1. Lakatos PL: Recent trends in the epidemiology of inflammatory bowel diseases: Up or down. World J Gastroenterol. 2006, 12: 6102-6108.
2. Karlinger K, Gyorke T, Mako E, Mester A, Tarjan Z: The epidemiology and the pathogenesis of inflammatory bowel disease. Eur J Radiol. 2000, 35: 154-167. 10.1016/S0720-048X(00)00238-2.
3. Eaden J: Colorectal carcinoma and inflammatory bowel disease. Aliment Pharmacol Ther. 2004, 20: 24-30. 10.1111/j.1365-2036.2004.02046.x.
4. Cashman KD, Shanahan F: Is nutrition an aetiological factor for inflammatory bowel disease. Eur J Gastroenterol Hepatol. 2003, 15: 607-613. 10.1097/00042737-200306000-00005.
5. Meister D, Ghosh S: Effect of fish oil enriched enteral diet on inflammatory bowel disease tissues in organ culture: differential effects on ulcerative colitis and Crohn's disease. World J Gastroenterol. 2005, 11: 7466-7472.
6. Aslan A, Triadafilopoulos G: Fish oil fatty acid supplementation in active ulcerative colitis: a double-blind, placebo-controlled, crossover study. Am J Gastroenterol. 1992, 87: 432-437.
7. Kim HS, Berstad A: Experimental colitis in animal models. Scand J Gastroenterol. 1992, 27: 529-537. 10.3109/00365529209000116.
8. Moyana TN, Xiang J, Qi Y, Kalra J: Development of the early mucosal lesions in experimental inflammatory bowel disease – implications for pathogenesis. Exp Mol Pathol. 1994, 60: 119-129. 10.1006/exmp.1994.1011.
9. Pricolo VE, Madhere SM, Finkelstein SD, Reichner JS: Effects of lambda-carrageenan induced experimental enterocolitis on splenocyte function and nitric oxide production. J Surg Res. 1996, 66: 6-11. 10.1006/jsre.1996.0364.
10. Bezabeh T, Somorjai RL, Smith IC, Nikulin AE, Dolenko B, Bernstein CN: The use of 1-H magnetic resonance spectroscopy in inflammatory bowel diseases: distinguishing ulcerative colitis from Crohn's disease. Am J Gastroenterol. 2001, 96: 442-448. 10.1111/j.1572-0241.2001.03523.x.
11. Casciani E, Polettini E, Bertini L, Masselli G, Emiliozzi P, Amini M, Pansadoro V, Gualdi GF: Contribution of the MR spectroscopic imaging in the diagnosis of prostate cancer in the peripheral zone. Abdom Imaging. 2007,
12. Narayana PA: Magnetic resonance spectroscopy in the monitoring of multiple sclerosis. J Neuroimaging. 2005, 46-57. 10.1177/1051228405284200. Suppl 15
13. Teksam M, Cakir B, Agildere AM: Proton MR spectroscopy in the diagnosis of early-stage subacute sclerosing panencephalitis. Diagn Interv Radiol. 2006, 12: 61-63.
14. Hahn P, Smith IC, Leboldus L, Littman C, Somorjai RL, Bezabeh T: The classification of benign and malignant human prostate tissue by multivariate analysis of 1H magnetic resonance spectra. Cancer Res. 1997, 57: 3398-3401.
15. Nicholson JK, Wilson ID: Opinion: understanding 'global' systems biology: metabonomics and the continuum of metabolism. Nat Rev Drug Discov. 2003, 2: 668-676. 10.1038/nrd1157.

16. Briere KM, Kuesel AC, Bird RP, Smith IC: 1H MR visible lipids in colon tissue from normal and carcinogen-treated rats. NMR Biomed. 1995, 8: 33-40. 10.1002/nbm.1940080108.
17. Kuesel AC, Kroft T, Saunders JK, Prefontaine M, Mikhael N, Smith IC: A simple procedure for obtaining high-quality NMR spectra of semiquantitative value from small tissue specimens: cervical biopsies. Magn Reson Med. 1992, 27: 349-355. 10.1002/mrm.1910270215.
18. Nikulin AE, Dolenko B, Bezabeh T, Somorjai RL: Near-optimal region selection for feature space reduction: novel preprocessing methods for classifying MR spectra. NMR Biomed. 1998, 11: 209-217. 10.1002/(SICI)1099-1492(199806/08)11:4/5<209::AID-NBM510>3.0.CO;2-5.
19. Somorjai RL, Alexander M, Baumgartner R, Booth S, Bowman C, Demko A, Dolenko B, Mandelzweig M, Nikulin AE, Pizzi N, Pranckeviciene N, Summers R, Zhilkin P: A data-driven, flexible machine learning strategy for the classification of biomedical data. Computational Biology, Artificial Intelligence Methods and Tools for Systems Biology. Edited by: Azuaje F, Dubitsky W. 2004, Boston: Kluwer Academic Publishers, 5: 67-85.
20. Bernstein CN, Shanahan F, Anton PA, Weinstein WM: Patchiness of mucosal inflammation in treated ulcerative colitis: a prospective study. Gastrointest Endosc. 1995, 42: 232-237. 10.1016/S0016-5107(95)70097-8.
21. Martin FP, Wang Y, Sprenger N, Holmes E, Lindon JC, Kochhar S, Nicholson JK: Effects of Probiotic Lactobacillus Paracasei Treatment on the Host Gut Tissue Metabolic Profiles Probed via Magic-Angle-Spinning NMR Spectroscopy. J Proteome Res. 2007,
22. Cullis PR, Hope MJ: Physical properties and functional roles of lipids in membranes. Biochemistry of lipids, lipoproteins and membranes. Edited by: Vance DE. 1992, Amsterdam: Elsevier Science publishers, 1-41.
23. White DA: The phospholipid composition of mammalian tissue. Form and function of phospholipids. Edited by: Ansell GB, Hawthorne JN, Dawson RMC. 1973, Amsterdam. Elsevier Scienctific publishing Co, 441-482.
24. Thyssen E, Turk J, Bohrer A, Stenson WF: Quantification of distinct molecular species of platelet activating factor in ulcerative colitis. Lipids. 1996, 31: 255-259. 10. 1007/BF02637086.
25. Hocke M, Richter L, Bosseckert H, Eitner K: Platelet activating factor in stool from patients with ulcerative colitis and Crohn's disease. Hepatogastroenterology. 1999, 46: 2333-2337.
26. Ende D, Rutter A, Russell P, Mountford CE: Chemical shift imaging of human colorectal tissue (ex vivo). NMR Biomed. 1996, 9: 179-183. 10.1002/(SICI)1099-1492 (199606) 9:4<179::AID-NBM407>3.0.CO;2-N.
27. Jockers-Wretou E, Giebel W, Pfleiderer G: Immunohistochemical localization of creatinkinase isoenzymes in human tissue. Histochemistry. 1977, 54: 83-95. 10.1007/BF00493332.
28. Graeber GM, Cafferty PJ, Wolf RE, Harmon JW: An analysis of creatine phosphokinase in the mucosa and the muscularis of the gastrointestinal tract. J Surg Res. 1984, 37: 376-382. 10.1016/0022-4804(84)90203-8.
29. Adosraku RK, Choi GT, Constantinou-Kokotos V, Anderson MM, Gibbons WA: NMR lipid profiles of cells, tissues, and body fluids: proton NMR analysis of human erythrocyte lipids. J Lipid Res. 1994, 35: 1925-1931.
30. Clandinin MT, Cheema S, Field CJ, Garg ML, Venkatraman J, Clandinin TR: Dietary fat: exogenous determination of membrane structure and cell function. FASEB J. 1991, 5: 2761-2769.
31. Simopoulos AP: Omega-3 fatty acids in inflammation and autoimmune diseases. J Am Coll Nutr. 2002, 21: 495-505.
32. Nishida T, Miwa H, Shigematsu A, Yamamoto M, Iida M, Fujishima M: Increased arachidonic acid composition of phospholipids in colonic mucosa from patients with active ulcerative colitis. Gut. 1987, 28: 1002-1007. 10.1136/gut.28.8.1002.
33. Nieto N, Giron MD, Suarez MD, Gil A: Changes in plasma and colonic mucosa fatty acid profiles in rats with ulcerative colitis induced by trinitrobenzene sulfonic acid. Dig Dis Sci. 1998, 43: 2688-2695. 10.1023/A:1026607428716.

CHAPTER 14

Serological and molecular investigation for brucellosis in swine in selected districts of Uganda

Joseph M. Kungu[2], Joyce Akol[2], Danilo Pezo[2], Hosny El Adawy[4], Falk Melzer[4], Mandy Elschner[4]

1 College of Veterinary Medicine, Animal Resources and Biosecurity, Makerere University, P. O. Box 7062, Kampala, Uganda

2 International Livestock Research Institute (ILRI), C/O Bioversity International, P. O. Box 24384, Kampala, Uganda

3 Mycobacteriology (BSL-3) Laboratory, College of Health Sciences, Makerere University, P. O. Box 7072, Kampala, Uganda

4 Friedrich-Loeffler-Institute, Federal Research Institute for Animal Health, Institute for Bacterial Infections and Zoonoses, Naumburger Str. 96a, 07743 Jena, Germany

5 International Livestock Research Institute, P. O. Box 30709, Nairobi, Kenya

ABSTRACT

Brucellosis is a notifiable zoonotic disease affecting livestock, humans, and wildlife in Uganda. Pigs can be infected with human pathogenic *Brucella suis* biovars 1 and 3 and can be a significant source of brucellosis for humans. Uganda has a rapidly growing pig population, and the pork consumption per capita is the highest in East Africa. The objective of this work was to determine the seroprevalence of brucellosis in Ugandan pigs. A cross-sectional serosurvey of pigs was conducted in three of the major pig-keeping districts in Uganda (Masaka ($n = 381$ samples), Mukono ($n = 398$), and Kamuli ($n = 414$)). In addition, pigs originating from these districts were sampled in the major pig abattoir in Kampala ($n = 472$). In total, 1665 serum samples were investigated by serological and molecular tests. Only three putative brucellosis-positive samples were detected serologically using indirect ELISA. These sera were found negative for *Brucella* antibodies by CFT; however, two had antibodies against *Yersinia enterocolitica* as determined by SAT. Presence of antibodies against *Yersiniae* was confirmed by *Y. enterocolitica* antibody-specific ELISA. The two *Yersiniae* ELISA-positive samples were brucellosis negative using real-time PCR. We tested additional 142 sera from the 1665 samples with real-time PCR. All tested negative. Under this type of production system, we expect a maximum *B. suis* prevalence of less than 1 % at 95 % confidence level, and therefore, the risk of acquiring brucellosis from the pigs or their products is negligible. However, pigs may harbor the zoonotic *Y. enterocolitica*. This is the first study to investigate the occurrence of brucellosis in pigs in Uganda and the first study to report *Y. enterocolitica* antibodies in swine in Uganda.

Keywords: Screening Porcine, brucellosis, Yersiniosis, Masaka, Mukono, Kamuli districts

1. INTRODUCTION

Brucellosis is a notifiable zoonotic disease affecting people, livestock, and wildlife globally (Perry and Sones 2007). It is widespread causing significant human suffering and serious economic losses in livestock (Nakavuma and Opuda-Asibo 1999; McDermott et al. 2013). Although national statistics are lacking, there is considerable concern about brucellosis in Uganda. A seroprevalence of 12 and 7 % was reported among beef abattoir workers in Kampala and Mbarara district, respectively (Nabukenya et al. 2013). In Mulago National Refererral Hospital, located in Kampala City, 652 cases of brucellosis were diagnosed between June 2004 and May 2006 alone (Makita et al. 2008).

Brucellosis, caused by *Brucella suis*, is a major disease of pigs causing infertility, production of small litters, and abortion in sows. *B. suis* is a notable occupational hazard particularly to abattoir workers, farmers, and veterinarians (Radostits et al. 2000; Swai and Schoonman 2012). Uganda has reported pig brucellosis to OIE between 1996 and 2000 (Scientific Opinion of the Panel on Animal Health and Welfare AHAW on a request from the Commission on porcine brucellosis Brucella suis 2009), but it is not clear how this was diagnosed and which strains or biovars were involved. Our literature review found no publications on porcine brucellosis in Uganda, but initial scoping studies with farmers revealed high levels of abortion and infertility (Ouma et al. 2014).

Pig production in Uganda can be categorized into three main systems namely (i) intensive, (ii) semi-intensive, and (iii) extensive (small-scale subsistence). Less than 10 % of pigs are kept in intensive or semi-intensive systems but extensive/tethered/small-scale pig-production system is common (Dione et al. 2014). The demand for pork in Uganda is growing rapidly, and the population of pigs has increased from 0.19 million in 1980 to 3.2 million in 2008 (Uganda Bureau of Statistics/Ministry of Agriculture and Fisheries 2009); currently, Uganda has the largest pig population in eastern Africa and ranks 3rd in sub-Saharan Africa (FAOSTAT/FAO 2011). The prevalence of porcine brucellosis in Uganda is unknown though the disease is present in sub-Saharan Africa (Godfroid et al. 2013; Ducrotoy et al. 2015).

The objectives of this work were to determine the seroprevalence of brucellosis in Ugandan pigs, to assess the brucellosis risk from pigs posed to humans, and to characterize the organisms involved. The work was part of Consultative Group for International Agricultural Research (CGIAR) research aiming at characterizing challenges and evaluating options that will allow emerging pig farmers to improve their productivity and livelihoods, while increasing the supply of critical nutrients to their communities and urban centers.

2. METHODOLOGY

2.1 Study area

The study was conducted from December 2012 to July 2013 in the only official pig abattoir in Kampala City and in pig farms in Masaka, Mukono, and Kamuli districts, Uganda. The three districts studied were selected using a process to identify district representative of high pork production potential (Ouma et al. 2014).

2.2 Abattoir sampling strategy

A cross-sectional abattoir survey was conducted on pigs brought from three districts of Uganda: Masaka, Mukono, and Kamuli. Visits were made twice a week for a period of 3 months when pigs from the target districts were scheduled for slaughter, and blood samples of at least 20 pigs were taken purposively. The blood was collected into sterile vacutainer tubes without anticoagulant, and metadata for each sample included the district of origin and date the animal was sampled. Blood samples were immediately safely transported in a cold box to the College of Veterinary Medicine, Animal Resources and Biosecurity, Makerere University, where sera were harvested and stored at −20 °C until analyzed.

2.3 On-farm sampling strategy

A cross-sectional farm survey was carried out in Masaka, Mukono, and Kamuli districts from April to July 2013. The pig population data from the government livestock census of 2008 was used to facilitate identification of locations within the selected districts where the pig value chain activities would take place. In each district, four to six sub-counties with high pig population were selected for further scrutiny of the existing value chain domains. Three value chain domains were considered: rural production for rural consumption (R-R), rural production for urban consumption (R-U), and urban/periurban production for urban consumption (U-U). In a particular district, two sub-counties were selected to represent each value chain domain. Within each selected sub-county, two to three villages were randomly selected for the pig value chain activities. In total, 35 villages were selected for the project value chain assessment activities (Ouma et al. 2014). For this study and due to financial limitations, 22 out of the 35 villages were purposively selected across the three districts. In the selected villages, a census of all pig farmers was provided by local governmental staff and farmer selection was performed using computer-generated, random numbers. The epidemiological unit was considered to be the farm. In each farm, one animal fitting the inclusion criteria was selected and included in the survey. A structured questionnaire was administered to the owner of the farm. The aim of the questionnaire was to assess the disease risk in the farming system. In each household, the selected pig was restrained and blood was collected from the jugular vein into a sterile vacutainer tube without anticoagulant and serum was harvested immediately upon return from the field.

2.4 Sample size determination

The sample size at each site was calculated considering an infinite population (no recent census data) using the formula adopted from Thrusfield (1995) as follows:

$$n = Z^2 P(1 - P) / d^2$$

where n is the required sample size; Z is the multiplier from normal distribution (1.96); P is the estimated prevalence which is 50 % considering the fact that there was no reliable prevalence data for brucellosis in Uganda; (1 – P) is the probability of having no disease and d is the desired precision (5 %). The level of confidence for the study was set at 95 % confidence interval. Based on this formula, a decision was made to sample a total of 384 farms in each district and was rounded to 400. Additional samples were purposively collected from village boars.

2.5 Criteria for exclusion

Piglets younger than 3 months old, pregnant sows and nursing sows with litters less than 2 months old, as well as weak pigs which might be adversely affected by bleeding were excluded from the survey.

2.6 Laboratory analysis

Serological detection of antibodies against Brucella

2.7 Indirect ELISAs

Specific detection of antibodies against *Brucella* in the porcine serum samples was carried out using two indirect ELISA kits: the Ingezim *Brucella* Porcine *Brucella*-Ab indirect ELISA kit (INGENASA, Madrid, Spain) and the ID. Vet Screen® (Grabels, France).

2.8 Ingezim *Brucella* Porcine *Brucella*-Ab indirect ELISA

A total of 472 serum samples from adult pigs slaughtered at the abattoir and 381 collected on farm from Masaka were screened for antibodies against *Brucella* at Makerere University, College of Veterinary Medicine, Animal Resources and Biosecurity, using Ingezim ELISA according to the manufacturer's instructions. The Ingezim *Brucella* Porcine *Brucella*-Ab indirect ELISA kit has a sensitivity of 75 %, is highly specific (98 %) (Muñoz et al. 2012), and is based on *Brucella* LPS antigen. The samples were run on *Brucella*-Ab indirect ELISA kit and the optical densities (ODs) were determined in a microplate spectrometer (BioTek EL 800, North Star Scientific, Ltd) at 450-nm wavelength. Positive and negative control serum samples were included in each test. Interpretation of the results was based on S/P calculation: S/P = OD sample/OD positive control. Samples with the S/P higher than or equal to 0.25 were considered as positive, and all those with S/P value less than 0.25 were considered negative.

2.9 ID. Vet Screen® ELISA

A total of 1193 sera from pigs aged 3 months and older, including village boars, were shipped to Friedrich-Loeffler-Institute (FLI) and screened for antibodies against *Brucella* using indirect ELISA. Of these sera, 381, 398, and 414 were collected from Masaka, Mukono, and Kamuli, respectively. The sera were screened for antibodies against *Brucella* using ID. Vet Screen® (ID. Vet Innovative Diagnostics, Grabels, France) according to the manufacturer's instructions. ID. Vet Screen® is a robust test that is based on the purified *Brucella* LPS antigen. The results by ID Screen® ELISA were validated by testing the field samples in parallel with archived *Brucella* antibody-positive (14RB6037 FLI, 10RB1462 EUPigBSS) controls, a negative control (13RB4440 FLI), and positive field samples (08RB3522, 08RB3538 FLI) from the 2008 outbreak of brucellosis in wild boars in Germany.

2.10 Complement fixation test

The complement fixation test (CFT) was performed as a confirmatory test on putative positive samples identified by indirect ELISA. The CFT was carried out at OIE Reference Laboratory for brucellosis (hosted by FLI) following a micromethod according to the Kolmer technique (fixation at 2–8 °C, overnight) and OIE Terrestial Manual 2012 guidelines. The serum samples and positive and negative controls were first diluted 1:5 with CFTB and incubated in a water bath at 60 °C for 30 min to inactivate the complement. Then, in 96-well, U-bottomed microplates, beginning from the second row, 25 µl of inactivated serum samples was serially diluted in CFTB from 1:5 to 1:320, after which 25 µl of brucellosis antigen (IDEXX Pourquier (France) (diluted 1:100 with CFTB)) was added to each well, followed by addition of 25 µl of complement (VIRION, Germany) to each well at working dilution 1:55 in CFTB. Control wells with diluent only, complement and diluent and antigen, and complement titrations 2, 1, 0.5, and 0.25, with diluent and antigen, were prepared to contain 75 µl in each well. The plates were shaken briefly and incubated for 16–20 h in a refrigerator. The plates after overnight incubation were pre-warmed in the incubator at 30 °C for 30 min. Then 50 µl of freshly prepared hemolytic system was added into each well and plates were shaken carefully. The plates were incubated at 37 °C for 15–30 min in a wet chamber. The plates were centrifuged for 5 min at 1000×*g*. The degree of hemolysis was compared with set standards corresponding to 0, 25, 50, 75, and 100 % lysis. The absence of anticomplementary activity was checked for each serum in the first row of the plates. Results were expressed in CFT international units (ICFTU) per millimeter calculated in line with those obtained in parallel titration with national standard serum calibrated against OIE International Standard Serum which contains 595 ICFTU/ml (OIE Terrestial Manual 2012). Sera giving a titer of 20 ICFTU/ml or more were considered to be positive.

2.11 Slow agglutination test

Slow agglutination to detect agglutinating antibodies against *Yersinia enterocolitica* was performed in U-bottomed microtiter plates as previously described (Burkhardt 1992). The serum samples were serially diluted twofold using 0.005 % solution of safranin in phosphate-buffered saline. Then an equal volume (50 µl) of *Y. enterocolitica* 0:9 standardized antigen (YAG 03/09) prepared from ATCC 55075 reference strain (FLI, Jena, Germany) was added to each well. The plates were shaken on a shaker and incubated in a wet chamber in the incubator at 37 °C for 20–24 h. Test sera were analyzed in parallel with *Brucella*-positive (KS Bruc 03/04) and *Brucella*-negative (KS Bruc 03/04) control sera and *Y. enterocolitica* weak and strong positive sera. After incubation, agglutination titers were determined and expressed as the reciprocal of the final dilution in which agglutination was observed. The results were evaluated as follows: for negative, occurrence of a clear red button/knob of antigen at the bottom; for positive, evaluated according to the degree of agglutination thus (1) film without red center—4+, (2) film with small red center—3+, (3) film with moderate red center—2+, (4) film with intensive/strong red center—1+, and (5) clear red button at bottom—0.

2.12 Indirect ELISA for *Y. enterocolitica* antibodies

Indirect ELISA using PIGTYPE® YOPSCREEN test kit (Qiagen® GmbH, Hilden, Germany) was used to detect antibodies to pathogenic *Yersiniae* in the pig sera that were positive by SAT. The test was performed according to the manufacturer's instructions. The results were validated by testing the Uganda samples in parallel with archived *Y. enterocolitica* antibody-positive field samples (13RB4901 FLI, 13RB4902 FLI, and 13RB4903FLI).

2.13 Molecular detection of *Brucella*

2.13.1 DNA preparation from sera

DNA was extracted and purified from porcine serum using DNeasy Blood and Tissue Kit (QIAGEN, Germany) according to the instructions of the manufacturer using QIAcube. The DNA concentration was determined photometrically using a Nano Drop ND-1000 UV-vis spectrophotometer (Nano-Drop Technologies, Wilmington, USA).

2.13.2 Real-time PCR

Multiplex real-time PCR was used for detection of the genus-specific *Brucella* cell surface salt extractable bcsp31 kDa protein gene, *Brucella abortus* alkB gene and *Brucella melitensis* BMEI1162 gene from extracted DNA (Probert et al. 2004). The *B. abortus* and *B. melitensis* TaqMan probes target the alkB and BMEI1162 gene, respectively. PCR reaction was performed using primer and probe set is shown in Table 1 (Jena Bioscience GmbH, Germany). The 25 μl multiplex PCR mixture consisted of 2 × TaqMan™ Environmental master mix (Applied Biosystems, NJ USA), 200 nM of each primer, 100 nM of each probe, and 5 μl of template DNA. Amplification and real-time fluorescence detection was performed on a Mx3000P thermocycler system (Stratagene, La Jolla, Canada) using the following reaction conditions: 2 min denaturation at 50 °C, polymerase activation step at 95 °C for 10 min, followed by 50 cycles of 95 °C for 25 s and 57 °C for 1 min. The samples scored positive were confirmed by visual inspection of the graphical plots showing cycle numbers versus fluorescence values. A sample with a fluorescence signal 30 times greater than the mean standard deviation in all wells over cycles 2 through 10 was considered a positive result, whereas a sample yielding a fluorescence signal less than this threshold value was considered a negative result. Cycle threshold values below 40 cycles were interpreted as positive.

Table 1 Oligonucleotide primers and probes used in the real-time multiplex PCR assay for the detection of *Brucella* spp., *B. abortus*, and *B. melitensis*

Target	Primer	
Brucella spp.	5′GCTCGGTTGCCAATATCAATGC 3′	Forward
	5′GGGTAAAGCGTCGCCAGAAG 3′	Reverse
	FAM-AAATCTTCCACCTTGCCCTTGCCATCA-BHQ1	Probe
B. abortus	5′GCGGCTTTTCTATCACGGTATTC 3′	Forward
	5′CATGCGCTATGATCTGGTTACG 3′	Reverse
	HEX-CGCTCATGCTCGCCAGACTTCAATG-BHQ1	Probe
B. melitensis	5′AACAAGCGGCACCCCTAAAA 3′	Forward
	5′CATGCGCTATGATCTGGTTACG 3′	Reverse
	CY5-CAGGAGTGTTTCGGCTCAGAATAATCCACA-BHQ2	Probe

FAM carboxyfluorescein, *HEX* hexachlorofluorescein, *BHQ1* Black Hole Quencher 1, *BHQ2* Black Hole Quencher 2

Since age has been alluded to as a significant risk factor for bovine brucellosis (Muñoz et al. 2012), we performed real-time PCR on almost all (86.57 %, $n = 58$) "village boar" samples. Additionally, ELISA positive and other sera, amounting to 12 % of 1193 on farm serum samples, were tested by real-time PCR.

2.14 Data analysis

Standard statistical analyses were performed using Microsoft Office Excel 2010. The disease prevalence was calculated by dividing the number of positive samples by the total number of samples tested. Test sensitivity (Se) of 75 % and specificity (Sp) of 98 % for i-ELISA (Muñoz et al. 2012) were used to calculate apparent prevalence of antibodies against *Brucella* infection according to Greiner and Gardner (2000).

3. RESULTS

3.1 Abattoir seroprevalence of porcine brucellosis

Of the 472 serum samples tested, 332, 2, and 138 sera were collected from pigs originating from Masaka, Mukono, and Kamuli districts, respectively. None of the serum samples was positive for *Brucella* antibodies, indicating 0 % apparent abattoir prevalence (95 % CI: 0.008) of porcine brucellosis in the three districts, suggesting absence of active *Brucella* exposure.

3.2 On-farm seroprevalence of porcine brucellosis

A total of 1 (0.28 %), 0 (0 %), and 1 (0.25 %) serum samples were putatively seropositive for *Brucella* in Masaka, Mukono, and Kamuli districts, respectively (Table 2). The positive sample from Masaka was detected by Ingezim ELISA and that from Kamuli by ID. Vet Screen®. There was no agreement between the results of the two tests on Masaka samples. The Ingezim ELISA kit was not employed for testing samples from Mukono and Kamuli districts.

Table 2 Seroprevalence of porcine brucellosis in Masaka, Mukono, and Kamuli districts as detected by indirect ELISA

District	Village	Sub-county	Number samples[a]	Number brucellosis positive[b]	
				Makerere (Ingezim porcine iELISA)	FLI (ID. Vet Screen iELISA)
Masaka	Kisoso	Kkingo	43 (2)	1	0
	Ssenya	Kkingo	34 (0)	0	0
	Lukindu	Kyanamukaka	37 (0)	0	0
	Kanoni-Bukunda	Kyanamukaka	50 (1)	0	0
	Senyange A	Ivan Kakembo	50 (0)	0	0
	Kyamuyimbwa Kikalala	Kabonera	35 (0)	0	0
	Butego	Katwe-Butego	27 (0)	0	0
	Kijjabwemi	Kimanya-Kyabakuza	43 (0)	0	0
	Kyabakuza B	Kimanya-Kyabakuza	35 (0)	0	0
Mukono	Kazo/Kalagala	Tenjeru	52 (1)	Nd	0
	Nsanja/Gonve	Tenjeru	43 (0)	Nd	0
	Bugoye/Kabira	Tenjeru	48 (0)	Nd	0
	Kyoga	Kyampisi	52 (0)	Nd	0
	Dundu	Kyampisi	61 (0)	Nd	0
	Kitete	Mukono TC	55 (0)	Nd	0
	Joggo	Mukono TC	64 (0)	Nd	0
Kamuli	Balubweneiwa	Bugulumbya	46 (0)	Nd	0
	Bukyonza B	Bugulumbya	30 (0)	Nd	0
	Butabala	Kitayunjwa	40 (0)	Nd	0
	Isingo A	Namwendwa	70 (0)	Nd	0
	Ntansi	Butansi	105 (0)	Nd	1
	Kantu zone	Butansi	110 (0)	Nd	0
Total	22		1130 (4)	1	1

Nd not done

[A]In brackets are the number of village boars sampled from each village as part of random sample

[b]Sites where samples were tested

3.3 Brucellosis status among purposively sampled village boars

Overall, 63 samples from purposively sampled village boars were analyzed; of these, 27, 25, and 13 were from Masaka, Mukono, and Kamuli, respectively. In total, 0 (0.0 %), 0 (0 %), and 1 (7.69 %) serum samples were putatively seropositive for *Brucella* in Masaka, Mukono, and Kamuli districts, respectively, as detected by ID. Vet Screen® (Table 3).

Table 3 Porcine brucellosis in purposively sampled village boars from Masaka, Mukono, and Kamuli districts as detected by indirect ELISA

District	Village	Sub-county	Number of village boar samples	Number brucellosis positive[a]	
				Makerere (Ingezim porcine iELISA)	FLI (ID. Vet Screen iELISA)
Masaka	Kisoso	Kkingo	2	Nd	0
	Ssenya	Kkingo	5	Nd	0
	Lukindu	Kyanamukaka	1	Nd	0
	Kanoni-Bukunda	Kyanamukaka	4	Nd	0
	Senyange A	Ivan Kakembo	3	Nd	0
	Kyamuyimbwa Kikalala	Kabonera	8	Nd	0
	Butego	Katwe-Butego	1	Nd	0
	Kijjabwemi	Kimanya-Kyabakuza	2	Nd	0
	Kyabakuza B	Kimanya-Kyabakuza	1	Nd	0
Mukono	Kazo/Kalagala	Tenjeru	8	Nd	0
	Nsanja/Gonve	Tenjeru	4	Nd	0
	Bugoye/Kabira	Tenjeru	0	Nd	0
	Kyoga	Kyampisi	2	Nd	0
	Dundu	Kyampisi	2	Nd	0
	Kitete	Mukono TC	3	Nd	0
	Joggo	Mukono TC	4	Nd	0

Kamuli	Balubweneiwa	Bugulumbya	2	Nd	0
	Bukyonza B	Bugulumbya	1	Nd	0
	Butabala	Kitayunjwa	1	Nd	1
	Isingo A	Namwendwa	0	Nd	0
	Ntansi	Butansi	9	Nd	0
	Kantu zone	Butansi	0	Nd	0
Total	22		63		1

Nd not done

[a]Sites where samples were tested

3.4 Testing of samples with complement fixation test

For confirmation, the two samples that presumptively tested positive by indirect ELISA (ID. Vet Screen®) were subjected to CFT analysis at FLI. Sera were tested in parallel with FLI *Brucella* antibody strong positive (14RB7378) and weak positive (14RB7379) samples. Both putatively positive samples from Ugandan swine tested negative for *Brucella* antibodies by CFT.

3.5 Ugandan pigs positive for antibodies against *Y. enterocolitica*

To check if the detected antibodies were due to cross-reactivity to *Y. enterocoltica*, the two putative positive samples were subjected to slow agglutination test for *Y. enterocolitica* 0:9 antibodies. Both samples from Uganda (14RB7097 and 14RB7324) tested positive for *Y. enterocolitica* antibodies by slow agglutination test, with titers of 40/2 and 40/3, respectively, indicating a 0.17 % prevalence of *Y. enterocolitica* antibodies in the pigs tested. To confirm their serological status, these samples were tested for *Y. enterocolitica* antibodies by indirect ELISA using PIGTYPE® YOPSCREEN kit (Qiagen® GmbH, Hilden, Germany). One sample (14RB7097) tested positive while the other (14RB7324) was negative.

3.6 Testing of serum samples with real-time PCR

To confirm the absence of *Brucella* infection, the *Y. enterocolitica*-positive sample and 143 other sera from Uganda [that comprised of 58 samples from boars (86.57 % of all boars), and 45 and 40 randomly selected sera that showed higher and low ODs in indirect ELISA, respectively, representing 12.07 % (n = 144) of the total serum samples] were tested for brucellosis using real-time PCR. The other *Y. enterocolitica*-positive sample could not be tested by real-time PCR due to lack of serum. No *Brucella*-specific DNA signal was detected in all samples tested.

3.7 Statistical analysis

Considering sample size to detect disease for infinite population: $n = ln(alpha) / ln(1 - p)$, where p = the minimum prevalence, and if assuming the p to be 5 % then $n = ln(0.05) / ln(0.95) = 58.4$. This work analyzed 1130 randomly collected samples, so reversing the equation $ln(alpha)/n = ln(1 - p)$ gives

$$\begin{aligned} ln(alpha)/n &= ln(1-p) \\ ln(1-p) &= ln(0.05)/1130 = -0.00265 \\ p &= 1-(exp(-0.00265)) = 1-0.9973 = 0.0027. \end{aligned}$$

Therefore, the sample size used in this study was sufficient to find the disease if it was present in 0.27 % of the population.

Assuming a perfect test (Se = 1), the prevalence in the population must be maximum of 0.27 % for us not to have found it in 1130 samples. The Ingezim Brucella Porcine Ab i-ELISA shows a sensitivity of 75 %; therefore, the prevalence will be no higher than 0.81 %.

4. DISCUSSION

To our knowledge, this is the first study to investigate the occurrence of brucellosis in pigs in Uganda or other East African countries. None of the sera tested was positive for porcine brucellosis. Under this type of production system, we expect a maximum *B. suis* prevalence of less than 1 % at 95 % confidence level. Since this study included pigs from Mukono and Masaka, two of the districts with the highest density of pigs in Uganda, and failed to detect disease there, it is likely to be absent.

Our findings that *Brucella* antibodies were absent from a large and representative sample of pigs in the major pig-producing districts of Uganda are consistent with earlier reports of 0 % (Cadmus et al. 2006) and 0.6 % (Onunkwo et al. 2011; Nwanta et al. 2011) prevalence of porcine brucellosis in Nigeria. A 0 % prevalence was also found in commercial piggeries in Zambia (Stafford et al. 1992). Our results however differed from those of Ngbede et al. (2013) who reported a high seroprevalence of porcine brucellosis (30.6 %) in Benue State, Nigeria. All the serological studies reporting the occurrence of porcine brucellosis in Nigeria used

RBT for testing whereas our study employed multiple tests. RBT reportedly has high sensitivity (100 %) but is less specific (84 %) (Akhtar et al.2010).

The finding of 0.17 % prevalence of *Y. enterocolitica* antibodies in our study population is of public health significance. *Y. enterocolitica* is an important zoonotic agent of global concern (Bhaduri et al. 2005; Ifeoma 2013) that can cause serious human suffering ranging from fever, abdominal pain, diarrhea to more serious complications such as appendicitis and erythema nodosum (Bari et al. 2011). Pigs are reported to be the main reservoir of these organisms (Adesiyun and Krishnan 1995; Bhaduri et al. 2005; Fredriksson-Ahomaa et al. 2010). To the best of our knowledge, this is the first report on the occurrence of *Y. enterocolitica* exposure in Ugandan pigs and calls for research on the infection and risks associated with this pathogen. Further studies are needed to investigate the prevalence and impact of *Y. enterocolitica* on Ugandan pigs and to investigate other causes for the high reported levels of abortion and infertility in pigs.

ACKNOWLEDGMENTS

The field research was funded by the Consultative Group for International Agricultural Research (CGIAR) Research Program on Agriculture for Nutrition and Health and the "Safe Food, Fair Food" project (funded by the Federal Ministry for Economic Cooperation and Development, Germany) in cooperation with the CGIAR Research Program on Livestock and Fish through the "Smallholder Pig Value Chains Development" project in Uganda (funded by IFAD-EU). Additional funding was provided by the German Academic Exchange Service (DAAD) through an in-region post-doctoral fellowship in partnership with the International Livestock Research Institute (ILRI), The World Academy of Sciences (TWAS)/German Research Foundation (DFG), and the Friederich-Loeffler-Institute (FLI), Germany. We would like to thank Dr Jane Poole (ILRI) and Dr Ian Dohoo (Prof. em. University of Prince Edward Island, Canada) for advice on the statistical analysis. Moreover, our appreciation goes to Uta Pfeil, Michaela Ganss, and Angella Musewa for technical assistance and Dr. Francis Mutebi and the veterinary officers at Wambizzi Cooperative Society Ltd. for their assistance and cooperation during abattoir sampling. We extend our utmost appreciation to the field veterinary staff and farmers in Masaka, Mukono, and Kamuli districts for the assistance and support during on-farm sample collection.

REFERENCES

1. Adesiyun, A.A. and Krishnan C., 1995. Occurrence of *Yersinia enterocolitica* 0:3, *Listeria monocytogenes* 0:4 and thermophilic *Campylobacter* spp. in slaughter pigs and carcasses in Trinidad. Food Microbiology, 12(2), 99-10
2. Akhtar, R., Chaudhry, Z.I., Shakoori, A.R., Mansoor ud Din, A. and Aslam, A., 2010. Comparative efficacy of conventional diagnostic methods and evaluation of polymerase chain reaction for the diagnosis of bovine brucellosis. Veterinary World, 3(2), 53-56
3. Bari, L., Hossain, M.A., Isshiki, K. and Ukuku, D.O., 2011. Behavior of *Yersinia enterocolitica* in Foods. Journal of Pathogens. doi: 10.4061/2011/420732.
4. Bhaduri, S., Wesley, I.V. and Bush, E.J., 2005. Prevalence of Pathogenic *Yersinia enterocolitica* Strains in Pigs in the United States. Applied and Environmental Microbiology, 71(11), 7117-7121
5. Burkhardt, F., 1992. Mikrobiologische Diagnostik(Georg Thieme Verlag Stuttgart–New York)
6. Cadmus, S.I.B., Ijagbone, I.F., Oputa, H.E., Adesokan, H.K. and Stack, J.A., 2006. Serological survey of Brucellosis in Livestock Animals and Workers in Ibadan, Nigeria. African Journal of Biomedical Research, 9, 163-168
7. Dione, M.M., Ouma, E.A., Roesel, K., Kungu, J., Lule, P. and Pezo, D., 2014. Participatory assessment of animal health and husbandry practices in smallholder pig production systems in three high poverty districts in Uganda. Preventive Veterinary Medicine, 117(3–4), 565-576
8. Ducrotoy, M., Bertu, W.J., Matope, G., Cadmus, S., Conde-Álvarez, R., A.M. Gusi, A.M., Welburn, S., Ocholi, R., Blasco, J.M. and Moriyón, I., 2015. Brucellosis in Sub-Saharan Africa: Current challenges for management, diagnosis and control. Acta Tropica. doi:10.1016/j.actatropica.2015.10.023
9. FAOSTAT/FAO, 2011. Food Balance/Food Supply - Livestock and Fish Primary Equivalent. Retrieved January 5, 2016, from http://faostat3.fao.org/browse/FB/CL/E.
10. Fredriksson-Ahomaa, M., Meyer, C., Bonke, R., Stüber, E. and Wacheck, S., 2010. Characterization of *Yersinia enterocolitica* 4/O:3 isolates from tonsils of Bavarian slaughter pigs. Letters in Applied Microbiology, 50(4), 412-418
11. Godfroid, J., Al Dahouk, S., Pappas, G., Roth, F., Matope, G., Muma, J., Marcotty, T., Pfeiffer, D. and Skjerve, E., 2013. A "One Health" surveillance and control of brucellosis in developing countries: moving away from improvisation. Comparative Immunology, Microbiology and Infectious Diseases, 36(3), 241-248
12. Greiner, M. and Gardner, I.A., 2000. Application of diagnostic tests in veterinary epidemiologic studies. Preventive Veterinary Medicine, 45(1–2), 43-59
13. Ifeoma, M.M., 2013. Incidence and Public Health Significance of *Yersinia enterocolitica* in Enugu Urban-Enugu State, Nigeria. Journal Of Environmental Science, Toxicology And Food Technology, 6(3), 5-56
14. Makita, K., Fèvre, E.M., Waiswa, C., Kaboyo, W., De Clare Bronsvoort, B.M., Eisler, M.C. and Welburn, S.C., 2008. Human brucellosis in urban and peri-urban areas of Kampala, Uganda. Annals of the New York Academy of Sciences, 1149, 309-311
15. McDermott, J., Grace, D. and Zinsstag, J., 2013. Economics of brucellosis impact and control in low-income countries. Revue scientifique et techniquecpcOffice International des Epizooties, 32(1), 249-261
16. Muñoz, P.M., Blasco, J.M., Engel, B., de Miguel, M.J., Marín, C.M., Dieste, L. and Mainar-Jaime, R.C., 2012. Assessment of performance of selected serological tests for diagnosing brucellosis in pigs. Veterinary Immunology and Immunopathology, 146(2), 150-158
17. Nabukenya, I., Kaddu-Mulindwa, D. and Nasinyama, G.W., 2013. Survey of *Brucella* infection and malaria among Abattoir workers in Kampala and Mbarara Districts, Uganda. BMC Public Health. 13:901. doi: 10.1186/1471-2458-13-901
18. Nakavuma, J.K. and Opuda-Asibo, J., 1999. Sero study of *Brucella abortus* in cattle and goat in central and southern Uganda. Uganda Journal of Agricultural Sciences, 4, 13-18

19. Ngbede, E.O., Momoh, A.H., Ruben Sylvester Bala, R.S., Madaki, B.D. and Maurice, N.A., 2013. An Abattoir-Based Study on Serodiagnosis of Swine Brucellosis in Makurdi, Benue State, North-Central Nigeria. Journal of Advanced Veterinary Research, 3(2), 57-59
20. Nwanta, J. A., Shoyinka, S.V.O., Chah, K.F., Onunkwo, J.I., Onyenwe, I.W., Eze, J.I., Iheagwam, C. N., Njoga, E.O., Onyema, I., Ogbu, K.I., Mbegu, E.C., Nnadozie, P.N., Ibe E.C. and Oladimeji, K.T. (2011). Production characteristics, disease prevalence, and herd-health management of pigs in Southeast Nigeria. Journal of Swine Health and Production, 19(6), 331-339
21. Onunkwo, J.I., Njoga, E.O., Nwanta, J.A., Shoyinka, S.V.O., Onyenwe, I.W. and Eze J.I., 2011. Serological Survey of Porcine *Brucella* Infection in South East, Nigeria. Nigerian Veterinary Journal, 32(1), 60-62
22. Ouma, E., Dione, M., Lule, P., Roesel, K. and Pezo, D., 2014: Characterization of smallholder pig production systems in Uganda: constraints and opportunities for engaging with market systems. Livestock Research for Rural Development, 26(3), Article #56. Retrieved January 5, 2016, from http://www.lrrd.org/lrrd26/3/ouma26056.htm
23. Perry, B. and Sones, K., 2007. Poverty reduction through animal health. Science, 315 (5810), 333—334, doi: 10.1126/science.1138614
24. Probert, W.S., Schrader, K.N., Khuong, N.Y., Bystrom, S.L. and Graves, M.H., 2004. Real-time Multiplex PCR Assay for Detection of *Brucella* spp., *B. abortus*, and *B. melitensis.* Journal of Clinical Microbiology, 42(3), 1290-1293
25. Radostits, O.M., Gay, C.C., Blood, D.C., and Hinchcliff, K.W., 2000. Veterinary Medicine: A textbook of the diseases of cattle, sheep, pigs, goats and horses. (W.B. Saunders Company Ltd, New York)
26. Scientific Opinion of the Panel on Animal Health and Welfare (AHAW) on a request from the Commission on porcine brucellosis (*Brucella suis*)., 2009. The EFSA Journal, 1144, 1-112. Retrieved January 5, 2016 from http://www.efsa.europa.eu/sites/ default/ files/scientific_output/files/main_documents/1144.pdf
27. Stafford, K., Tafford, Y., Paton, D. and Gamble, P., 1992. Antibodies to some swine disease in commercial piggeries in Central Zambia. Revue d'élevage et de médecine vétérinaire des pays tropicaux, 45(3–4), 229-230
28. Swai, E.S. and Schoonman, L., 2012. A survey of zoonotic diseases in trade cattle slaughtered at Tanga city abattoir: a cause of public health concern. Asian Pacific Journal of Tropical Biomedicine, 2(1), 55-60
29. OIE Terrestial Manual, 2012. Chapter 2.4.3. Bovine brucellosis, 616-650
30. Thrusfield, M., 1995. Veterinary Epidemiology. Blackwell Science Ltd, Oxford.
31. Uganda Bureau of Statistics/Ministry of Agriculture, Animal Industry & Fisheries, 2009. A summary report of the livestock census, 2008.

CHAPTER 15

Effects of Chronic D-Serine Elevation on Animal Models of Depression and Anxiety-Related Behavior

David-Marian Otte[1], Maria Luisa Barcena de Arellano[1], Andras Bilkei-Gorzo[1], O¨ nder Albayram1, Sophie Imbeault[1], Haang Jeung[2], Judith Alferink[1,2], Andreas Zimmer[1*]

1 Institute of Molecular Psychiatry, University of Bonn, Bonn, Germany

2 Department of Psychiatry, University of Bonn, Bonn, Germany

ABSTRACT

NMDA receptors are activated after binding of the agonist glutamate to the NR2 subunit along with a co-agonist, either L-glycine or D-serine, to the NR1 subunit. There is substantial evidence to suggest that D-serine is the most relevant co-agonist in forebrain regions and that alterations in D-serine levels contribute to psychiatric disorders. D-serine is produced through isomerization of L-serine by serine racemase (Srr), either in neurons or in astrocytes. It is released by astrocytes by an activity-dependent mechanism involving secretory vesicles. In the present study we generated transgenic mice (SrrTg) expressing serine racemase under a human GFAP promoter. These mice were biochemically and behaviorally analyzed using paradigms of anxiety, depression and cognition. Furthermore, we investigated the behavioral effects of long-term administration of D-serine added to the drinking water. Elevated brain D-serine levels in SrrTg mice resulted in specific behavioral phenotypes in the forced swim, novelty suppression of feeding and olfactory bulbectomy paradigms that are indicative of a reduced proneness towards depression-related behavior. Chronic dietary D-serine supplement mimics the depression-related behavioral phenotype observed in SrrTg mice. Our results suggest that D-serine supplementation may improve mood disorders.

1. INTRODUCTION

The ionotropic N-methyl-D-aspartic acid (NMDA) receptor is activated by glutamate [1], if an allosteric co-activation site is occupied by either glycine or D-serine [2]. While glycine is probably the main co-agonist in the spinal cord and in the hindbrain, several lines of evidence indicate that D-serine is the main co-agonist in the forebrain [3]–[8]. Thus, glycine concentrations at forebrain glutamatergic synapses are low, most probably due to the presence of high-affinity glycine transporters in closely apposed astrocytes [9]–[14]. In contrast, the forebrain contains relatively high levels of D-serine [6], [7], [15], [16], particularly in regions with high NMDA receptor expression [6], [7], [16]. For instance, D-serine is found at high concentrations in hippocampal astrocytes located in close proximity to CA1 apical dendrites, which contain a high NMDA receptor density [6]. D-serine is also a more potent activator of NMDA receptors than glycine when examined in heterologous expression systems [17], [18], or in brain slices [9]. Recent studies demonstrated that the NMDA receptor glycine/D-serine site is most probably not fully saturated in many brain regions (i.e. hippocampus, thalamus, neocortex, brain stem and retina) [9], [10], [13], [14], [19]–[22]. Fine-tuning of D-serine levels may therefore serve as a mechanism to modulate NMDA receptor activity. Most recently, it has been shown that synaptic and extrasynaptic NMDARs are gated by D-serine and glycine, respectively. The regionalized availability of the coagonists matches the favored affinity of synaptic NMDARs for D-serine and extrasynaptic NMDARs for glycine. Interestingly, Papouin et al. demonstrated that long-term potentiation and NMDA-induced neurotoxicity rely on synaptic NMDARs only while long-term depression requires both [23]. In the mammalian brain, D-serine is produced from L-serine by the enzyme serine racemase [3], [24]–[26]. Neurons are probably the main source of serine racemase and D-serine in mammals, because serine racemase is predominantly localized in this cell type in the brain. Serine racemase has also been found to a lesser extent in astrocytes [27]). Interestingly, despite being less expressed, astrocytic D-serine seems to be essential for hippocampal LTP [28].

Alterations in glutamatergic signaling through NMDA receptors have been implicated in the pathology of psychiatric disorders and thus a tight regulation of D-serine levels may be important [29], [30]. Indeed, it has been shown that patients with mood disorders display increased post mortem brain glutamate levels [31] and schizophrenic patients exhibit reduced D-serine levels in the cerebrospinal fluid (CSF) [32].

Glycine site agonists were found to be beneficial in several chronic neurological disorders, including Alzheimer's disease and schizophrenia [33]–[36]. Additionally, D-cycloserine, a partial agonist at the strychnine-insensitive glycine site of the NMDA-receptor [37], facilitates declarative learning and hippocampal activity in healthy humans [38]. A role of D-serine in the modulation of affective behaviors was recently demonstrated using a serine racemase knockout mouse model with dramatically reduced D-serine levels. These mice exhibited elevated levels of anxiety [39] and a decline of cognitive functions [39]. Furthermore, it has been shown that an acute treatment with D-serine produces antidepressant effects in rodents [40].

We have now generated a mouse strain with chronically increased D-serine levels by transgenic expression of serine racemase in astrocytes (SrrTg). These animals were analyzed in behavioral paradigms of depression and cognition to determine the behavioral effects of increased D-serine levels. SrrTg mice showed a reduced depression-related behavior in the ultrasonic vocalization, the forced swim and the novelty-suppressed feeding test. Higher brain serine racemase expression did not alter the animals' learning ability, as evidenced by the results of the operant learning paradigm and the water maze test. Interestingly, D-serine added to the drinking water in wild type (Wt) mice produced behavioral changes that were similar to the SrrTG phenotype.

2. MATERIALS AND METHODS

2.1 Generation of transgenic mice

For the generation of the expression construct, we obtained a vector containing the complete murine serine racemase ORF (BC011164-NCBI) under the control of the CMV promoter (CMV-SPORT6 backbone; RZPD, Germany, order nr. IRAKp961M0624Q). A 2.4 kb *Dra*III/*Eco*RV fragment harboring the serine racemase ORF, the SV40 intron and polyA were cut out and cloned into a *NotI/BamHI* restricted and blunted pGFAP-GFP vector [41]. The resulting GFAP-serine racemase vector was cut with BglII/DraIII to release a 5.1 kb fragment containing the GFAP promoter, the serine racemase ORF and the SV40 intron and polyA (Fig 1A). This fragment was gel-purified, diluted in oocyte injection buffer (5 mM Tris, pH 7.4, 0.2 mM EDTA, 100 mM NaCl) to 3 ng/µl and injected into pronuclei of fertilized oocytes from CD1 mice. The transgenic founder mice were crossed to CD1 Wt mice. Transgenic lines were maintained and analyzed as hemizygous lines (Srr2 and Srr12) using CD1 Wt mice for breeding. Genotyping was done by PCR with genomic DNA obtained from mouse tail biopsies using the primers SrrTg F (5′TTCGAGGTGCCCTTAATGCC′3) and SrrTg R (5′AGAGCTTGGC CATGGTTTCC′3).

2.2 Total RNA preparation and RT PCR

Mouse brain tissues from Srr12, Srr2 and Wt mice were rapidly dissected, snap frozen in isopentane and stored at −80°C. Total RNA was prepared using the Trizol method (Invitrogen, Germany). Five µg RNA and 0.5 µg oligo dT (20) primers (Invitrogen) were heated at 70°C for 4 min, chilled on ice and reverse transcribed in a total volume of 20 µl containing 4 µl first strand buffer (Invitrogen), 2 µl 0.1 M DTT, 1 µl 10 mM dNTPs, 0.5 µl RNase OUT (Invitrogen), and 200 U Superscript II reverse transcriptase (Invitrogen) at 42°C for 50 min.

2.3 Taqman analysis

Real-time quantitative PCR was performed using an ABI 7900 sequence detector (Perkin Elmer, MA, USA) on cDNA samples. The PCR reaction was carried out at 50°C for 2 min, 95°C for 10 min followed by 40 cycles of 95°C for 15 s, then 60°C for 1 min using the manufacturers Universal PCR Master Mix (Perkin Elmer, MA, USA). Taqman primer and probe sets were purchased from Applied Biosystems: Mm00489125_m1 for serine racemase and for glyceraldehyde-3-phosphate-dehydrogenase (GAPDH). GAPDH was used to normalize for the amount of sample in a given reaction. Results are expressed as fold change calculated relative to CD1 Wt after normalization.

2.4 Western blot analysis

Brains were homogenized in 10 mM Tris HCl pH 8, 150 mM NaCl, 5 mM EDTA, 1% NP-40, 0.5% sodium deoxycholate, and 0.1% SDS containing Complete Mini protease inhibitor cocktail (Roche). Protein concentration was determined using the BCA Protein Assay (Pierce). 20 µg of the protein extracts were separated by 12% SDS-PAGE and probed with mouse anti-serine racemase polyclonal antibody (1:500; BD, overnight at 4°C) followed by rabbit anti-mouse peroxidase-conjugated antibody (1:10,000; Jackson ImmunoResearch, 1 h at room temperature) and then exposed to enhanced chemiluminescence substrate (ECL; Pierce) for 5 minutes. After stripping, the blot was labeled with a mouse anti-β-actin antibody (Sigma A2228, 1:10000, 2 h at room temperature) as a housekeeping gene followed by a goat □nti-mouse peroxidase-conjugated antibody (Jackson Lab, 1:3000, 1 h at room temperature) and again incubated with ECL for 5 minutes. The blots were analyzed using the ChemiDoc™ MP Imagine System from Bio-Rad and the ImageLab 4.01 software.

2.5 Immunohistochemistry

All procedures were performed at room temperature and all solutions were dissolved in 1X PBS (Gibco 18912-014) unless otherwise indicated. All chemicals were purchased from Sigma-Aldrich unless otherwise specified. Tissues were obtained from 8–9 week old mice transcardially perfused with PBS followed by 4% paraformaldehyde and post-fixed between 6–16 hours at 4°C. Organs were cryoprotected in 20% sucrose and 20 μm-thick slices were cut using a cryostat (Leica CM3050S), mounted onto glass slides (FisherBrand Superfrost Plus) and stored at −80°C until further use.

Sections were thawed for 20 minutes in PBS and permeabilized using 0.2% Triton-X 100 for 20 minutes before they were washed in PBS for 5 minutes. Blocking of unspecific binding sites was done with 5% goat serum (Abcam) for 10 minutes followed by another PBS wash for 5 minutes. Rabbit anti-serine racemase (1:4000, Abcam ab45434; lot434164) and mouse anti-GFAP (1:800, Sigma G3893; lot080M4838) were applied and sections stored in a humid chamber overnight at 4°C. Sections were rinsed 3×5 minutes in PBS prior to incubation in a humid chamber with secondary antibodies (goat anti-rabbit Cy3-conjugated, 1:600; goat anti-mouse Alexa488-conjugated, 1:400; both from Invitrogen) for 1 hour in the dark. Sections were rinsed 3×5 minutes in PBS prior to coverslipping and mounting with Fluoromount G (Southern Biotech). Images were obtained using a Zeiss Imager.M2 epifluorescence microscope equipped with an Axiocam MRm camera and AxioVision software (rel. 4.8). Sections from both genotypes were always processed in parallel and imaged using the same exposure times. ZVI files were converted to TIFF format, deconvolved using the Iterative Deconvolution plugin by Bob Dougherty, and merged in Image J (version 1.44o, NIH). Figures were produced with Photoshop CS4 (v10.0.1, Adobe). Background levels were adjusted identically for both genotypes. Since there is no serine racemase expression in the spleen [42], [43] we used wild type spleen slices to confirm antibody specificity for the rabbit anti-serine racemase antibody (data not shown).

2.6 High performance liquid chromatography analysis (HPLC)

Tissue samples were weighed after extraction and homocysteine as an internal standard was added. After that the samples were homogenized in 10 volumes of 5% trichloroacetic acid (TCA) and the homogenates were centrifuged at 18000×*g* at 4°C for 30 min. To remove TCA, the supernatants were washed three times with water-saturated diethyl ether. The resultant samples were used for HPLC analysis. Amino acid enantiomers were separated by HPLC using a carbon 18 reverse-phase column (250 mm) (Knauer, Advanced scientific instruments, Germany) with fluorimetric detection after derivatization with N-isobutyryl-L-cysteine and O-phthalaldehyde (OPA) as described in [44]. The N-isobutyryl-L-cysteine/ OPA derivatives were immediately applied to the HPLC system (Knauer, Advanced scientific instruments, Germany). Mobile phase was 8% MeCN in 0.1 M sodium acetate buffer (pH 6.0). Amino acids were separated isocratically for 37 minutes and then the column was eluted with 50% H_2O/50% MeCN for 5 minutes. After that, the column was reequillibrated for 15 minutes under the initial conditions. The flow-rate was 0.125 ml/min. Fluorescence detection of each amino acid derivative was carried out at 443 nm with excitation at 344 nm [44]. The absolute L-serine and D-serine levels referring to the internal standard were calculated and data were related to the wet weight of the initial tissue samples.

2.7 Behavioral tests

Experiments were performed with 8–10 weeks old SrrTg from two independent founders (Srr12 and Srr2) and Wt littermates on a CD1 background. Animals received water and food *ad libitum* except during the operant conditioning period and the novelty-suppressed feeding paradigm. Mice were housed in groups of 3–5 animals and kept in a reversed light–dark cycle (with a dark period between 9:00 am and 6:00 pm). Experiments were conducted during the active phase of the animals in a dimly lit, low-noise environment. The experimenter was blind to the genotype. SrrTg mice were always compared to Wt littermates pooled from both transgenic lines. Care of the animals and performance of all experiments followed the guidelines of the German Animal Protection Law and were approved by the legal authorities (LANUV NRW, Permit number: 50.203.2-BN 34 2/06). The animals were left undisturbed for at least one week between testing when used in multiple models. The sequence of testing was as follows: Open-field test, Y-maze test, novelty-induced suppression of feeding and forced swim test.

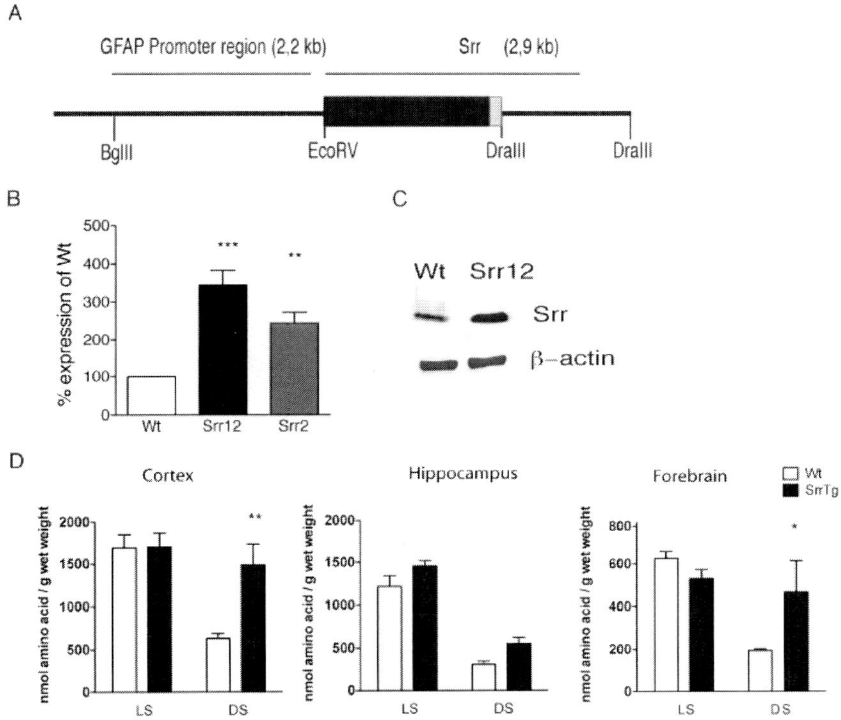

Figure 1. Serine racemase expression and brain D-serine levels in transgenic mice.

(A) Schematic representation of the construct for serine racemase overexpression in glial cells under a GFAP promoter. The GFAP promoter is represented as a black line, the serine racemase ORF as a black bar and the SV40 intron/polyA signal as a grey bar. The cutting sites for the endonucleases BglII, EcoRV and DraIII are listed. (B) Quantitative analysis of serine racemase expression in adult mouse brains by TaqMan® Assay. Serine racemase expression was calculated as ΔCT value, normalized to GAPDH. Shown are the average values of 4 animals, expressed as mean ±SEM. (C) Western blot analysis of serine racemase in whole brain of a Srr12 mouse and a Wt littermate. Images of Western blot analysis of serine racemase and β-actin were performed with specific antibodies using the same extract (n=4). All values are expressed as mean ±SEM of 5 animals; *p<0.05 and **p<0.01. (D) Quantitative analysis of absolute brain D-serine and L-serine levels per wet weight in 5 Srr12 mice and corresponding Wt littermates by HPLC. D-serine levels of cortex, hippocampus and forebrain were measured. Mice were treated with D-serine (D), L-serine (L), glycine (G) and water (W).

In the ultrasonic vocalisation test, bulbectomy-induced hypermotility paradigms and in the operant learning and Morris water maze models we used separate groups of animals that were tested in only one model, respectively. The home cages were brought into the test room at least 30 min before each experiment.

2.8 Forced swim test.

The forced swim test is a commonly used procedure of preclinical screening of drugs for the antidepressant activity [45]. Animals were placed in a Plexiglas cylinder (10 cm inner diameter, 50 cm high) filled with 22–23°C water (20 cm height). The duration of the experiment was 6 min. The immobility time of the animals was evaluated between the 2nd and 6th minute for 4 min. A mouse was judged to be immobile when it remained floating in the water, making only movements necessary to keep its head above the water [45]–[47]. Means and SEM were calculated for each group.

2.9 Open-field test.

Mice were placed into the centre of a dimly lit (20–30 lux) chamber of the open-field apparatus (44×44×30 cm). Movements of the animals were tracked by an automatic monitoring system (TSE Systems, Bad Homburg, Germany) for 30 min. Horizontal motor activity was evaluated by calculating the distance that the animals travelled in the arena. For each group, the mean value and SEM was calculated.

2.10 Novelty-suppressed feeding.

For this test we used 20 Wt and 20 transgenic mice that were food deprived for 24 hr before testing. During testing, animals were placed individually close to a wall of an open-field apparatus lit by normal house lighting (light intensity 300–400 lux in the centre of the box). Six pellets (2.8–3.3 mg) of standard mouse chow were placed in the center of the open field. The latency to eat (nt sniffing or manipulating the pellets) was recorded by an observer who was unaware of the genotype [48].

2.11 Olfactory bulbectomy-induced hyperactivity.

The bulbectomy was performed as described previously [49]. Mice were anesthetized with avertin (2.5%, 20 ml/kg). An incision was made in the skin overlying the skull, and a 2 mm hole was drilled into the skull through the frontal suture. Olfactory bulbs were removed by gentle aspiration. The same procedure was performed on sham-operated animals, except that the olfactory bulbs were not removed. Each animal was housed singly after surgery. Locomotor activity of the bulbectomized and sham-operated animals was studied 14 d after the surgery in a dimly lit, sound-attenuated room. The animals were placed in an open-field arena (45×45×22 cm) and the distance traveled was recorded for 5 min (Actimeter, TSE). After completion of the behavioral experiments, mice were decapitated, brains were removed, and the success of the OBX was evaluated visually.

2.12 Ultrasonic vocalization.

Five to seven day-old pups were removed from their mother, placed individually into a glass jar, and put into a sound isolated box kept at room temperature (23±1°C). These pups emitted characteristic ultrasonic calls, which are known to be reduced by antidepressants [50]–[54]. The duration and number of these calls was registered automatically using the Ultravox system (Noldus, The Netherlands) for 5 min.

2.13 Y- maze.

The Y-maze was composed of three equilaterally intersecting Plexiglas arms (58 cm long ×19 cm wide ×38 cm high). Mice were placed in the apparatus and arm visits were analyzed for 10 minutes using the Videomot 2 (TSE Systems, Bad Homburg, Germany) video observation system. Mean values and SEM were calculated for each group and expressed as percent spontaneous arm visits and the total number of arm visits.

2.14 Operant conditioning test.

The day before and during the operant conditioning test the animals received only 80% of the food consumed under *ad libitum* conditions (Bilkei-Gorzo et al, 2005). Operant conditioning was developed by B.F. Skinner and is a type of learning in which organisms learn to voluntarily respond in a certain way depending on the consequences e.g. rewards [55]. Test cages (17×17×17 cm) were made from transparent plastic material and placed into a larger (50×40×70 cm) wooden box. Each cage contained a nose-poke sensor, a feeder and signal lamps for visual cues. The cages were connected to a computer-controlled central unit (TSE GmbH, Germany). Animals were placed individually into the cages, and number of nose-pokes into the sensor hole was registered for twenty minutes. Each nose-poke resulted in a delivery of a 50 mg food-pellet. The timeout period was 5 seconds and was signaled with a yellow lamp. The animals were tested daily until the variation between the responses on three consecutive days was less than 30%, but for a maximum of 24 days. Means and standard deviation of nose pokes were calculated daily for each group. Throughout the testing period the animals were single housed.

2.15 Morris Water maze test.

In the Morris water maze we assessed hippocampal dependent spatial learning and memory abilities. One week before the experimental period, male mice (9–10 weeks old) were transferred to standard single-mouse cages, maintained at a 12:12-h inverted cycle (light on 07:00 pm–07:00 am), and tested during the dark period. The home cages were brought into the test room at least 30 min before each experiment. Water maze test was conducted with hemizygous transgenic animals, which were compared to their Wt littermates.

A circular pool (diameter 120 cm and height 30 cm) made of black PVC was used in a dimly illuminated room. The pool was filled to a depth of 15 cm with opacified water using non-toxic white paint (24°C). Four orthogonal starting positions were situated around the perimeter of the pool, dividing its surface into four quadrants. A platform in form of a transparent Plexiglas cylinder (15 cm tall and 8 cm diameter) covered with a white aluminum perforated plate (14 cm diameter) was placed in the center of one quadrant, approximately 1.5 cm under the water level and served as an escape platform. The pool was located in a room containing numerous extra-maze visual cues. A camera was fixed on the ceiling above the water maze.

2.16 Water maze procedure.

Each mouse was tested for four consecutive sessions daily over 5–6 days. The hidden platform remained at a fixed spatial location for the entire acquisition period and each mouse was assigned a different escape sector. The mice were released, facing the wall of the maze at each trial session. During the first two days, animals were put into the same starting points (N) for each session. From day 3rd to day 5th, animals were released into the four consecutive positions for each session (N, E, S, W).

A trial ended when the mouse reached the hidden platform and managed to remain there for 5 sec. If a successful escape did not take place within 70 sec, the mouse was guided to the platform and the trial was recorded as an escape failure with an arbitrary latency of 70 sec. The tested mouse was left for a 15 sec inter-trial interval in a dark dry container.

2.17 D-serine supplementation

D-serine (Sigma), L-serine (Sigma) and glycine (Sigma) were dissolved in water and the pH was adjusted to 7.0. Wt mice consumed water, or water supplemented with 350 mg/L D-serine, L-serine, or glycine for 35 days. This resulted in a daily dose of 58 mg/kg D-serine per body weight (average weight: 30 g; average drinking volume: 5 ml/day). Preliminary studies using a two bottle choice test revealed no differences in preference for water alone or water supplemented with the amino acids used here. Although D-serine is more brain penetrant than L-serine and glycine, we used the latter amino acids to control for possible unspecific effects of the food supplementation. The liquid consumption was documented every second day before the liquid was replaced with freshly prepared solution. After the behavioral analyses all mice were sacrificed and the brains prepared, dissected and submitted to HPLC analysis.

2.18 Statistical analysis

Behavioral data analysis for males and females are provided separately even though there was no effect of sex nor a sex x genotype interaction. Data were analyzed, where appropriate, by paired or unpaired t-test, one-way, two-way/three-way or repeated measures analysis of variance (ANOVA) followed by either Dunnett's- or Bonferroni *post hoc* tests. If the data were not normally distributed, we applied non-parametric tests for analysis (Kruskal–Wallis test). Analysis was done by using the GraphPad Prism version 4.00 for MAC or by Statistica version 7.1. Unless otherwise indicated, data are expressed as mean ±SEM, and statistical significance was considered when $p<0.05$.

3. RESULTS

3.1 Generation of GFAP-Srr transgenic mouse strains

Fig. 1A). Quantitative real-time PCR with mRNA from whole brain extracts showed that both lines expressed the transgene, with higher levels in the Srr12 strain (Fig. 1B, $p<0.001$). *Western* blot analysis of whole brain protein lysates from Srr12 transgenic and Wt animals also revealed a higher serine racemase protein content (Fig. 1C) and concomitantly increased absolute D-serine levels in different brain regions, except for the forebrain of Srr12 mice (Fig. 1D, two-way ANOVA: hippocampus: genotype $F(1,16)=8.89$, $p=0.0088$; cortex: genotype $F(1,16)=6.48$, $p=0.0216$; amino acid

$F(1,16)=14.10$, $p=0.0017$, interaction: $F(1,16)=6.08$, 0.0254; forebrain: genotype $F(1,13)=1.41$, $p=0.2556$; amino acid $F(1,13)=11.19$, $p=0.0053$, interaction: $F(1,13)=6.22$, $p=0.0269$).

Immunofluorescence analysis of the Srr12 mouse line revealed stronger immunoreactivity in the cortex of Srr12 mice, especially within the neuropil (Fig. 2A, main panels). Although $GFAP^+$ cells express serine racemase in wildtype animals, the expression is much higher in Srr12 mice as demonstrated for the cortex (Fig. 2A/B). This is quite noticeable in the Srr12 cortex where serine racemase immunoreactivity localizes to $GFAP^+$ astrocytes (arrows) whereas Wt astrocytes seemed to be at least partially devoid of serine racemase expression (asterisks).

We observed no gross abnormalities, reproductive deficits or behavioral abnormalities in transgenic animals by routine handling and therefore conducted a series of behavioral tests to determine if the elevated D-serine levels affect distinct mouse behaviors. Although we examined both lines in parallel in many behavioral tests, we conducted some experiments only with Srr12 mice, because they generally showed a slightly more pronounced phenotype.

3.2 Mood-related behaviors in SrrTg mice

In the Porsolt forced swim test, we observed a significantly reduced immobility time in both transgenic lines compared to Wt animals (Fig. 3A, Kruskal–Wallis test=28.29, $p<0.0001$). We did not find any sex (two-way ANOVA $F(1,56)=0.15$, $p=0.7$) or sex genotype interaction (two-way ANOVA $F(2,56)=0.32$, $p=0.72$,). Furthermore, we assessed exploratory behavior of Srr12 and Srr2 mice in the open field test. Both transgenic lines exhibited a similar exploratory activity and habituation during the 30 minutes observation time as Wt mice (three-way ANOVA genotype effect, $F(2,80)=0{,}312$, $p>0.77$; time effect $F(5,395)=52{,}435$, $p>0.01$; Fig. 3B). In the novelty-suppressed feeding (NSF) test, Srr12 animals showed a highly significant reduction in the latency to approach the food compared to Wt mice (Fig. 3C, one-way ANOVA $F(1,28)=16.89$, $p<0.01$), whereas the general food intake in both transgenic mouse strains was not changed (data not shown) Again, we did not observe any sex (two-way ANOVA $F(1,50)=3.86$, $p=0.055$) or sex genotype interaction (two-way ANOVA $F(1,50)=2.78$, $p=0.10$). We also analyzed the effects of a bilateral olfactory bulbectomy (OBX), an animal model for depression, in Wt and Srr12 transgenic mice. We found a significant genotype x surgery interaction for both sexes (two-way ANOVA male $F(1,26)=6.39$. $p<0.05$; two-way ANOVA female $F(1,26)=12.89$ $p<0.05$). OBX produced a characteristic hyperactivity, indicated by an increased distance traveled during the first 6 minutes in the open field (Fig. 3D) in Wt mice, but not in Srr12 transgenic mice. Finally, we analyzed the SrrTg mice in the ultrasonic vocalization test (USV). Transgenic pups of the Srr12 line displayed a significant reduction in the number (Fig. 3E left, one-way ANOVA $F(2,44)=3.31$, $p<0.05$) and in the duration of calls (Fig. 3E right, $p<0.01$, Kruskal–Wallis test). We did not notice any sex (number of calls: two-way ANOVA $F(1,73)=0.03$, $p=0.90$; duration of calls: two-way ANOVA $F(1,73)=0.25$, $p=0.62$) or sex genotype interaction (number of calls: two-way ANOVA $F(2,73)=0.14$, $p=0.87$; duration of calls: two-way ANOVA $F(2,73)=1.53$, $p=0.22$). The behavior of Srr2 pups in this paradigm was not different from Wt pups.

3.3 Cognition and memory in SrrTg mice

The Y-maze test assesses spatial working memory as percentage of spontaneous alternations, calculated as the number of alternations (entering into three different arms consecutively) divided by the number of possible alternations (total arm visits minus 2) and multiplied with 100. Neither the spontaneous alternation (Fig. 4A, left) or the total number of arm visits (Fig. 4A, right) differed significantly between Srr12, Srr2 and Wt littermates (Fig. 4A left, one-way ANOVA, same: $F(2,69)=1.204$, $p>0.05$; Fig. 4A right: $F(2,69)=0.8257$, $p>0.05$).

The Morris water maze is widely used to study spatial memory and learning [56]. Repeated measures ANOVA for the latency to find the platform during acquisition (day 1–5) revealed that overall both Wt and Srr12 mice learned to find the hidden platform, which was indicated by a significant time effect for both strains ($F(5,85)=11.01$, $p<0.0001$, Fig. 4B). Srr12 mice did not perform differently from their Wt littermates ($F(1,17)=2,87$, $p=0.11$) and there was no genotype x time interaction ($F(5,85)=1.35$, $p=0.25$). The swim speed was neither affected by genotype ($F(1,17)=0.00$, $p=0.9$) nor by time ($F(4,68)=0.75$, $p=0.8$). Additionally, no significant genotype x time interaction was found ($F(4,68)=2.04$, $p=0.10$).

We also studied cognitive functions in an operant learning paradigm (Fig. 4C). Repeated measures ANOVA revealed that Srr12 mutant mice did not show altered operant learning abilities compared to Wt mice. There was a significant effect of time ($F(11,308)=29.22$, $p<0.001$), but no genotype effect ($F(1,28)=2.162$, $p>0.05$) and no time x genotype interaction ($F(11,308)=1.177$, $p>0.05$).

3.4 D-serine supplementation

We investigated the consequences of a dietary D-serine drinking supplementation in Wt mice. Accordingly, we added 350 mg/kg D-serine to the drinking water for a 5-week period.

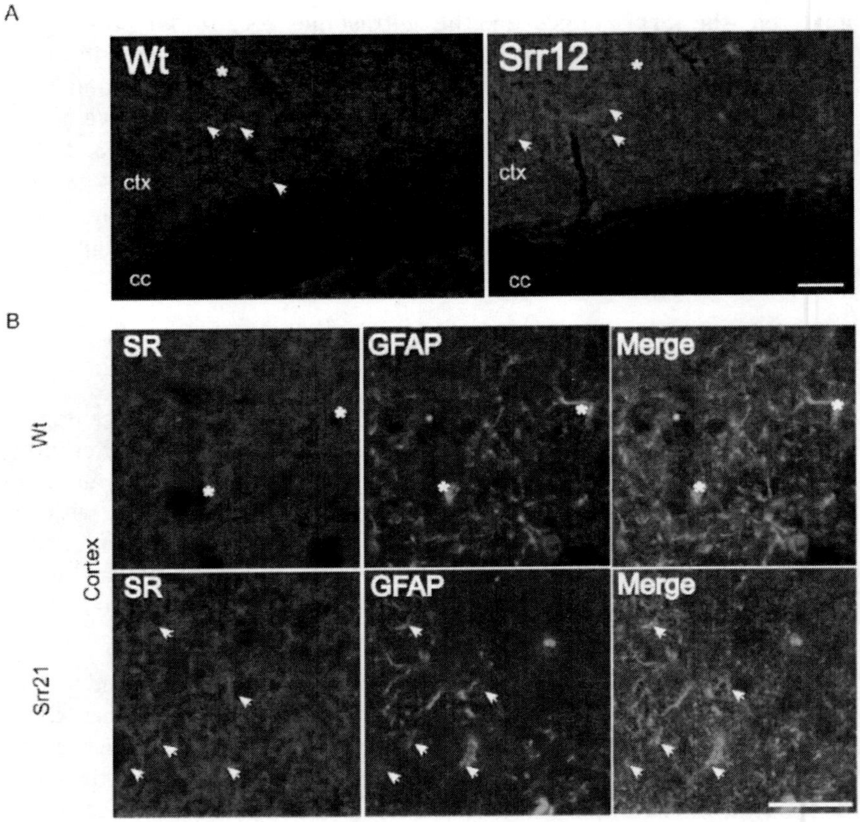

Figure 2. Specific serine racemase immunofluorescence in Srr12 mice. (A) Immunoreactivity for serine racemase is higher in transgenic animals than in Wt controls. Arrows - indicate serine racemase positive cell soma. * - indicates immunoreactivity of the neuropil which is much higher in Srr12 transgenic mice than Wt controls, possibily due to higher expression in astrocytic processes. (B) Serine racemase expression is specifically increased in astrocytes as labeled by the astrocyte marker glial fibrillary acidic protein (GFAP). Arrowheads - indicate GFAP+ processes bearing SR+ puncta, although both Wt and Srr12 astrocytes express serine racemase, immunoreactivity and colocalization is more pronounced in Srr12 mice. Asterisks indicate GFAP positive cells which seemed to be devoid of serine racemase expression in WT mice. All images were acquired using the same exposure times. Scale bar - 100 µm top panels and 50 µm all other panels. ctx - cortex, cc - corpus callosum.

Control mice either received water, L-serine or glycine. One-way ANOVA revealed that mice receiving D-serine had considerably higher brain D-serine levels compared to the control groups in all brain regions studied (one-way ANOVA, Fig 5A, cortex $F(3,22)=18.17$, $p<0.0001$; hippocampus $F(3,22)=14.15$, $p<0.0001$; forebrain $F(3,22)=8.872$, $p<0.0004$). Additionally, D-serine treatment led to significantly increased L-serine levels in the cortex and L-serine treatment significantly increased the L-serine levels in the forebrain (Fig 5A). Behavioral testing showed that mice receiving D-serine had a significantly reduced immobility time in the Porsolt forced swim test when compared to the water control group (one-way ANOVA, $F(3,46)=3,747$, $p<0.05$, Fig. 5B). L-serine or glycine administration had no effect. The latency to approach food in the novelty-suppressed feeding test was also significantly decreased in mice receiving D-serine (one-way ANOVA, $F(4,59)=2,812$, $p<0.05$, Fig. 5C). Again, there was no effect of L-serine or glycine supplementation. The open field test revealed no differences in the distance travelled between the different groups indicated that D-serine does not influence open field activity (one-way ANOVA Fig. 5C) The operant learning test revealed that a chronic D-serine treatment had no influence on learning behavior in this paradigm. Both groups, D-serine treated and controls, performed equally (two-way repeated ANOVA, time effect: $F(13,494)=54.86$, $p<0.0001$; treatment effect: $F(1,38)=0.19$, $p<0.66$; Fig. 5E). In the water maze test, D-serine treated mice behaved similarly compared to the control group (two-way ANOVA repeated measurers, time effect: $F(4,72)=158.83$, $p< <0.0001$; treatment effect: $F(1,18)=1.45$, $p<0.24$; Fig. 5F. Thus, the dietary and genetic elevation of D-serine levels produced similar behavioral effects in depression- related and cognitive paradigms as those observed in SrrTg animals.

4. DISCUSSION

In this study we investigated the consequences of genetically or pharmacologically increased brain D-serine levels on animal behavior. Both the astrocytic expression of serine racemase in transgenic mice or the supplementation of the drinking water with D-serine resulted in significantly elevated brain D-serine levels in the cortex and the hippocampus. This affected emotional behaviors but not the cognitive performance of the mice. SrrTg mice displayed increased serine racemase protein expression shown by *Western* blot and immunofluorescence analyses. This increased serine racemase expression co-localized with GFAP expression, which is an astrocytic marker in the brain [42], [57]–[60]. This co-localization indicated that SrrTg mice indeed have augmented astrocytic serine racemase expression.

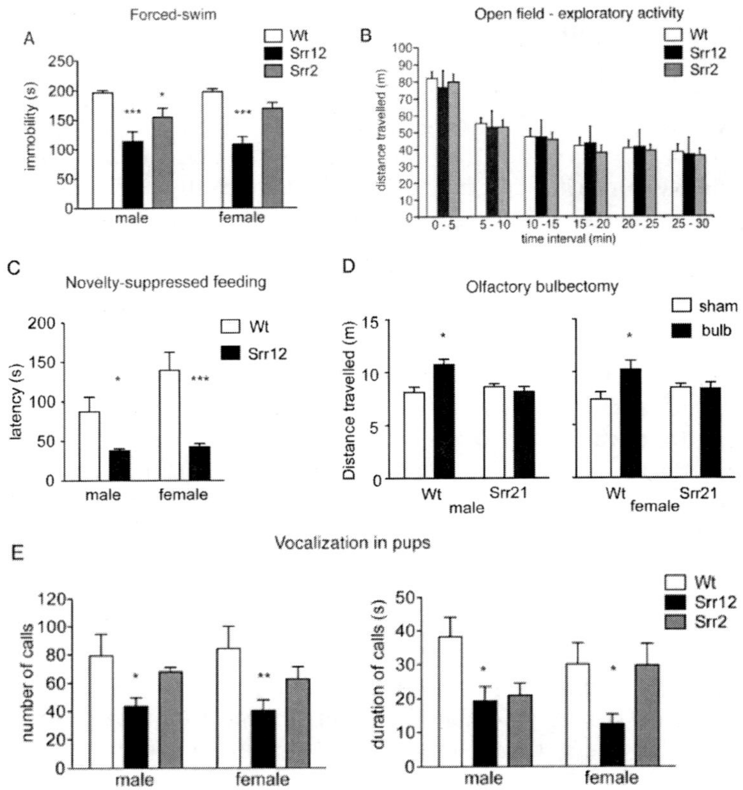

Figure 3. Depression-related behavior in SrrTg mice. (A) Forced-swim test. Immobility time, defined as a lack of activity aside from small movements needed to keep the body floating, was measured throughout the last 4 min of the session. (Wt: n=21 (10 males, 11 females); Srr2: n=17 (8 males, 9 females); Srr12: n=24 (9 males, 15 females)). (B) Locomotor activity was recorded over a 30 min time period and expressed as distance traveled. (n=15 per group). (C) Novelty suppressed feeding test. Srr12 mice displayed a significantly reduced latency to approach and eat food compared to Wt mice (Wt: n=26 (14 males, 12 females); Srr12: n=28 (15 males, 13 females)). (D) Horizontal (distance traveled) activity 2 weeks after surgery of sham-operated and bulbectomized SrrTg mice and their sham operated littermates in the open-field apparatus (Wt: female 16 (sham 10, bulb 6), male 18 (sham 14, bulb 4); Srr12: female 15 (sham 8, bulb 7), male 13 (sham 5, bulb 7). (E) Ultrasonic vocalization. Measurements of social isolation-induced ultrasonic vocalizations (USV) in SrrTg and Wt pups revealed a significant increase in Srr12 mice in number (left) and duration (right) of calls (Wt: n=22 (12 males, 10 females); Srr2: n=23 (9 males, 13 females); Srr12: n=35 (20 males, 15 females)). All values are expressed as mean ±SEM. *$p<0.05$ and ***$p<0.001$.

4.1 Chronic D-serine elevation results in antidepressive behavior

The Porsolt forced swim test evaluates behavioral despair in mice. SrrTg mice and animals chronically treated with D-serine showed a reduced immobility time in this paradigm, thus behaving like animals treated with antidepressant drugs. This behavior was not due to a generally elevated activity in these mice, because they showed no changes in locomotor activity in the open field test. Hence, it is more likely that the elevated D-serine levels altered affective or depression-related behaviors in these animals. These findings are in line with emerging evidence that glutamatergic signaling may be involved in the pathogenesis of major depressive disorders [61], although the mechanisms are still not completely understood. Accordingly, a single application of the NMDA-receptor selective antagonist ketamine, a widely used anesthetic drug, can produce a long-lasting antidepressant effect. This effect seems to be mediated through an up-regulation of mammalian target of rapamycin (mTor) signaling [62], [63]. There is also evidence that ketamine enhances glutamatergic signaling in the prefrontal cortex, at least at low doses [64], [65],[66]. D-cycloserine (DCS), a partial agonist at the strychnine-insensitive glycine site of the NMDA receptor [37], also showed antidepressant effects in animal models [67] and in humans [68], which is in line with our observation of antidepressant-like behavior in the transgenic mice with elevated brain D-serine levels. The influence of low D-serine levels on depression-like behavior has not been tested so far as serine racemase-deficient mice have not been examined in behavioral paradigms of depression. They display an anxiety-related behavioral phenotype with cognitive deficits [39].

We also investigated the consequences of an olfactory bulbectomy (OBX) in SrrTg mice. Removal of the *bulbus olfactorius* in rodents is considered as an animal model of depression with high predictive validity, because it triggers physiological and behavioral changes that are sensitive to antidepressant treatment, but not to anxiolytic drugs [69]. Behavioral deficits in bulbectomized mice are accompanied by NMDA receptor-mediated dysfunctions in the hippocampus and amygdala [70]. It has been suggested that these dysfunctions are due to a disconnection of projections from the olfactory bulb to the olfactory cortex, including the olfactory tubercle, amygdala, and entorhinal cortex after bulbectomy [71]. While Wt mice displayed an elevated locomotor activity in the open field after OBX, we could not observe such behavioral effects in SrrTg mice. Together these findings strongly suggest that chronically elevated D-serine levels protect against the behavioral changes in the OBX mouse model of depression.

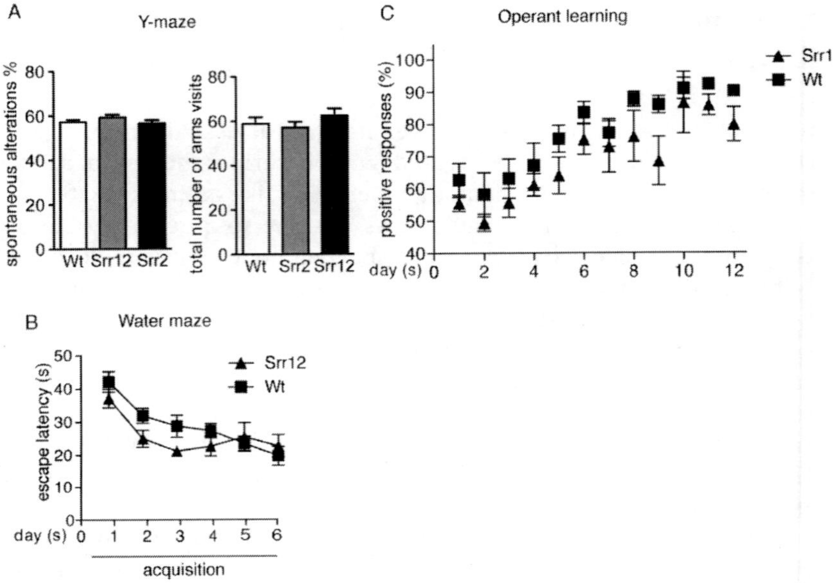

Figure 4. Cognition in SrrTg mice. (A) In the Y-maze task, SrrTg mice displayed no significant differences in the percentage of spontaneous alternations (left) and total number of arm visits (right) compared to Wt mice (Wt: n=24; Srr2: n=19; Srr12: n=27). (B) Morris water maze test. Average water maze escape latencies across 5 training days of Wt and Srr12 mice (n 9–10 male mice per group). (C) Operant conditioning test. The learning performance of SrrTg mice did not differ compared to Wt mice (Wt: n=16; Srr12: n=16). All values are expressed as mean ±SEM indicated animals.

The USV test evaluates distress-like ultrasonic calls that pups emit after separation from their mothers. This behavior is considered as an expression of affect and is used for the preclinical evaluation of anxiolytic drugs [72]–[74]. SrrTg pups showed a significantly lower number and shortened duration of USV, again indicating an effect of elevated D-serine levels on affective responses. Glutamatergic signaling has been implicated in this behavior, as USVs were enhanced after administration of low to moderate doses of the NMDA receptor antagonists memantine or neramexane, while higher doses reduced USVs [75]. The high-affinity NMDA receptor antagonist dizocilpine reduced USVs, without showing biphasic dose-response effects [75]. Additionally, 1-aminocyclopropanecarboxylic acid (ACPC), a partial agonist on the NMDA receptor, reduces USV in rats, whereas glycine does not alter USV [76]. Despite these results the mechanism by which D-serine alters USV in SrrTg mice remains unclear.

We also evaluated SrrTg mice and D-serine-treated mice in the NSF test, a conflict paradigm with a food reward being the positive reinforcer and fear in a novel environment the negative reinforcer. This test is sensitive to chronic antidepressant treatment, as well as to anxiolytic drugs [77]. Again, SrrTg and D-serine-treated mice showed altered emotionality, as they approached the food significantly faster than control animals.

Interestingly, we observed a more pronounced behavioral phenotype in serine racemase overexpressing mice, when compared to animals that received D-serine orally, although the latter animals exhibited higher brain D-serine levels, albeit not measured at the same time. This is probably due to the fact that astrocytic D-serine is more efficient in enhancing NMDA receptor signaling than D-serine provided via drinking water. Astrocytic D-serine is released by a vesicle-dependent mechanism that is closely related to glutamate release and calcium influx and thus regulated by neuronal activity [78].

4.2 Mild chronic D-serine elevation and cognition

350 mg/L D-serine added to the drinking water increased D-serine levels in all brain regions studied. That D-serine treatment at least acutely increases D-serine content in mice and rats has already been demonstrated [79], [80]. Moreover, D-serine treatment also increases L-serine levels in the cortex indicating that D-serine is possibly converted to L-serine by the serine racemase in this brain region [3], [24]. Studies investigating the effects of D-serine administration on cognitive performances showed different results. Thus, an acutely applied dose of 1000 mg/kg D-serine to rats led to cognitive enhancement in the Morris water maze test, while a dose of 100 mg/kg showed no effect [81]. A single administration of 800 mg/kg D-serine also enhanced social memory in rats [82]. These studies suggested that only higher doses of D-serine enhance the cognitive performance of rats. However, functional magnetic resonance imaging demonstrated that administration of 50 mg/kg of D-serine is enough to enhance hippocampal activity in rats [83]. Therefore, Bado et al., examined a 50 mg/kg dose of D-serine applied prior to cognitive test situations and showed that this low dose effectively enhanced cognitive performances in several tasks [79]. In this study, we investigated the effects of a chronic low dose (58 mg/kg/d) D-serine treatment on the cognitive performance in mice. This treatment had no effect on the animals' performances in the operant conditioning paradigm or the Morris water maze test. Likewise, the performance of SrrTg animals was also not significantly different from Wt littermates in these tests.

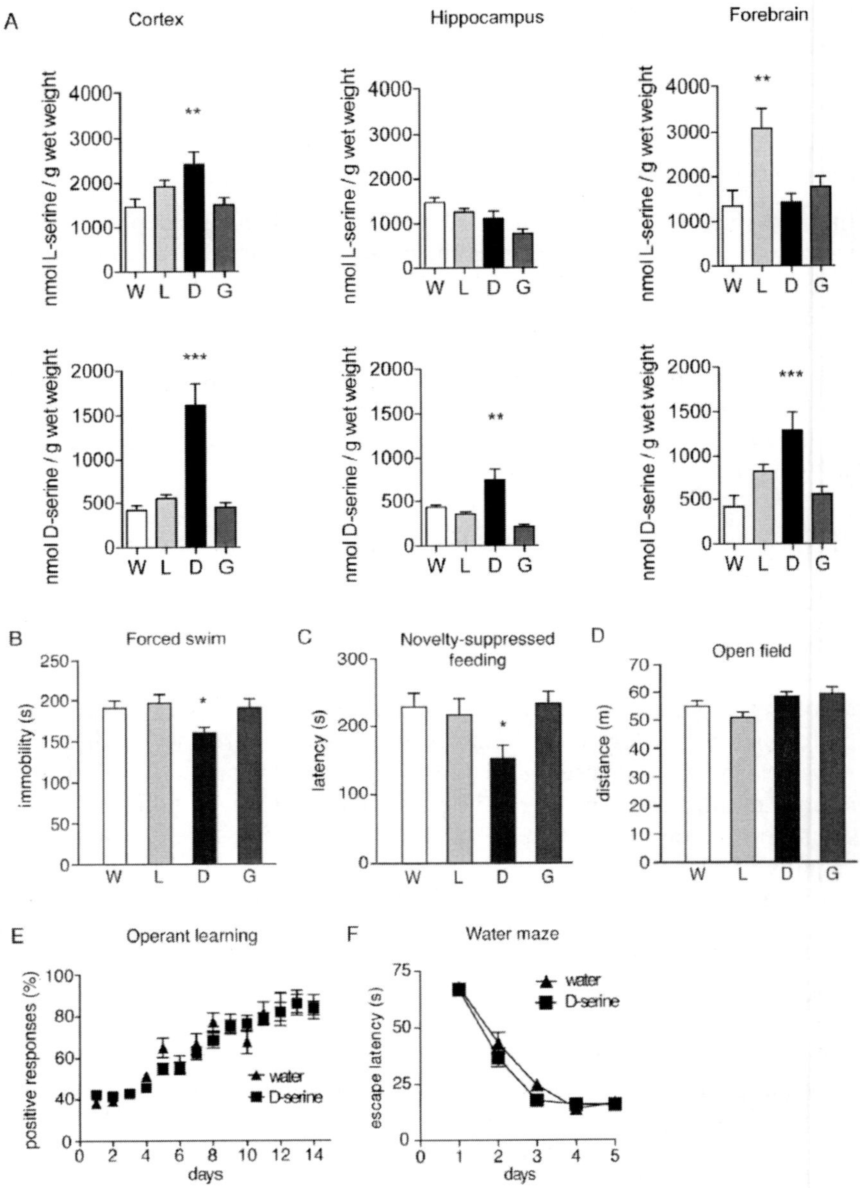

Figure 5. Brain D-serine levels and behavior of Wt mice after D-serine administration. (A) Quantitative analysis of brain D-serine and L-serine levels per wet weight in 4–8 dietary-treated Wt mice by HPLC. D-serine levels of cortex, hippocampus and forebrain were measured. Mice were treated with D-serine (D), L-serine (L), glycine (G) and water (W). (B)

Forced-swim test. Immobility time, defined as a lack of activity aside from small movements needed to keep the body floating, was measured throughout the last 4 min of the session. D-serine treated mice spent less time immobile compared to controls. (C) Novelty suppressed feeding test. The mice that were treated with D-serine displayed a significantly reduced latency to approach the food and to start eating. (D) Open field test. Distance traveled in the open field arena of Wt mice. Behavior in this paradigm was not changed after amino acid treatment. (E) Operant conditioning test. The Hill equation indicates learning improvement during the training days. Data of all groups resulted in similar learning curves. (F) Water maze test. Escape latencies across 5 training days of D-serine treated and control mice were recorded and did not differ between the groups. *$p<0.05$ and **$p<0.01$. Each error bar represents the mean ±SEM of 10 animals.

It has recently been demonstrated that the GFAP promoter is also present in slowly cycling, stationary cells of the brain that develop into intermediate progenitors [84]. We therefore cannot exclude that the serine racemase is also expressed in these neuronal progenitors of SrrTg mice. However, D-serine treatment also produced the same behavioral changes observed in SrrTg mice and, therefore, it is unlikely that the observed behavioral changes are caused by developmental effects.

Our present data show that a chronic elevation of D-serine levels, either by overexpression of serine racemase or by drinking supplementation, improves the emotional behavior of mice. It should be noted that the total brain D-serine levels were only significantly changed under our experimental conditions, with up to 3-fold elevation in the SrrTg mice and after food supplementation of Wt mice. The D-serine increase was most pronounced in the cortex area of both models. We suggest D-serine as useful for the pharmacotherapy of mood disorders.

ACKNOWLEDGMENTS

The authors thank Kerstin Michel and Hanna Schrage for technical support.

REFERENCES

1. Li F, Tsien JZ (2009) Memory and the NMDA receptors. N Engl J Med 361: 302–303.
2. Kemp JA, Leeson PD (1993) The glycine site of the NMDA receptor–five years on. Trends Pharmacol Sci 14: 20–25.

3. Wolosker H, Blackshaw S, Snyder SH (1999) Serine racemase: a glial enzyme synthesizing D-serine to regulate glutamate-N-methyl-D-aspartate neurotransmission. Proc Natl Acad Sci U S A 96: 13409–13414.
4. Baranano DE, Ferris CD, Snyder SH (2001) Atypical neural messengers. Trends Neurosci 24: 99–106.
5. Mothet JP, Parent AT, Wolosker H, Brady RO Jr, Linden DJ, et al. (2000) D-serine is an endogenous ligand for the glycine site of the N-methyl-D-aspartate receptor. Proc Natl Acad Sci U S A 97: 4926–4931.
6. Schell MJ, Brady RO Jr, Molliver ME, Snyder SH (1997) D-serine as a neuromodulator: regional and developmental localizations in rat brain glia resemble NMDA receptors. J Neurosci 17: 1604–1615.
7. Schell MJ, Molliver ME, Snyder SH (1995) D-serine, an endogenous synaptic modulator: localization to astrocytes and glutamate-stimulated release. Proc Natl Acad Sci U S A 92: 3948–3952.
8. Kim PM, Aizawa H, Kim PS, Huang AS, Wickramasinghe SR, et al. (2005) Serine racemase: activation by glutamate neurotransmission via glutamate receptor interacting protein and mediation of neuronal migration. Proc Natl Acad Sci U S A 102: 2105–2110.
9. Berger AJ, Dieudonne S, Ascher P (1998) Glycine uptake governs glycine site occupancy at NMDA receptors of excitatory synapses. J Neurophysiol 80: 3336–3340.
10. Martina M, Krasteniakov NV, Bergeron R (2003) D-Serine differently modulates NMDA receptor function in rat CA1 hippocampal pyramidal cells and interneurons. J Physiol 548: 411–423.
11. Lim R, Hoang P, Berger AJ (2004) Blockade of glycine transporter-1 (GLYT-1) potentiates NMDA receptor-mediated synaptic transmission in hypoglossal motorneurons. J Neurophysiol 92: 2530–2537.
12. Kinney GG, Sur C, Burno M, Mallorga PJ, Williams JB, et al. (2003) The glycine transporter type 1 inhibitor N-[3-(4′-fluorophenyl)-3-(4′-phenylphenoxy)propyl]sarcosine potentiates NMDA receptor-mediated responses in vivo and produces an antipsychotic profile in rodent behavior. J Neurosci 23: 7586–7591.
13. Chen L, Muhlhauser M, Yang CR (2003) Glycine tranporter-1 blockade potentiates NMDA-mediated responses in rat prefrontal cortical neurons in vitro and in vivo. J Neurophysiol 89: 691–703.
14. Bergeron R, Meyer TM, Coyle JT, Greene RW (1998) Modulation of N-methyl-D-aspartate receptor function by glycine transport. Proc Natl Acad Sci U S A 95: 15730–15734.
15. Hashimoto A, Nishikawa T, Hayashi T, Fujii N, Harada K, et al. (1992) The presence of free D-serine in rat brain. FEBS Lett 296: 33–36.
16. Hashimoto A, Nishikawa T, Oka T, Takahashi K (1993) Endogenous D-serine in rat brain: N-methyl-D-aspartate receptor-related distribution and aging. J Neurochem 60: 783–786.
17. Matsui T, Sekiguchi M, Hashimoto A, Tomita U, Nishikawa T, et al. (1995) Functional comparison of D-serine and glycine in rodents: the effect on cloned NMDA receptors and the extracellular concentration. J Neurochem 65: 454–458.
18. Priestley T, Laughton P, Myers J, Le Bourdelles B, Kerby J, et al. (1995) Pharmacological properties of recombinant human N-methyl-D-aspartate receptors comprising NR1a/NR2A and NR1a/NR2B subunit assemblies expressed in permanently transfected mouse fibroblast cells. Mol Pharmacol 48: 841–848.
19. Thomson AM, Walker VE, Flynn DM (1989) Glycine enhances NMDA-receptor mediated synaptic potentials in neocortical slices. Nature 338: 422–424.
20. Yang Y, Ge W, Chen Y, Zhang Z, Shen W, et al. (2003) Contribution of astrocytes to hippocampal long-term potentiation through release of D-serine. Proc Natl Acad Sci U S A 100: 15194–15199.
21. Stevens ER, Esguerra M, Kim PM, Newman EA, Snyder SH, et al. (2003) D-serine and serine racemase are present in the vertebrate retina and contribute to the physiological activation of NMDA receptors. Proc Natl Acad Sci U S A 100: 6789–6794.
22. Baptista V, Varanda WA (2005) Glycine binding site of the synaptic NMDA receptor in subpostremal NTS neurons. J Neurophysiol 94: 147–152.

23. Papouin T, Ladepeche L, Ruel J, Sacchi S, Labasque M, et al. (2012) Synaptic and extrasynaptic NMDA receptors are gated by different endogenous coagonists. Cell 150: 633–646.
24. Wolosker H, Sheth KN, Takahashi M, Mothet JP, Brady RO Jr, et al. (1999) Purification of serine racemase: biosynthesis of the neuromodulator D-serine. Proc Natl Acad Sci U S A 96: 721–725.
25. De Miranda J, Santoro A, Engelender S, Wolosker H (2000) Human serine racemase: moleular cloning, genomic organization and functional analysis. Gene 256: 183–188.
26. De Miranda J, Panizzutti R, Foltyn VN, Wolosker H (2002) Cofactors of serine racemase that physiologically stimulate the synthesis of the N-methyl-D-aspartate (NMDA) receptor coagonist D-serine. Proc Natl Acad Sci U S A 99: 14542–14547.
27. Benneyworth MA, Li Y, Basu AC, Bolshakov VY, Coyle JT (2012) Cell selective conditional null mutations of serine racemase demonstrate a predominate localization in cortical glutamatergic neurons. Cell Mol Neurobiol 32: 613–624.
28. Henneberger C, Papouin T, Oliet SH, Rusakov DA (2010) Long-term potentiation depends on release of D-serine from astrocytes. Nature 463: 232–236.
29. Coyle JT (2006) Glutamate and schizophrenia: beyond the dopamine hypothesis. Cell Mol Neurobiol 26: 365–384.
30. Coyle JT, Tsai G, Goff D (2003) Converging evidence of NMDA receptor hypofunction in the pathophysiology of schizophrenia. Ann N Y Acad Sci 1003: 318–327.
31. Hashimoto K, Sawa A, Iyo M (2007) Increased levels of glutamate in brains from patients with mood disorders. Biol Psychiatry 62: 1310–1316.
32. Bendikov I, Nadri C, Amar S, Panizzutti R, De Miranda J, et al. (2007) A CSF and postmortem brain study of D-serine metabolic parameters in schizophrenia. Schizophr Res 90: 41–51.
33. Duncan EJ, Szilagyi S, Schwartz MP, Bugarski-Kirola D, Kunzova A, et al. (2004) Effects of D-cycloserine on negative symptoms in schizophrenia. Schizophr Res 71: 239–248.
34. Heresco-Levy U, Kremer I, Javitt DC, Goichman R, Reshef A, et al. (2002) Pilot-controlled trial of D-cycloserine for the treatment of post-traumatic stress disorder. Int J Neuropsychopharmacol 5: 301–307.
35. Laake K, Oeksengaard AR (2002) D-cycloserine for Alzheimer's disease. Cochrane Database Syst Rev CD003153.
36. Kantrowitz JT, Malhotra AK, Cornblatt B, Silipo G, Balla A, et al. (2010) High dose D-serine in the treatment of schizophrenia. Schizophr Res 121: 125–130.
37. Lopes T, Neubauer P, Boje KM (1997) Chronic administration of NMDA glycine partial agonists induces tolerance in the Porsolt swim test. Pharmacol Biochem Behav 58: 1059–1064.
38. Onur OA, Schlaepfer TE, Kukolja J, Bauer A, Jeung H, et al. (2010) The N-methyl-D-aspartate receptor co-agonist D-cycloserine facilitates declarative learning and hippocampal activity in humans. Biol Psychiatry 67: 1205–1211.
39. Basu AC, Tsai GE, Ma CL, Ehmsen JT, Mustafa AK, et al. (2009) Targeted disruption of serine racemase affects glutamatergic neurotransmission and behavior. Mol Psychiatry 14: 719–727.
40. Malkesman O, Austin DR, Tragon T, Wang G, Rompala G, et al. (2011) Acute d-serine treatment produces antidepressant-like effects in rodents. Int J Neuropsychopharmacol 1–14.
41. Nolte C, Matyash M, Pivneva T, Schipke CG, Ohlemeyer C, et al. (2001) GFAP promoter-controlled EGFP-expressing transgenic mice: a tool to visualize astrocytes and astrogliosis in living brain tissue. Glia 33: 72–86.
42. Wang LZ, Zhu XZ (2003) Spatiotemporal relationships among D-serine, serine racemase, and D-amino acid oxidase during mouse postnatal development. Acta Pharmacol Sin 24: 965–974.
43. Horio M, Kohno M, Fujita Y, Ishima T, Inoue R, et al. (2011) Levels of D-serine in the brain and peripheral organs of serine racemase (Srr) knock-out mice. Neurochem Int 59: 853–859.

44. Grant SL, Shulman Y, Tibbo P, Hampson DR, Baker GB (2006) Determination of d-serine and related neuroactive amino acids in human plasma by high-performance liquid chromatography with fluorimetric detection. J Chromatogr B Analyt Technol Biomed Life Sci 844: 278–282.
45. Bilkei-Gorzo A, Michel K, Noble F, Roques BP, Zimmer A (2007) Preproenkephalin knockout mice show no depression-related phenotype. Neuropsychopharmacology 32: 2330–2337.
46. Porsolt RD, Le Pichon M, Jalfre M (1977) Depression: a new animal model sensitive to antidepressant treatments. Nature 266: 730–732.
47. Porsolt RD, Bertin A, Jalfre M (1977) Behavioral despair in mice: a primary screening test for antidepressants. Arch Int Pharmacodyn Ther 229: 327–336.
48. Rochford J, Beaulieu S, Rousse I, Glowa JR, Barden N (1997) Behavioral reactivity to aversive stimuli in a transgenic mouse model of impaired glucocorticoid (type II) receptor function: effects of diazepam and FG-7142. Psychopharmacology (Berl) 132: 145–152.
49. Bilkei-Gorzo A, Racz I, Michel K, Zimmer A (2002) Diminished anxiety- and depression-related behaviors in mice with selective deletion of the Tac1 gene. J Neurosci 22: 10046–10052.
50. Noirot E (1972) Ultrasounds and maternal behavior in small rodents. Dev Psychobiol 5: 371–387.
51. Brunelli SA, Shair HN, Hofer MA (1994) Hypothermic vocalizations of rat pups (Rattus norvegicus) elicit and direct maternal search behavior. J Comp Psychol 108: 298–303.
52. Gardner CR (1985) Distress vocalization in rat pups. A simple screening method for anxiolytic drugs. J Pharmacol Methods 14: 181–187.
53. Miczek KA, Weerts EM, Vivian JA, Barros HM (1995) Aggression, anxiety and vocalizations in animals: GABAA and 5-HT anxiolytics. Psychopharmacology (Berl) 121: 38–56.
54. Miczek KA, Yap JJ, Covington HE 3rd (2008) Social stress, therapeutics and drug abuse: preclinical models of escalated and depressed intake. Pharmacol Ther 120: 102–128.
55. Morse WH, Skinner BF (1958) Some factors involved in the stimulus control of operant behavior. J Exp Anal Behav 1: 103–107.
56. Morris R (1984) Developments of a water-maze procedure for studying spatial learning in the rat. J Neurosci Methods 11: 47–60.
57. Papageorgiou IE, Gabriel S, Fetani AF, Kann O, Heinemann U (2011) Redistribution of astrocytic glutamine synthetase in the hippocampus of chronic epileptic rats. Glia 59: 1706–1718.
58. Suarez I, Bodega G, Fernandez B (2002) Glutamine synthetase in brain: effect of ammonia. Neurochem Int 41: 123–142.
59. Martinez-Hernandez A, Bell KP, Norenberg MD (1977) Glutamine synthetase: glial localization in brain. Science 195: 1356–1358.
60. Hallermayer K, Harmening C, Hamprecht B (1981) Cellular localization and regulation of glutamine synthetase in primary cultures of brain cells from newborn mice. J Neurochem 37: 43–52.
61. Skolnick P, Popik P, Trullas R (2009) Glutamate-based antidepressants: 20 years on. Trends Pharmacol Sci 30: 563–569.
62. Hashimoto K (2010) Role of the mTOR signaling pathway in the rapid antidepressant action of ketamine. Expert Rev Neurother 11: 33–36.
63. Welberg L (2010) Psychiatric disorders: Ketamine modifies mood through mTOR. Nat Rev Neurosci 11: 666.
64. Breier A, Malhotra AK, Pinals DA, Weisenfeld NI, Pickar D (1997) Association of ketamine-induced psychosis with focal activation of the prefrontal cortex in healthy volunteers. Am J Psychiatry 154: 805–811.
65. Rowland LM, Bustillo JR, Mullins PG, Jung RE, Lenroot R, et al. (2005) Effects of ketamine on anterior cingulate glutamate metabolism in healthy humans: a 4-T proton MRS study. Am J Psychiatry 162: 394–396.

66. Moghaddam B, Adams B, Verma A, Daly D (1997) Activation of glutamatergic neurotransmission by ketamine: a novel step in the pathway from NMDA receptor blockade to dopaminergic and cognitive disruptions associated with the prefrontal cortex. J Neurosci 17: 2921–2927.
67. Trullas R, Skolnick P (1990) Functional antagonists at the NMDA receptor complex exhibit antidepressant actions. Eur J Pharmacol 185: 1–10.
68. Heresco-Levy U, Javitt DC, Gelfin Y, Gorelik E, Bar M, et al. (2006) Controlled trial of D-cycloserine adjuvant therapy for treatment-resistant major depressive disorder. J Affect Disord 93: 239–243.
69. Kelly JP, Wrynn AS, Leonard BE (1997) The olfactory bulbectomized rat as a model of depression: an update. Pharmacol Ther 74: 299–316.
70. Adamec RE, Burton P, Shallow T, Budgell J (1999) Unilateral block of NMDA receptors in the amygdala prevents predator stress-induced lasting increases in anxiety-like behavior and unconditioned startle–effective hemisphere depends on the behavior. Physiol Behav 65: 739–751.
71. Heimer L (1968) Synaptic distribution of centripetal and centrifugal nerve fibres in the olfactory system of the rat. An experimental anatomical study. J Anat 103: 413–432.
72. Rex A, Voigt JP, Voits M, Fink H (1998) Pharmacological evaluation of a modified open-field test sensitive to anxiolytic drugs. Pharmacol Biochem Behav 59: 677–683.
73. Rodgers RJ, Boullier E, Chatzimichalaki P, Cooper GD, Shorten A (2002) Contrasting phenotypes of C57BL/6JOlaHsd, 129S2/SvHsd and 129/SvEv mice in two exploration-based tests of anxiety-related behaviour. Physiol Behav 77: 301–310.
74. Merali Z, Levac C, Anisman H (2003) Validation of a simple, ethologically relevant paradigm for assessing anxiety in mice. Biol Psychiatry 54: 552–565.
75. Minkeviciene R, Banerjee P, Tanila H (2008) Cognition-enhancing and anxiolytic effects of memantine. Neuropharmacology 54: 1079–1085.
76. Winslow JT, Insel TR, Trullas R, Skolnick P (1990) Rat pup isolation calls are reduced by functional antagonists of the NMDA receptor complex. Eur J Pharmacol 190: 11–21.
77. Santarelli L, Saxe M, Gross C, Surget A, Battaglia F, et al. (2003) Requirement of hippocampal neurogenesis for the behavioral effects of antidepressants. Science 301: 805–809.
78. Mothet JP, Pollegioni L, Ouanounou G, Martineau M, Fossier P, et al. (2005) Glutamate receptor activation triggers a calcium-dependent and SNARE protein-dependent release of the gliotransmitter D-serine. Proc Natl Acad Sci U S A 102: 5606–5611.
79. Bado P, Madeira C, Vargas-Lopes C, Moulin TC, Wasilewska-Sampaio AP, et al. (2011) Effects of low-dose D-serine on recognition and working memory in mice. Psychopharmacology (Berl) 218: 461–470.
80. Hashimoto A, Chiba S (2004) Effect of systemic administration of D-serine on the levels of D- and L-serine in several brain areas and periphery of rat. Eur J Pharmacol 495: 153–158.
81. Zhang Z, Gong N, Wang W, Xu L, Xu TL (2008) Bell-shaped D-serine actions on hippocampal long-term depression and spatial memory retrieval. Cereb Cortex 18: 2391–2401.
82. Shimazaki T, Kaku A, Chaki S (2010) D-Serine and a glycine transporter-1 inhibitor enhance social memory in rats. Psychopharmacology (Berl) 209: 263–270.
83. Panizzutti R, Rausch M, Zurbrugg S, Baumann D, Beckmann N, et al. (2005) The pharmacological stimulation of NMDA receptors via co-agonist site: an fMRI study in the rat brain. Neurosci Lett 380: 111–115.
84. Platel JC, Gordon V, Heintz T, Bordey A (2009) GFAP-GFP neural progenitors are antigenically homogeneous and anchored in their enclosed mosaic niche. Glia 57: 66–78.

CHAPTER 16

Bovine diseases causing neurological signs and death in Mexican feedlots

Rafael Ramírez-Romero, Cecilia Ramírez-Hernández, Luis Jorge García-Márquez, Rafael Julio Macedo-Barragán, Julio Martínez-Burnes, Alfonso López-Mayagoitia

ABSTRACT

The number of large feedlot operations, similar to that of USA and Canada, has notably increased in Mexico in the last three decades. Clinical and laboratory diagnoses of neurological diseases in feedlot cattle are crucial in Mexico and America because of the high incidence of bovine paralytic rabies (BPR). Because of its zoonotic potential, BPR must be promptly diagnosed and differentiated from other bovine neurological diseases such as thrombotic meningoencephalitis (TME), polioencephalomalacia (PEM) and botulism. More recently, BPR and botulism have been diagnosed with increasing frequency in Mexican feedlots. Neither BPR nor botulism has relevant gross lesions, thus post-mortem diagnosis without laboratory support is impossible. Herein, we describe five outbreaks of neurological diseases in Mexican feedlots in which BPR, botulism and PEM were diagnosed either independently or in combination. A diagram illustrating the most conspicuous pathologic findings and ancillary laboratory test required to confirm the diagnoses of these neurological diseases in feedlot cattle is proposed.

Keywords: Feedlot, cattle Bovine, paralytic, rabies Botulism, Thrombotic, meningoencephalitis Polioencephalomalacia Mexico

1. INTRODUCTION

Bovine paralytic rabies (BPR) is endemic in many geographical areas of Mexico, particularly in those which are normal habitat for vampire bat (*Desmodus rotundus*). These high-risk areas include zones along the Gulf of Mexico and the Pacific Coast (Fig. 1) (FAO/SAGARPA 2007; Lee et al. 2012). Several outbreaks of vampire-transmitted rabies have been described in Aldama, Mexico (Fig. 1, Tamaulipas) (Martinez-Burnes et al. 1997; Ramírez Romero et al. 2011).

Mexico has extensive grass lands which are suitable to raise beef cattle (Huntsinger and Starrs 2006). At 2007, Mexico registered an inventory of 23 million bovines of which almost 19 million were beef cattle (Peel et al. 2011). Nonetheless, the inventory is actually declining. Over a million heads are exported yearly to the USA, mostly from the northern Mexico close to the Mexican-American border (Huntsinger and Starrs 2006). Crossbreed beef cattle with minimal zebu blood are preferred by USA importers, while zebu crosses are typically used for domestic consumption in Mexico (Peel et al. 2011). The vast majority of animals fattened in northern Mexico feedlots originate from tropical and subtropical regions of Mexico and America where vampire bats and BPR are endemic (Peel et al. 2011; Lee et al. 2012).

In 2011, 148 confirmed cases of animal rabies were reported in Mexico, most of them (82 %) occurred in cattle (Blanton et al. 2012). Interestingly, a cow imported from Mexico was diagnosed with rabies in Texas, USA, and the virus involved was identified as vampire bat strain probably originated from the border state of Tamaulipas, Mexico (Fig. 1, Tamaulipas) (Blanton et al. 2011). Three confirmed cases of BPR have also been reported in a feedlot in the border town of Mexicali in Baja California, Mexico, a geographical location far distant from the rabies-endemic areas (Fig. 1). Baja California is an arid territory unsuitable to be inhabited by vampire bats (Lee et al. 2012).

The risk of moving cattle from endemic to BPR-free areas has been underestimated, and it is not until recently that the importance of BPR in feedlots of northern Mexico has been documented.

Figure. 1 Map of Mexico. BPR control zones are depicted in *darker colour*; these zones harbour vampire bats. Those states included have been referred in the text. *Black dots* are confirmed with BPR outbreaks (Note: the outbreaks in Aldama and Mexicali are not described in text but included in map). The *white dot* corresponded to a confirmed botulism outbreak (outbreak 4, Sabinas Hidalgo, Nuevo León). The *black dot with a white outline* refers to a BPR outbreak (outbreak 2, Tamuin, San Luis Potosi) and to other different outbreak occurring in the same area in which there were both BPR and botulism (outbreak 3 Tamuin, San Luis Potosi). The *gray dot with a black outline* corresponds to the last outbreak, which is a concomitance of both PEM and BPR (outbreak 5, General Escobedo, Nuevo León). The map was modified from the original source: FAO/SAGARPA 2007

1.1 Representative outbreaks

During 2012 and 2013, several outbreaks of feedlot mortality in cattle showing neurological signs were investigated.

1.2 First outbreak

The first outbreak occurred in Escarcega, Campeche (Fig. 1), where a mortality of 32.4/1,000 was reported. Prior to death, one of the affected animals showed progressive hindlimb ataxia, aggressive behaviour, prostration and tenesmus. Post-mortem examination revealed no significant gross findings, but the brain of this bullock subsequently tested positive for rabies by fluorescence antibody test (FAT). Another animal also tested positive for rabies by FAT but full necropsy was not done. Similar cases occurred in animals maintained in the feedlot and in animals on prairie. In this system, animals are grazing for at least 45 days (preconditioning), before being moved to the feedlot (finishing). Although this particular geographical area is endemic for BPR, rabies vaccination was not done until the BPR diagnosis had been made. In this outbreak, BPR most likely occurred locally where the animals were kept.

1.3 Second outbreak

Another outbreak occurred in Tamuin, San Luis Potosi (Fig. 1), in a feedlot that also had extensive grassland areas for preconditioning. Detailed clinical and post-mortem studies were done in one animal, and the brain was collected and submitted to the laboratory from another one. These two animals were representative of the outbreak with a mortality of 28.6/1,000; most of the affected animals were recently introduced to grass (less than 12 days before). The majority of animals had been purchased in Chiapas (Fig. 1). Clinically, these animals developed progressive hindlimb ataxia and posterior hypoaesthesia, became prostrated, showed trismus and tenesmus and died. Both animals tested positive for rabies by FAT. Although vampire bats are present in this zone, the infection most likely occurred in southern Mexico prior the animals been moved to the feedlot. It is worthwhile noting that in these outbreaks of Tamuin and Escarcega, the presumptive diagnosis was originally thrombotic meningoencephalitis (TME), an infectious disease caused by the bacterium *Histophilis somni*. Hence, treatments and management of affected animals were carried out ignoring the possible risk of rabies. Veterinarians and farm personnel received post-exposure treatment after the rabies diagnosis was confirmed.

1.4 Third outbreak

This outbreak occurred also in Tamuin, San Luis Potosi (Fig. 1), but in a different feedlot where the local veterinarian provided information regarding a previously confirmed case of BPR in a heifer that died few days after being put on a prairie. The outbreak continued, and by the time the problem was properly investigated, there were many feedlot and grazing cattle prostrated (mortality 75.2/1,000). Because of the neurological signs and previous diagnostic history in the premises, BPR was suspected in spite that it was unusual for BPR to cause recumbence in many animals at the same time (Fig. 2). Typically, cattle with BPR become ataxic and fall prostrate at different time during an outbreak, presumably due to variations in the incubation period. Most of the animals in this outbreak had no fever, continued conscious, and showed a tendency to remain in sternal recumbence with the head turned to the flank, but skin sensation remained. Necropsy was performed in two animals, and in both cases, the brain tested negative for rabies by FAT. Neither post-mortem nor histopathological examinations of the brain resulted in significant microscopic changes. Further field investigation revealed that poultry litter had been recently added to the feed. By this reason alone, botulism became the top differential diagnosis. Blood serum collected from these animals was inoculated intraperitoneally to adult mice causing the death of the rodents within 36 h. This finding was considered positive for botulism toxin, but the specific toxin type could not be identified. Additional laboratory tests confirmed that poultry litter which was added to the diet contained botulism toxin. Mortality in this feedlot stopped when the poultry litter was eliminated from diet.

1.5 Fourth outbreak

This outbreak of feedlot mortality (62.5/1,000) appeared in Sabinas Hidalgo, Nuevo Leon (Fig. 1), where affected animals showed signs of paralysis and prostration. Sabinas Hidalgo is in an arid region unsuitable for vampire bat life, and in the state of Nuevo Leon, BPR had not yet been reported (Lee et al. 2012) (Fig. 1). The affected animals had arrived from various locations of and south-eastern Mexico. The neurological signs were similar to those registered in the previous outbreak, but in addition, several animals lost tongue tone and showed poor tongue retraction. Post-mortem examination of three animals failed to reveal any significant gross changes. Since the FAT for rabies was negative and there were no microscopic lesions in the brain, botulism was suspected. This disease was subsequently confirmed by botulinum toxin bioassay in mice. Similar to the previous outbreak, poultry manure had been added to the diet few days before the presentation of clinical signs. The poultry litter also tested positive for botulism toxin, and mortality stopped 16 days after the poultry litter was excluded from diet.

Figure. 2 Prostrated animals from the same pen showing signs consistent with botulism. However, in this ranch, a BPR-positive case confirmed by FAT occurred previously in a grazing heifer. The zone is within the risk areas for BPR (outbreak 3, Tamuin, San Luis Potosi)

1.6 Fifth outbreak

The last outbreak with low mortality (18/1,000) occurred in November, 2013, in General Escobedo, Nuevo León (Fig. 1), in animals that had arrived 2 weeks earlier from the south-eastern Mexico. The clinical signs in affected animals consisted of ataxia, opisthotonus, collapse and death. The presumptive diagnosis was polioencephalomalacia (PEM), and the veterinarian in charge decided that rabies testing was unnecessary. In this feedlot, animals are received with forage during first day and are abruptly changed to a ration with 75 % processed grain and other by-products and supplement (molasses, cottonseed hulls, dried distillers' grain, fat, and others), leaving a proportion of 25 % or less, with bad quality fibre. Formalin-fixed tissues were submitted to our diagnostic laboratory, and microscopic examination of brain revealed neuronal necrosis and severe oedema in cerebral cortex, changes that were consistent with PEM. Interestingly, one of the animals had mild non-suppurative encephalitis with gliosis and some neurons, particularly Purkinje's cells in cerebellum showed large, cytoplasmic, and eosinophilic inclusions which were considered by the pathologist pathognomonic for rabies (Negri bodies). The veterinarian and personnel handling these animals received post-exposure vaccination and became aware that testing for rabies in cattle with neurological signs should be done even in places like Nuevo Leon state, where vampire bats are not present.

2. DISCUSSION

Since most of the cattle fattened in northern Mexico originate from south-eastern regions of the country where BPR is endemic, it is likely that the sick animals that tested positive for rabies were already infected at the time of arrival. Similar scenario has been reported in Texas, USA, where a cow imported from Mexico developed rabies after arrival (Blanton et al. 2011). Therefore, BPR should always be suspected, either in feedlot or in prairie, if cattle exhibit ataxia, knuckling of the fetlocks and ascending hind leg paralysis that progress to prostration, hypoaesthesia and death. This is particularly true for animals that recently have been moved from BPR-endemic areas such as south-eastern Mexico or America. Likewise, BPR should be suspected in neurological cases occurring at the late phases of fattening, weather in feedlot or prairie, particularly in zones known to be the normal habitat for vampire bats. In these feedlots, animals should be vaccinated for BPR on arrival as part of the preventive and health management programme. Only because BPR has not been previously diagnosed in a bovine exploitation is not a valid reason for not suspecting or vaccinating for BPR. Outbreaks of BPR occur suddenly when otherwise healthy vampire colonies become infected and easily spread the virus (Martinez-Burnes et al. 1997; Lee et al. 2012).

Animals with BPR tend to show neurological signs and die one by one while in botulism, several animals develop clinical signs almost simultaneously. If an outbreak of neurological disease affects many animals simultaneously and poultry manure has been added to the diet, botulism should be suspected. Botulism should always be considered in cattle with neurological signs after FAT has ruled out rabies and microscopic examination of the brain had not revealed lesions. In Mexico, outbreaks of botulism typically occur when poultry litter is added to the diet. This form of botulism is much common in the Americas than that of the classic botulism seen in Africa and Australia where phosphorous-deficient cattle develop pica and eat botulism-contaminated material (Bienvenu et al. 1990; Jean et al. 1995; Fitzpatrick 2006). Bovine diets containing poultry litter have been widely used in Mexico despite of the occasional botulism outbreaks registered since the 1980s. The main reason is the low price and availability of this non-protein nitrogenous source as diet supplement for cattle and small ruminants.

TME should also be properly investigated in outbreaks of bovine neurological diseases. For instance, two fattening heifers that died after a brief presentation of neurological signs in an endemic area were tentatively diagnosed as BPR in spite the fact that the herd had been vaccinated. After FAT came back negative for rabies, TME was suspected and confirmed when microscopic examination revealed a severe suppurative meningoencephalitis with widespread vasculitis and thrombosis, the typical lesion of this disease (García-Márquez et al. 2013).

It is not possible to diagnose or differentiate clinically BPR from botulism. For instance, during an outbreak of neurological disease in Trinidad, botulism was first suspected, yet BPR was subsequently proven to be the cause (Carneiro 1954). Conversely, in an outbreak of bovine neurological disease in Turkey, rabies was initially considered when, in fact, the problem was botulism (Senturk and Cihan 2007); it should be noted that BPR and botulism can occur simultaneously in a feedlot, as in the third outbreak described here. Also, BPR may occur as co-morbidity with PEM as described in the fifth outbreak. The clinical signs for BPR, botulism, TME and PEM are summarized in Table 1. (Radostits et al. 2000; Ramírez Romero et al. 2011; Headley et al. 2013). Additionally, an algorithm for pathologic diagnosis and some complementary procedures is proposed in Fig. 3. All of these neurological diseases, including hypovitaminosis A, are common in feedlot cattle (Glock 1998; Ramírez Romero et al. 2011; Headley et al. 2013).

The presence of vampire wound bites in cattle with neurological signs should point out towards a possible BPR problem (Martinez-Burnes et al. 1997; Ramírez Romero et al. 2011).

Table 1 Clinical comparison among common diseases in feedlot characterized by neurological manifestations

Clinical signs	Disease			
	Bovine paralytic rabies (BPR)	Botulism	Thrombotic meningoencephalitis (TME)	Polioencephalomalacia (PEM)
Ataxia	Yes, progressive; most obvious in hind legs	Yes, progressive; most obvious in hind legs	Yes	Yes
Amaurosis	No	No	No	Yes
Fever	No	No	Yes	No
Loss of tongue tone	No	Yes	No	No
Trismus	Yes	No	No	No
Hypoaesthesia	Yes, hind quarters	No	No	No
Bellowing	Yes	No	No	No
Opisthotonus	No	No	Yes	Yes
Tenesmus	Yes	No	No	No
Depression	No	No	Yes	Yes
Alertness	Yes	Yes	No	No
Ptosis	No	Yes	No	No
Thiamine response	No	No	No	Yes (except in sulphur-induced polio)

Antibiotic response	No	No	Yes (early treatment)	No
Gross and microscopic lesions	No gross lesion in the brain but skin lesions caused by vampire bats; microscopically, non-suppurative encephalitis with inclusion bodies	No gross or microscopic lesions	Hemorrhages and infarcts in the brain and microscopic suppurative meningoencephalitis with vascular thrombosis	Mild brain oedema, cerebellar herniation, yellow discoloration of the cerebral cortex (only in severe cases), and microscopic necrosis of cortical neurons
Hematologic abnormalities	No	No	Yes, severe neutrophilia	No
Other laboratory tests	FAT, mice inoculation test and immunohistochemistry and virus isolation	Mice inoculation, bioassay, and demonstration of botulism toxin in feed	Bacterial culture of *Histophilus somni*	Autoflorescence of gray matter under UV light

Modified from Radostits et al. 2000; Ramírez Romero et al. 2011; Glock 1998; Headley et al. 2013

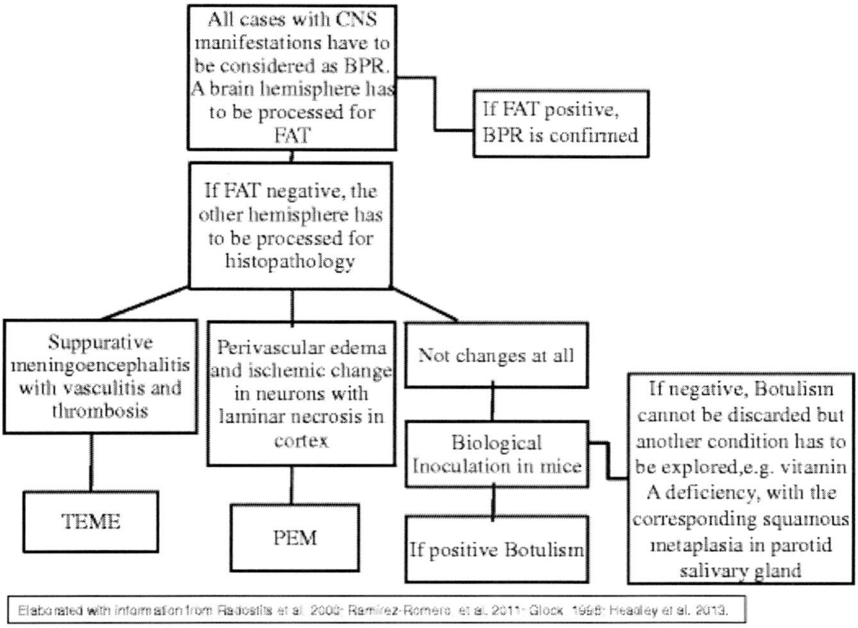

Figure. 3 The diagram illustrates the diagnosis of BPR and ancillary laboratory tests for differential diagnosis of bovine neurological diseases

These wounds or scars are more commonly seen in dorsum of the ears and cervical region, among other anatomical sites (Fig. 4). In botulism, perhaps the most reliable clinical indicator is to pull and release the tongue where slow retraction suggests flaccid paralysis which is typically seen in botulism but not in BPR (Fitzpatrick 2006; Senturk and Cihan 2007).This clinical procedure proved to be useful in the diagnosis of botulism in the fourth outbreak reported here. It should be noted however that tongue retraction test is problematic to do for BPR because rabid cattle often develop trismus ("lockjaw"). Wearing gloves and taking precautions when handling animals with neurological signs are imperative since saliva is a good source of rabies virus. Contact with saliva from a rabid bovine or from tissues at a post-mortem examination of a rabid animal is an occupational risk for veterinarians, animal technicians and farmers. A fatality was reported in a veterinarian who handled a rabid ruminant in Brazil (Brito et al. 2011).

In conclusion, neurological diseases in feedlot cattle should always be properly investigated. BPR should be suspected even in northern Mexico since cattle may arrive to feedlot cattle already sick or incubating the disease. BPR should be considered in areas known to be the normal habitat for vampire bats. Preventive (pre-exposure) rabies vaccination should be mandatory for veterinarians and personnel working in Mexican feedlots. In cases where the rabies FAT results negative, other bovine neurological diseases such as PEM, TME and botulism should be investigated with the help of diagnostic laboratory. Botulism does not produce gross or microscopic lesions; thus, diagnostic is based on clinical signs, history and thoroughly investigating possible sources of botulism toxin such as poultry manure, decayed tissues or bones. Brain histopathology is a reliable diagnostic test for PEM and TME.

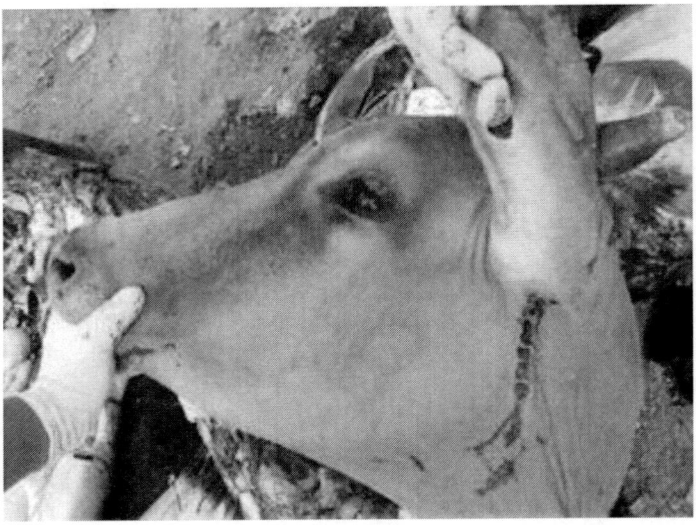

Figure. 4 Bleeding bites inflicted by vampire bats in a confirmed case of BPR (outbreak 2, Tamuin, San Luis Potosi)

ACKNOWLEDGMENTS

This study was supported by the FUNDACION PRODUCE, Nuevo Leon, A.C. project 1349, 2013–2014. Additionally, PROMEP SEP offered partial financial support through the extended 2011 project: Red de estudios sobre enfermedades en fauna silvestre y su papel como reservorios.

REFERENCES

1. Bienvenu, J.-G., Morin, M., Forget, S., 1990. Quebec. Poultry litter associated botulism (type C) in cattle. The Canadian Veterinary Journal, 31, 711.

2. Blanton, J.D., Palmer, D., Dyer, J., Rupprecht, C.E., 2011. Rabies surveillance in the United States during 2010. Journal of the American Veterinary Medical Association, 239, 773–783.
3. Blanton J.D., Dyer, J., McBrayer, J., Rupprecht, C.E., 2012. Rabies surveillance in the United States during 2011. Journal of the American Veterinary Medical Association, 241, 712–722.
4. Brito, M.G., Chamone, T.L., Silva, F.J., Wada, M.Y., Miranda, A.B..., Castilho, J.G., Carrieri, M.L., Kotait, I., Lemos, F.L., 2011. Antemortem diagnosis of human rabies in a veterinarian infected when handling an herbivore in Minas Gerais, Brazil. Revista do Instituto de Medicina Tropical de São Paulo, 53, 39–44.
5. Carneiro, V., 1954. Transmission of rabies by bats in Latin America. Bulletin of the World Health Organization, 10, 775–780.
6. FAO SAGARPA, 2007. Evaluación Alianza para el Campo 2006. Informe de Evaluación Nacional. Subprograma de Salud Animal. Secretaría de Agricultura Ganadería, Desarrollo Rural, Pesca y Alimentación (SAGARPA), México.
7. Fitzpatrick, S., 2006. Botulism poisoning in cattle in the northern territory. Department of Regional Development, Primary Industry, Fisheries and Resources, Northern Territory Government 2006, Serial No. 651, Agdex No. 420/654
8. García-Márquez, L.J., Ramírez-Romero, R., Macedo-Barragán, R.J., Prado-Rebolledo, O.F., 2013. Memorias del XXXVIII Congreso Nacional de Buiatría, Acapulco, México (Asociación Mexicana de Médicos Veterinarios Especialistas en Bovinos, ocasional publication, 38), 306–309.
9. Glock, R.D., 1998. Diagnostic and feedlot pathology. Veterinary Clinics of North America. Food Animal Practice 14, 315–324.
10. Headley, S.A., Oliveira, V.H.S., Figueira, G.H., Bronkhorst, D.E., Alfieri, A.F., Okano, W., Alfieri, A.A., 2013. *Histophilus somni*-induced infections in cattle from Southern Brazil. Tropical Animal Health and Production, doi 10.1007/s11250-013-0402-7
11. Huntsinger, L., Starrs, P.F., 2006. Grazing in arid North America: a biogeographical approach. Sécheresse, 17, 219–233.
12. Jean, D., Fecteau, G., Scott, D., Higgins, R., Quessy, S., 1995. Clostridium botulinum type C intoxication in feedlot steers being fed ensiled poultry litter. The Canadian Veterinary Journal, 3, 626–628.
13. Lee, D.N., Papes, M., Van Den Bussche, R.A., 2012. Present and potential future distribution of common vampire bats in the Americas and the associated risk to cattle. PLoS ONE 7: e42466. doi:10.1371/journal.pone.0042466
14. Martinez-Burnes, J., Lopez, A., Medellin, J., Haines, D., Loza, F., Martinez, M., 1997. An outbreak of vampire bat-transmitted rabies in cattle in northeastern Mexico. The Canadian Veterinary Journal, 38:175–177.
15. Peel, D.S., Mathews Jr., K.H., Johnson, R.J., 2011. Trade, the expanding Mexican beef industry, and feedlot and stocker cattle production in Mexico. LDP-M-206-01. Economic Research Service USDA, 2011.
16. Radostits, O.M., Gay, C.C., Blood, D.C., Hinchcliff, K.W., 2000. Veterinary medicine. Saunders/Elsevier, St. Louis Missouri, pp. 1582–1584.
17. Ramírez Romero, R., González Báez, A., Nevárez Garza, A.M., Rodríguez Tovar, L.E., 2011. Informe de tres casos de rabia paralítica y babesiosis bovina en el municipio de Aldama, Tamaulipas. Veterinaria México, 42, 331–338.
18. Senturk, S., Cihan, H., 2007. Outbreak of botulism in a dairy herd in Turkey. Irish Veterinary Journal, 60, 481–484.

CHAPTER 17

Memory Effects on Movement Behavior in Animal Foraging

Chloe Bracis[1,2], Eliezer Gurarie[3], Bram Van Moorter[4,5], R. Andrew Goodwin[6]

[1]*Quantitative Ecology and Resource Management, University of Washington, Seattle, Washington, United States of America,*

[2]*Senckenberg Biodiversity and Climate Research Centre (BiK-F), Senckenberg Gesellschaft für Naturforschung and Goethe Universität Frankfurt, Frankfurt (Main), Germany,*

[3]*Department of Biology, University of Maryland, College Park, Maryland, United States of America,*

[4]*Centre for Conservation Biology, Norwegian University of Science and Technology, Trondheim, Norway,*

[5]*Norwegian Institute for Nature Research (NINA), Trondheim, Norway*

6 Environmental Laboratory, US Army Engineer Research and Development Center, Portland, Oregon, United States of America

ABSTRACT

An individual's choices are shaped by its experience, a fundamental property of behavior important to understanding complex processes. Learning and memory are observed across many taxa and can drive behaviors, including foraging behavior. To explore the conditions under which memory provides an advantage, we present a continuous-space, continuous-time model of animal movement that incorporates learning and memory. Using simulation models, we evaluate the benefit memory provides across several types of landscapes with variable-quality resources and compare the memory model within a nested hierarchy of simpler models (behavioral switching and random walk). We find that memory almost always leads to improved foraging success, but that this effect is most marked in landscapes containing sparse, contiguous patches of high-value resources that regenerate relatively fast and are located in an otherwise devoid landscape. In these cases, there is a large payoff for finding a resource patch, due to size, value, or locational difficulty. While memory-informed search is difficult to differentiate from other factors using solely movement data, our results suggest that disproportionate spatial use of higher value areas, higher consumption rates, and consumption variability all point to memory influencing the movement direction of animals in certain ecosystems.

1. INTRODUCTION

Sustaining animal populations in fragmented landscapes often depends on understanding and managing their movements, but a major challenge comes from individuals changing their response to a stimulus depending on the context of their decision [1, 2]. An event can therefore elicit variable animal responses as an individual's context changes based on past experiences, even in seemingly static or unchanging environments. Learning and memory are key factors in the emergence of decision context. The unique experience of an individual can change its perceived value of an environmental stimulus over time, which can shape the decisions that underlie an observed movement trajectory [1, 3].

A range of taxa, from insects to primates, have been shown to exhibit spatial memory during foraging. Bees, for example, learn reward values [4]. When displaced into familiar territory, they can choose between two goals and navigate to that goal, demonstrating a map-like memory [5], although path integration and learned landmarks may be sufficient to accomplish these tasks [6]. Fish on coral reefs use learned landmarks to navigate between food patches, including the possibility of novel routes indicating a cognitive map [7]. Birds repeatedly return to previously visited sites [8]. An analysis of visit length and time between visits suggests that monkey movements are non-random due to the use of memory and visitation patterns are driven by resource availability [9].

Studies of foraging behavior in the field have provided evidence for learning and memory across a variety of taxa, complementing the study of foraging and memory in controlled laboratory studies (reviewed in [10]). Memory may be key to understanding patterns observed in animal foraging, yet memory-informed search is difficult to differentiate from other sensory-driven search behaviors [11, 12]. Unlike physiological attributes such as energy reserves or hormone levels, memory is an internal state of the animal that cannot be measured directly. This is especially true in ecological contexts when the history of the animal's experiences may be unknown. In fact, formulating clear behavioral criteria to infer cognitive processes is a particular challenge [13]. Simulation models are therefore a key analytical tool to investigate hypotheses that involve direct measures or manipulation of memory.

Prior work suggests that in foraging, memory-informed movement is advantageous in predictable landscapes [14] and therefore negatively correlated with the rate of environmental change [15]. Incorporating memory has been suggested as a future direction for improving movement models [16, 17]. Methods based on random walks or area-restricted search are often used to interpret movement [11, 18, 19]. When memory is incorporated, the models are often system-specific [20, 21] or use an homogeneous landscape [22, 23], ignoring that heterogeneity in habitat quality can also affect revisit probability [9]. Other studies suggest memory is important to home range formation, which emerges from animals learning their environment as an intermediate state between dispersive and site-fidelious tendencies [24–26]. Returning to a location is a function of information decay and distance: either the balance between a repulsive working memory and an attractive reference memory decaying at different rates and weighted by distance [24], or when foragers benefit from site-specific information after a time delay, such that there is a dynamic equilibrium between resource consumption and regeneration [25].

We explored conditions under which the added cognitive complexity of maintaining memory is advantageous to foraging by creating a flexible, continuous model incorporating memory for an animal moving through a heterogeneous landscape. The main strategic objectives in constructing the model were (1) to build a nested sequence of increasingly more complex models from a random walk null model, (2) to have a straightforward currency for comparing outcomes that, in this case, is resource consumption, and (3) to make the model continuous in space and time to explore properties across scales without concern for the effect of discretization. To illustrate the model, we compare an animal using three nested movement processes: (1) a simple correlated random walk, (2) kinesis, in which correlated random walk parameters, i.e., searching and feeding behaviors, change as a function of the consumption rate, and (3) memory-informed movement, in which directional bias in the searching behavior is informed by previous experiences. We compare the three movement processes across a range of landscape characteristics (patchiness, spatial correlation, and regeneration rate).

To facilitate comparisons of the benefit of the memory process at various interactions of scales, particularly the spatial scales of the landscape with the spatial scale of foraging and memory, we use continuous processes with similar underlying Gaussian kernels for the movement model, the landscape representation, and the foraging and memory processes. This makes the prototypical scenario examined here an animal grazing across stationary resources that deplete and regenerate, rather than a predator consuming discrete prey.

These nested movement processes (random walk, kinesis, and memory) allowed us to evaluate the landscape characteristics that result in the memory model outperforming other behaviors, with implications for the evolution of memory [27]. We were also interested in how the spatial and temporal scales of the memory process relate to the scale of patterns in the landscape, particularly the resource regeneration rate that affects the preferred timing of patch revisits. To better understand the reasons underlying differences among movement processes, we analyzed the corresponding behavioral differences in terms of habitat use and time allocation among the movement processes, and what they suggest for patterns one is likely to observe for an animal using memory.

2. METHODS

In our model, a simulated individual moves through a habitat of variable resource quality, consuming those resources. The individual's movements may be random (random walk), informed by its current consumption rate (kinesis), or informed by its memory of resource quality and consumption rate (memory). In the framework of Nathan et al. [28], finding food resources motivates the animal's movement, and its internal goal is to maximize consumption. We vary only the animal's directional tendency; all other model attributes are similar across movement processes. While the random walk model has no external influence, the current resource quality influences the kinesis movement process. In memory-based movement, both current resource quality and spatial memory influence movement, and navigation towards previously learned habitat areas is possible. The model is two-dimensional in space $z \in R^2$ with time t (e.g., $f(z, t)$, though the dependent variables z and t are omitted in some equations for clarity). Parameters and symbols are listed in Table 1.

Table 1. Parameters used in the foraging model and values for simulations. Because units are arbitrary in the simulations, L is used for generic length units and T is used for generic time units.

Parameter	Definition	Units	Values
Simulations			
Δt	model time step	T	0.1
T	simulation length (time steps)		1000
Landscapes			
μ_Q	patch concentration (GRF mean)		-1.5, -1, -0.5, 0, 1
γ_Q	patch size (GRF scale)		2, 10
Consumption			
β_R	regeneration rate	1/T	0.005, 0.01, 0.05
β_C	consumption rate	1/T	1
γ_C	consumption spatial scale	L	1
Memory[a]			
ψ_M	short-term memory factor		2, 5, 10
β_L, β_S	learning rates	1/T	1
ϕ_L, ϕ_S	decay rates	1/T	0, 1e-5, 1e-4, 0.001, 0.01, 0.1, 0.5
γ_L, γ_S	learning spatial scale	L	1
Movement[b]			
τ_S, τ_F	autocorrelation time scale	T	4, 2
ν_S, ν_F	length of μ	L/T	6, 1
γ_Z	memory spatial scale	L	1, 5, 10
λ	mean time to update θ	T	1

[a] L = long-term memory, S = short-term memory
[b] S = searching, F = feeding

2.1 Habitat quality and consumption

The habitat is modeled as a continuous scalar measure of resource quality that varies across the landscapes. The intrinsic quality, $Q_0(z)$, is constant through time, meaning that we do not consider transient or moving patches. The instantaneous habitat quality, $Q(z, t)$ depends on both consumption, $C(z, t)$, by the animal and a logistic regeneration, $R(z, t)$,

$$\frac{\partial Q}{\partial t} = (R - C)Q \tag{1}$$

Consumption and regeneration are defined as

$$C = \beta_C f_C(|z - Z|) \tag{2}$$

$$R = \beta_R\left(1 - \frac{Q}{Q_0}\right) \tag{3}$$

where β_R is the regeneration rate, and consumption occurs at animal's position, $z = Z$, described by an isotropic spatial kernel, $fC(|z - Z|)$, and consumption rate, β_C. We use a bivariate normal distribution for f_C, and the variance parameter $\gamma^2{}_c$ (length scale) characterizes how widely the animal consumes about its location (**Table 2**). This conception of resource consumption, high in the animal's immediate vicinity and decaying to zero at greater distances, is a good fit for grazing animals, with the length scale representing how far an animal can reach for food as it moves or fine-scale movements on a smaller scale than the trajectory [29]. Within our framework, however, any kind of kernel, including those with limited spatial extents such as discrete top-hat kernels, can be implemented as straightforwardly as others.

Table 2. Spatial kernels used in the foraging model. $\mathcal{N}_2$ is the bivariate normal distribution and I is the 2 × 2 identity matrix.

Description	Equation	Form
Consumption kernel	f_C	$\mathcal{N}_2(\mathbf{0}, \gamma_C^2 I)$
Short-term memory learning kernel	f_S	$\mathcal{N}_2(\mathbf{0}, \gamma_S^2 I)$
Long-term memory learning kernel	f_L	$\mathcal{N}_2(\mathbf{0}, \gamma_L^2 I)$
Memory distance-weighting kernel	f_Z	$\mathrm{Exp}(\gamma_Z)$

2.2 Memory map

As the animal moves across the landscape observing habitat quality, it builds up a memory map, $M(z, t)$ made up of two streams. Two memory streams have been used to detect changes in the environment [30], combine short-term tactical and longer-term strategic behaviors in foraging [31], give rise to stable non-territorial home ranges [24], and represent distal and proximal expectations of reward in conditioning experiments [32]. Multiple memory layers have been suggested for modeling memory decay [11]. The memory map includes a long-term stream, $L(z, t)$, which decays slowly and attracts the animal to high quality habitat, and a short-term stream, $S(z, t)$, which decays more quickly and repels the animal from depleted habitat it has recently occupied. The two memory streams combine linearly to form the memory map, so that positive values are attractive, zero indicates neutrality, and negative values are repulsive:

$$M = L - \psi_M S \tag{4}$$

Because L and S have the same maximum value (Q_0) and S decays faster than L, the short-term memory factor, ψ_M, ensures that the value at a just-visited location will initially be negative, or repulsive, with $\psi_M > 1$. As L and

S decay, the value eventually turns positive and thus attractive for good quality habitat.

Each memory component (L and S) is a mixture of two parts, learning and forgetting,

$$\frac{\partial L}{\partial t} = \beta_L f_L(|z - Z|)(Q_0 - L) - \phi_L L \qquad (5)$$

$$\frac{\partial S}{\partial t} = \beta_S f_S(|z - Z|)(Q_0 - S) - \phi_S S \qquad (6)$$

where β_L and β_S are the learning rates of the long- and short-term memory streams, f_L and f_S are spatial kernels describing learning (Table 2), and ϕ_L and ϕ_S are the decay rates ($\phi_L < \phi_S$).

2.3 Movement model

Movement process. An animal's movements through a landscape are described by a continuous trajectory, $Z(t)$. Taking velocity, $V(t)$, the animal's position is thus $Z(t) = \int_0^t V(t')dt' + Z(0)$, where $Z(0)$ is the animal's initial position. An autocorrelated, directed, continuous movement process is

$$dV = \frac{1}{\tau}(\mu(t) - V)dt \qquad (7)$$

This is similar to the Ornstein-Uhlenbeck process, but without the white noise component. Instead, stochasticity is introduced through the bias vector, described by its magnitude and angle as $\mu(t) = (\nu, \angle\theta)$. The movement process is parameterized by τ, the time scale of autocorrelation, and $\nu = ||\mu(t)||$, the magnitude of the bias vector which controls the average speed of the process. The angle θ is set probabilistically, either from a uniform circular distribution resulting in a random walk or from a probability distribution computed from the memory map. A Poisson process with rate parameter λ is used to update θ, and $\mu(t)$ is constant between updates.

Behavior states.Three nested versions of the movement process are compared: random walk, kinesis, and memory. For the single-state random walk model, a single set of speed and time scale parameters determine the movement process. With the kinesis and memory models, the animal switches between searching and feeding states (**Table 3**). These three movement processes have the advantage of being nested, facilitating comparisons. Thus, random movement is a special case of kinesis where both behaviors have identical parameter values, and kinesis in a special case of memory with either zero learning rates or immediate decay rates. Kinesis is a movement model that performs well in a variety of environments and avoids the strong assumptions of perceptual abilities of area-restricted search [19], which is also known to perform sub-optimally in very patchy environments [33]. Random walk provides a useful null model to compare against.

Table 3. Behavioral states in the model and corresponding movement process parameters.

State	Direction ($\angle\mu$)	Speed ($\lVert\mu\rVert$)	Time scale (τ)
Searching	memory: $\angle\mu \sim g(\theta)$ kinesis: $\angle\mu \sim U(0, 2\pi)$	v_S (fast)	τ_S (large)
Feeding	$\angle\mu \sim U(0, 2\pi)$	v_F (slow)	τ_F (small)

For both kinesis and memory models, movement in the feeding state is tortuous and slow as the animal seeks to exploit the local high quality habitat. Movement is undirected, so the angle θ of the bias term is drawn from a uniform distribution with update times determined by the Poisson process. The searching state, on the other hand, is characterized by more linear movements and directional persistence, and the animal moves with a faster speed. Bias angles still change randomly for kinesis, but with memory, the animal seeks productive patches with the direction determined from the memory map. The angular probability distribution is computed by integrating transects of the memory map radiating out from the forager's location with the memory value at each point weighted by distance. The integrated transects are then normalized by the integrated value for the whole memory. The angular probability density function is given by

$$g(\theta) = \frac{\int_0^r M(r,\theta) f_Z(r)\,dr}{\int_0^{2\pi}\int_0^r M(r,\theta') f_Z(r)\,dr\,d\theta'} \tag{8}$$

where $r = |z - Z|$ and $f_Z(r)$ is a kernel function (e.g., exponential with length scale parameter γ_Z) that weights according to distance, such that closer resources are preferred all else being equal (**Table 2**).

Behavioral state transitions. The animal begins in the searching state, transitions to the feeding state when consumption increases, and transitions back to searching whenever consumption drops. Following the marginal value theorem of optimal foraging theory, an animal should leave a patch when the foraging rate in the patch drops below the average foraging rate in the environment [34]. Thus, in our simulations, the switch between states occurs when the instantaneous consumption rate, $C(t)$, crosses the average consumption rate, C. The average consumption rate is based on the intrinsic habitat quality, $\bar{C} = \beta_C f_C(|z - Z|)\overline{Q_0}$ where Q0 is the average habitat quality over all space.

2.4 Simulations

We ran tens of thousands of simulations of the foraging process based on all three movement models across a range of landscape properties according to the parameter values in **Table 1**. The parameter space was explored systematically, with simulations run for all combinations of specified parameter values. For each movement model, we ran twenty replicate simulations for each combination of the landscape parameters of patch concentration, patch size, and regeneration rate. For the memory model, each combination of memory parameters were run across all landscape replicates.

To make the simulations as comparable as possible, we used the same speed and timescale autocorrelation parameters for both the memory and kinesis models, with the random walk model using the faster, more linear searching behavior speed and timescale autocorrelation, i.e., the random movement parameterization which leads to the highest foraging efficiency as per encounter theory [35].

To examine a stationary scenario (i.e., after the animal is familiar with its environment), simulations begin with the long-term attractive memory stream initialized to the habitat quality, Q_0, and the short-term repulsive stream initialized to zero. Similarly, for the kinesis and memory models, the animal is assumed to know the average consumption rate,C. This assumption corresponds to a forager that is broadly familiar with its environment of regenerating resources, in contrast to a completely naive (e.g., introduced) forager.

The continuous time model (Eqs 4–8) was implemented in Java, with time discretized with small regular intervals Δt approximating dt. If an event from the Poisson process that updates $\mu(t)$ occurred during the interval, then a new angle θ was selected at that interval. The differential equations described above were approximated using the Euler forward method [36]. The time step was selected to satisfy the requirements of the Courant–Friedrichs–Lewy criterion [37] that the time step be smaller than the spatial resolution divided by the magnitude of the fastest velocity to ensure the accuracy and stability of the approximation and also to satisfy $\Delta t < \tau$ (**Table 1**).

Landscapes. To simulate the model, the continuous variation in habitat quality is discretized onto a grid. Landscapes were generated with a Gaussian random field (GRF) using the RandomFields R package [38] to be 50 × 50 in size using an exponential covariance function with variance = 1, nugget = 0, and a set of mean (patch concentration, μ_Q) and scale (patch size, γ_Q) values (**Fig 1**, **Table 1**). Differing levels of patchiness were simulated by truncating all negative values to zero, resulting in more areas of no resources for smaller mean values. Landscapes were normalized to sum to one so that the total resources are the same across all landscapes. For each combination of μ_Q and γ_Q, twenty landscape replicates were generated.

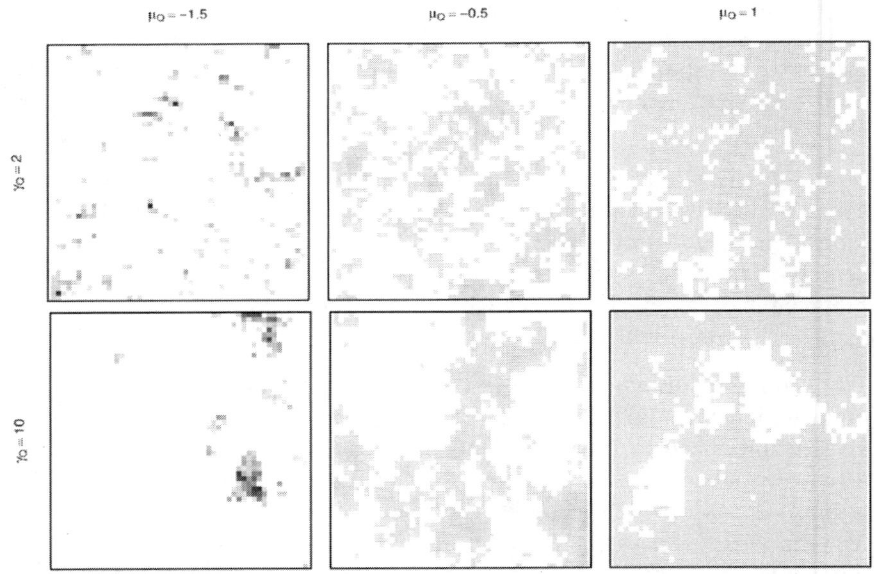

Figure 1. Sample generated landscapes for different combinations of patch concentration from patchy to smooth and patch size from small to large. Color indicates resource quality from none (white) to low (light green) to high (dark green). Total resources in each landscape are the same.

The GRF mean parameter, or patch concentration, controls how concentrated resources are in space along a gradient from patchy (sparsely located high-value resources) to smooth (widespread lower-value resources). The GRF scale parameter, or patch size, controls the relative size and continuity of patches in the generated landscape via the spatial autocorrelation in the generated field. Note that the landscape parameters are not completely orthogonal: patchier landscapes have smaller patches than smooth landscapes. These parameters allowed us to explore a range of landscape types, though the behavioral model is not tied to these generated landscapes.

The initial location of foragers was in all cases the midpoint of each landscape, and the boundaries were reflective. In the memory case, the μ vector was reset on hitting a boundary, either randomly for kinesis or according to the angular probability density function for memory, but restricted to the quadrants away from the boundary. While the memory model can lead to home range behavior [24], the kinesis and random walk models are dispersive, and thus using a bounded landscape makes these processes more comparable while providing a conservative estimate for any advantage from memory.

Metrics. The primary metric used to evaluate simulations was total consumption, i.e., the sum of the forager's consumption (Eq 2) over the fixed duration of each simulation (1000 time steps). Additional metrics were habitat usage (time spent in areas of zero resources and the four quartiles of resources for areas with positive quality) and time spent in each behavioral state, searching and feeding, for kinesis and memory movement processes.

The analysis of model outputs was done with R [39]. To compare the three movement processes we performed the Approximative K-Sample Permutation test [40] where post-hoc comparisons were made with the Nemenyi-Damico-Wolfe-Dunn (NDWD) test using the coin R package [41, 42] and a p-value <0.05 considered significant. Comparisons were made in aggregate, pooling both memory and landscape parameterizations. We also compared the movement processes for each memory model parameterization across all the landscape paramterizations. In these cases we adjusted the p-values to control the false discovery rate using the Benjamini–Hochberg (BH) procedure [43]. To evaluate the contribution of different parameters in the memory model, we used random forests to compute a statistic of relative importance for each parameter [44] using the party R package [45]. An advantage of random forests is their robustness to nonlinearity and complex interaction effects [45]. The method first permutes one of the predictors (thereby removing the potential association between that predictor and the response), then generates random forests (a set of classification trees fit to bootstrap samples drawn from the original), and compares the prediction accuracy of the permuted and unpermuted predictor.

3. RESULTS

3.1 Foraging efficiency

Differences in simulation trajectories and space use, which translate to differences in consumption, were apparent (Fig 2). Averaging over all memory parameterizations and landscapes, the memory model (consumption mean = 0.43, s.e. = 0.29, n = 90720) outperformed the kinesis model (consumption mean = 0.27, s.e. = 0.11, n = 360) and the random walk model (consumption mean = 0.18, s.e. = 0.03, n = 360) and was significant across all groups (Approximative K-Sample Permutation test: maxT = 30.67, p <0.0001; NDWD post-hoc tests: p <0.0001 for all group comparisons). When examining each memory model parameterization separately against the kinesis and random walk models for each landscape parameterization, the total consumption for the three models was nearly always significantly different (98.7% Approximative K-Sample Permutation tests significant with BH p-value adjustment; S1 Appendix). Every memory model parameterization had higher mean consumption than the random walk model across all landscape parameterizations, and 98.7% of these were significantly different (NDWD post-hoc tests with BH p-value adjustment; S1 Appendix). Of the 67.2% of memory model parameterizations that were significantly different from the kinesis model (NDWD post-hoc tests with BH p-value adjustment; S1 Appendix), the memory model had higher average consumption in all cases. Thus, while there was more variability across simulations for the memory model, total consumption was also higher, meaning the increased variability with memory was towards improved performance.

Landscape characteristics affected the degree of benefit memory provides (Fig 3). In more patchy environments (negative μ_Q) with larger patches (large γ_Q), the memory model (using the best overall parameterization across all landscapes) strongly outperformed the kinesis model, which in turn outperformed the random walk model, although the effect weakened as the landscapes become smoother (positive μ_Q) and patches smaller (small γ_Q). Similarly, the percent of memory parameterizations for which consumption was significantly different from the kinesis model was highest in the best performing landscapes (S1 Fig). The pattern held across regeneration rates, but was strongest for higher regeneration rates. Mean consumption values remained constant across landscape parameterizations for the random walk model, increased with only increasing patch concentration for the kinesis model, and increased with both increasing patch concentration and size (γ_Q) for memory. Variability across simulations also increased for all movement processes with increasing patch concentration and size (i.e., landscapes with fewer larger patches).

Even with increased variability in landscapes with high-value and/or large patches, the minimum memory model consumption was higher than the maximum random walk model and generally higher than or close to the mean kinesis model value.

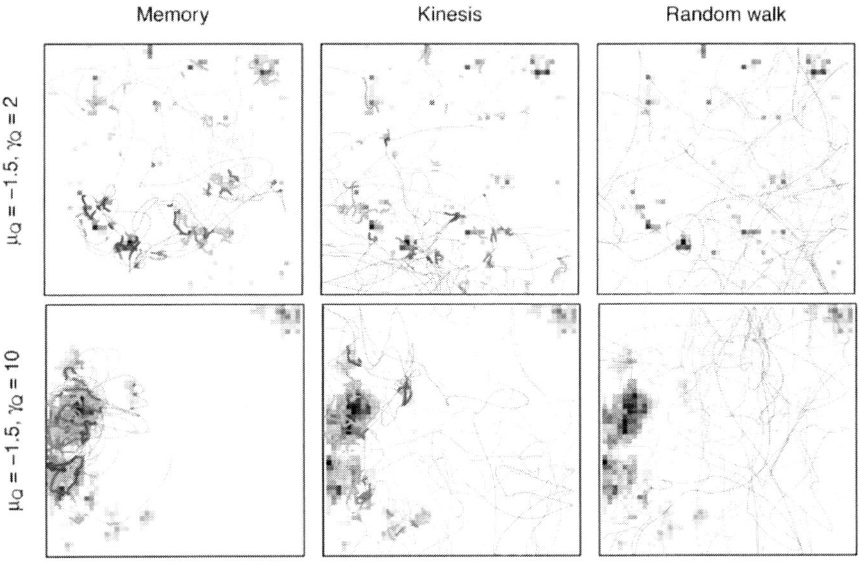

Figure 2. Sample trajectories for three movement models (left to right) on two different landscapes (top to bottom). Trajectories start at the center with color changing through time (from green to light blue to dark blue). For memory and kinesis, thin lines indicate searching and thick lines indicate feeding behavior. Resources are shown at their undepleted level at the beginning of the simulation. Memory is parameterized with best overall parameters, $\phi_L = 1e - 05$, $\phi_S = 0.01$, $\psi_M = 2$, $\gamma_Z = 10$.

3.2 Memory parameters

To assess the best performing memory parameter combination for each landscape parameter combination, we compared the total consumption averaged over landscape replicates (Table 4). While there was less variability within memory parameterizations than across movement processes, patterns still emerged. The best performing short decay rate varied consistently with regeneration rate. There was less consistency in the long decay rate. Interestingly, a long decay rate of 0 was rarely selected, though resource stability means there is no obvious advantage to decaying the long-term attractive memory stream. Little pattern was apparent with the short memory factor in relation to the landscape parameters, although there appeared to be an interaction between the short decay rate and the short memory factor. Larger short memory factors were associated with relatively faster short decay rates. Thus the short-term repulsive memory stream may similarly adjust the overall memory by either a lower weighting of a slowly decaying memory or a higher weighting of a faster decaying memory (Eqs 4 and 6). Finally, a large value was favored for the memory spatial scale, i.e., giving more weight to distant patches.

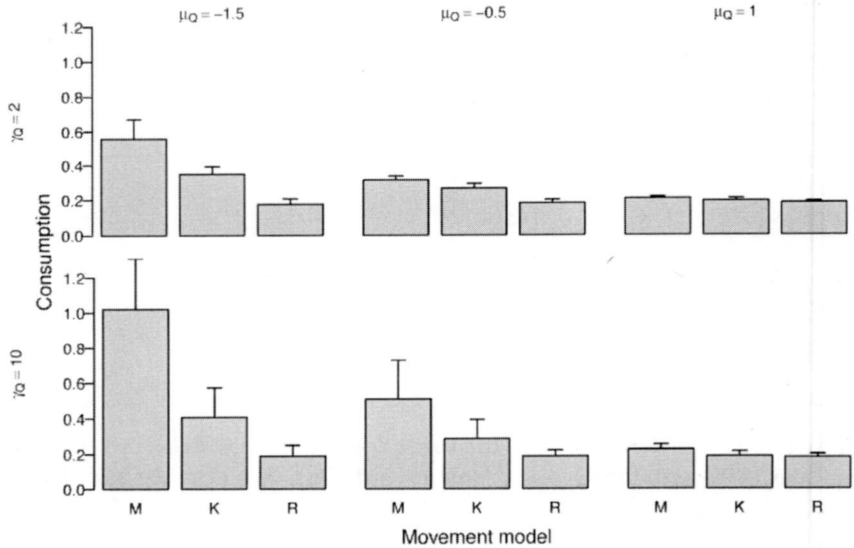

Figure 3. Consumption for the three movement models across different landscape parameters patch concentration μ_Q and size γ_Q for medium regeneration rate β_R = 0.01. Bars show mean consumption values across replicates of landscape parameters while lines show minimum and maximum. Memory is parameterized with best overall parameters, $\phi_L = 1e - 05$, $\phi_S = 0.01$, $\psi_M = 2$, $\gamma_Z = 10$. In the figure, M = memory, K = kinesis, R = random walk.

Comparing variable importance to explain the differences in consumption, landscape parameters dominated memory parameters (Table 5). A conservative rule of thumb to interpret variable importance values is that a variable is informative if its value is greater than the absolute value of the lowest negative value, as irrelevant predictors will randomly vary around zero [46]. All parameters had positive variable importance values.

Table 4. Best performing memory parameters (ϕ_L, ϕ_S, ψ_M, γ_Z) for each landscape environment, a combination of regeneration rate (β_R), patch concentration (μ_Q), and patch size (γ_Q).

		$\beta_R = 0.005$				$\beta_R = 0.01$				$\beta_R = 0.05$			
μ_Q	γ_Q	ϕ_L	ϕ_S	ψ_M	γ_Z	ϕ_L	ϕ_S	ψ_M	γ_Z	ϕ_L	ϕ_S	ψ_M	γ_Z
-1.5	2	1e-05	0.01	2	10	1e-04	0.1	5	10	0.01	0.5	10	5
-1.5	10	1e-05	0.01	2	10	0.001	0.5	5	10	1e-04	0.5	2	10
-0.5	2	1e-04	0.01	10	5	1e-04	0.1	10	5	0.01	0.5	5	5
-0.5	10	1e-05	0.01	2	10	0	0.1	5	5	1e-05	0.1	5	10
1	2	0.001	0.1	10	5	0.01	0.1	10	10	0.01	0.1	5	10
1	10	0.01	0.1	5	10	0.001	0.1	2	10	0.001	0.1	10	10

Of the landscape parameters, patch concentration was the most important, followed by patch size and regeneration rate at a similar order of magnitude. For the memory parameters, the short decay rate was most important, followed by memory spatial scale at a similar order of magnitude, then short memory factor and long decay rate. The long decay rate may only be important as a threshold (i.e., slow enough), and the bulk of the observations were for smaller long decay rates.

Table 5. Permutation importance scores (mean decrease in accuracy) calculated using random forests for the memory model. Results shown treat parameters as continuous variables. Results were similar when parameters were treated as categorical.

Parameter		Variable importance (e-3)
μ_Q	landscape patch concentration	96.3
γ_Q	landscape patch size	27.4
β_R	regeneration rate	6.32
ϕ_S	short decay rate	2.65
γ_Z	memory spatial scale	2.47
ψ_M	short memory factor	0.310
ϕ_L	long decay rate	0.0267

As another approach to evaluating memory, we examined different parameterizations for how often they were significantly different from the kinesis model (S2 Fig) and how much improvement over kinesis (S3 Fig) they provided (memory parameterization nearly always outperformed the random walk model). The clearest pattern emerged with memory spatial scale, γ_Z, with larger values outperforming smaller values almost exclusively, where larger values mean less discounting of distant resources in memory. For the short decay rate, ϕ_S, larger values, and thus faster decay, were generally better, while the opposite was true for the long decay rate, ϕ_S. There was some clustering evident with the short memory factor, ψ_M, as well. For specific landscape parameterizations, the general patterns in parameter values still held but the magnitude of difference between the kinesis and memory models mirrored Fig 3.

3.3 Behavior

The differences in habitat usage observed across the movement processes for different landscape parameterizations (Fig 4) mirrored the differences in spatial concentration for the individual trajectories (Fig 2). The amount of habitat with zero resources varied with patch concentration, from most areas having zero resources ($\mu_Q = -1.5$) to some positive amount of resources nearly everywhere (μ_Q = 1). As expected, the habitat usage of the random walk model matched the distribution of habitat on the landscape. For the kinesis model, habitat usage was markedly biased towards areas of higher quality, though this was not affected by patch size, γ_Q. For the memory model, on the other hand, habitat usage was more skewed towards better quality areas than the kinesis model and also differed by patch size as well as concentration.

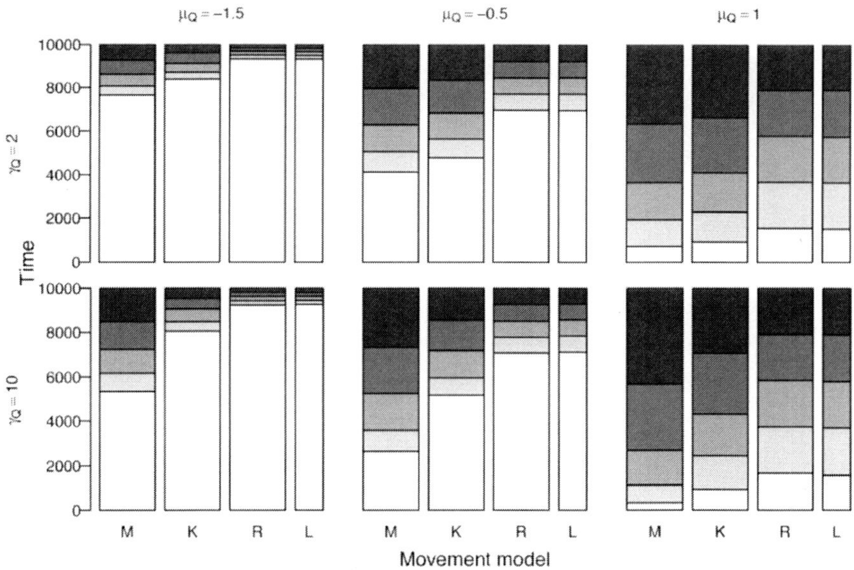

Figure 4. Time spent in areas of different resource quality across different landscape parameters patch concentration μ_Q and size γ_Q for medium regeneration rate β_R = 0.01 compared to the distribution of resources on the landscape. White represents zero resources while shades of gray from light to dark show quartiles of increasing quality. The memory model is parameterized with best overall parameters, ϕ_L = 1e − 05, ϕ_S = 0.01, ψ_M = 2, γ_Z = 10. In the figure, M = memory, K = kinesis, R = random walk, L = landscape.

The memory-informed foragers spent less time searching on average than the kinesis-driven foragers (Fig 5). In the kinesis model, time searching decreased with smoother landscapes where most areas contained some resources. Search time for the memory model was generally consistent across differing patch concentration values, increasing only slightly in patchier landscapes. On the other hand, the kinesis model showed increasing search time variability with increasing patch size, but the median time searching was similar. For the memory model, time searching decreased substantially with increasing patch size.

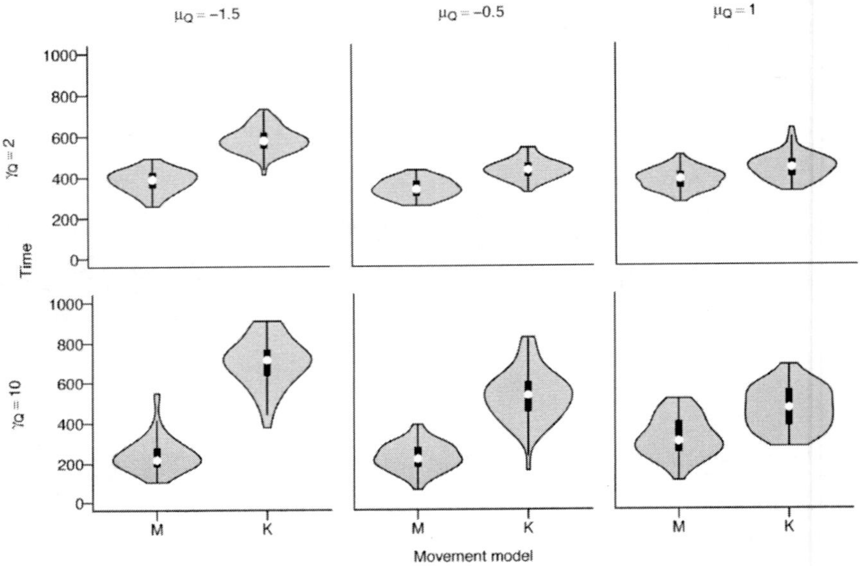

Figure 5. Time spent searching (as opposed to feeding) for memory and kinesis models across different landscape parameter values for patch concentration μ_Q and size γ_Q for medium regeneration rate $\beta_R = 0.01$ as a violin plot showing median values and kernel density plot. The memory model is parameterized with best overall parameters, $\phi L = 1e - 05$, $\phi_S = 0.01$, $\psi_M = 2$, $\gamma_Z = 10$. In the figure, M = memory, K = kinesis.

4. DISCUSSION

4.1 Comparative analysis of model implementation and outcomes

We developed a continuous-time, continuous-space foraging model that incorporates movement with directional preference based on a memory of habitat quality that allowed us to examine the effect of landscape characteristics on the performance of different movement processes, particularly memory. Our model simulation exercises suggest that while the different movement processes perform similarly in smooth landscapes, more complex processes perform better in patchier landscapes. The concentrated resources in patchier landscapes are both harder to locate and of higher value once located, leading to the memory model outperforming the random walk model and the kinesis model. Similarly, foragers using memory also receive higher rewards with a faster regeneration rate and larger high-value patches, which also provide a stronger signal in memory. In general, landscapes that favor memory are those with resources that are higher value or more difficult to encounter, and these landscapes are the places best-suited to look for evidence of memory-informed foraging behavior.

The three movement processes modeled here are nested, so the same underlying movement model is used. Additional behavioral complexity, like switching states or using memory-informed directional biases, can be enabled or disabled within the same framework. This allows for better comparisons between movement processes for simulation studies and also for hypothesis testing if the model is fit to data. The memory implementation is both simple and conforms to current knowledge in animal cognition. Two memory streams are a common feature of cognitively-based ecological models [24, 30–32] and reflect evidence for multiple parallel memory systems that interact [47, 48].

We utilize a continuous space and time architecture in the the model to compare processes that occur across scales, e.g., a single lever can transition between highly tortuous and highly linear movement, between highly patchy and highly homogenous landscapes, and between highly local and highly distributed spatial kernels of foraging and memory. For the movement process particularly, a continuous framework mimics the actual continuous movement of an animal across a continuous landscape, even though sampled discretely. Recent work has focused on parameterizing continuous movement processes from trajectories [49–51] and using continuous-space formulations in spatial population models to represent spatial heterogeneity from multiple sources and at multiple scales [1]. While discrete correlated random walks (CRW) are a common modeling approach (e.g., [14, 21, 24]), CRWs are highly scale-dependent with difficulties in both reformulating from one interval to another and selecting the interval [52, 53]. Discrete movements between patches is one clearly observable and common behavior. This process is modeled, approximately, by the behavioral changes inherent in the behavioral switches of the memory model. Indeed, a discrete movement between known patches under certain conditions is an emergent property of the memory model. However, the model as currently formulated does not accommodate many features of animal movement, such as diurnal behavior. Additionally, we acknowledge that there are problems with scaling at the extremes with both very large and infinitesimal scales.

Using spatially continuous landscape generated from Gaussian random fields allows us to explore a continua of resource patchiness and heterogeneity. However, the behavioral model is completely independent of the landscape, and it can be run on a real landscape based on sampled or remotely sensed data (e.g., [20, 54]), or a landscape of discrete patches (e.g., [24, 26, 35]). An advantage to a continuous landscape representation is that field data are virtually always continuous in some way, while patch-based approaches often make unrealistic assumptions about the shape of the patches and are thus harder to apply to field data. However, this representation is not necessarily applicable to all situations, such as fruiting trees [21] or mobile prey [55]. However, other representations for the landscape and foraging process could be used with the continuous movement model, including discrete models more appropriate to some systems that can almost always be generated as special cases of the continuous models we present. Additionally, the model simplifies the environment into a single gradient, while real animals nearly always have to balance tradeoffs between different requirements and therefore do not move on a single gradient landscape. Moreover, most ecological applications do not directly measure the distribution of an animal's resources on the landscape but instead use remotely-sensed proxies for those resources (e.g., the normalized difference vegetation index (NDVI) as a proxy for forage quality [56]).

Our use of continuous Gaussian kernels in the foraging and memory processes are perhaps more limited in their applicability. The local distance-dependent kernel for learning is analogous to the detected local environment (e.g., via vision or smell), while the kernel of foraging can be thought of as local foraging movements that occur on a finer scale than the displacements, though these analogies break down at infinitesimal scales. Nonetheless, in either case, any kind of kernel can be implemented as straightforwardly as others within our framework, including smaller, non-infinite, or discrete (top-hat) kernels. Other limitations of the model include the exclusion of processes such as bioenergetics, including starvation due to the temporal pattern of energy consumption, stochasticity in resource acquisition or handling time, and diurnal or resting behavior. While all these would be possible to incorporate in the model, we chose to omit them for simplicity. Instead, we focused on the movement, foraging, and memory processes necessary to address our main question of how landscape characteristics affect the performance of the memory model compared to other movement processes.

Our findings corroborate prior work, even where the conception of memory is different. Grove's [27] conclusion that memory is more useful for lower patch density is analogous to the result here for patchier landscapes (with less area containing resources). Our finding that larger patches decrease the utility of memory is less directly comparable due to differences in landscape construction, but the qualitative conclusion that memory is less useful when patches are more easily discoverable is supported. In models that investigate learning and memory of consumption and encounter rates, rather than the spatial location of resources, patch heterogeneity is also an important predictor of the forager's consumption, with memory most useful in conditions of resource heterogeneity and a long-term memory allowing the forager to concentrate effort in the highest-value patches [57–59]. Avgar et al. [15] suggest only physiological constraints should limit memory in an unchanging environment, but that forgetting is adaptive in a temporally changing environment, with memory capacity negatively correlated to the rate of change. Here, the available amount of resources varied temporally, and the rate of forgetting (the short-term memory decay rate) tracked the rate of environmental change (the regeneration rate). The utility of memory in high complexity landscapes may be reduced due to the high cost of tracking sufficient information [11]. An interesting extension of this work would be to model memory at different spatial resolutions and determine if an increasing cost for more detailed memory would produce the dome-shaped relationship between memory utility and resource complexity predicted by Fagan et al. [11].

4.2 Landscape characteristics

Landscape characteristics (Fig 1) mediate the magnitude of consumption variation among the three movement processes. In smooth, continuous landscapes with a relatively even distribution of low resources, the movement processes perform similarly. In patchier landscapes with more area devoid of resources, the kinesis model generally outperforms the random walk model and the memory model outperforms the kinesis model. Resources are both more difficult to locate and of higher value in patchier landscapes, as we held the total amount of resources constant across landscapes. Patchy landscapes favor memory as the benefit of finding resources is high and memory directs movement towards hard-to-find patches. Any tendency to slow down in the few higher-valued patches is an advantage, however, as seen by the kinesis model outperforming the random walk model. The kinesis model has perfect knowledge of the average consumption rate, while if the consumption rate must be learned, the ability to exploit high-value patches increases with memory length [58].

Patch size is also important to consumption rates, as it informs the likelihood that nearby locations also have resources. As patch size increases, memory-based movement dramatically outperforms the kinesis and random walk models, though larger patches can also lead to higher consumption in perceptually-guided foragers [60]. Differences in search times highlight these differences: the memory model uniquely spends less time searching with increasing patch size and exploits the larger, more contiguous patches more efficiently, spending less time traveling between them. Travel time between patches is expected to differ between perceptually-guided and omniscient foragers [27], and this is an emergent property of our model. Larger patches give a stronger signal to the probabilistic direction (Eq 8). The memory-guided forager can easily return to a large patch after wandering off, but the forager using kinesis depends on randomly encountering patches again. Only consumption variability increases with patch size for the kinesis and random walk models, while both consumption and consumption variability increase with patch size for memory. Models struggle to capture the variability present in natural systems, and it has been suggested that spatially explicit models are necessary for foraging behavior to account for the effects of heterogeneous resource distribution and to allow for non-random walk strategies [59]. Although memory-informed movement can give rise to more predictable movement paths such as home range behavior [24, 25] and repeated routes [21], we suggest that it can also lead to increased variability in consumption and thus forager condition.

Performance of memory-based movement also depends on the interaction between the spatial scale of memory and patch sizes in the landscape. Larger memory spatial scales allow the forager to make movements towards distant locations [18]. The better performance of large memory spatial scales show the benefit of considering food farther away. Exceptions occur with small patch sizes, where small spatial ranges may perform better (Table 4). This may be due to small patches having smaller signals in memory, making it advantageous to weigh close resources higher. Small, distant patches may not be worth the travel time. Including perception [14] would likely amplify this effect, as more discontinuous patches would be harder to detect and exploit. Some results suggest that intermediate perceptual abilities perform best with heterogeneous resource distribution as foragers must balance habitat exploration with tradeoffs between time spent searching versus feeding [61].

Lastly, landscape regeneration impacts the performance of the alternative movement processes, something that has been considered in some models [24, 25], but is frequently omitted even in models that explicitly include memory [14, 59]. Differences become readily apparent with high regeneration rates. When patches regenerate quickly there is a benefit to returning soon, favoring memory. This has been demonstrated in other models, with slow regeneration rates resulting in the forager exploring the entire environment and fast regeneration rates resulting in home range behavior [25, 26].

Among the landscape characteristics, regeneration has the largest effect on the memory model's optimal parameterization (Table 4). Faster short-term memory decay is preferred as the regeneration rate increases, returning the animal sooner to productive patches. Animals may track an assessment of the lag time to return to a patch, either inherited or learned based on environmental conditions. However, memory generally outperforms the other movement processes at non-optimal decay rates in patchy landscapes, suggesting that the performance of the memory model could be robust to regeneration rates varying seasonally or among resource types without tracking an assessment of patch return intervals.

4.3 Potential applications

Our model accounts for environmental change through the regeneration rate and depletion by foragers, suggesting its potential application to systems with temporally varying heterogeneous resources, though our reliance on consumption as a metric provides challenges to model calibration given the difficulty of collecting such ancillary data to animal movements. Scales could vary spatially from a nectarivore foraging on flowering plants to a fish feeding in seagrass beds to a cervid browsing forest clearings. Scales could vary temporally from a plant producing additional flowers or fruit from one day to the next to more seasonal regeneration. For example, dugongs (Dugong dugon) forage on seagrass meadows that are patchily located in coastal waters, and the meadows themselves are spatially heterogeneous in quality based on biomass density and species composition [62]. Dugongs have been observed to revisit grazing locations with the return time correlating to regeneration time [63]. As evidence that memory can shape spatial population processes, dugong grazing pressure appears to drive community structure by creating favorable conditions for preferred species, a process known as cultivation grazing [63, 64]. Another candidate application is the Mongolian gazelle (Procapra gutturosa), a wide-ranging species [65], whose habitat preferences correlate to satellite-derived measures of habitat quality [56]. Gazelle movements are highly nomadic through a temporally dynamic landscape, leaving open the question of how much movements are driven by memory, perception, or randomness.

While evaluated separately here, the differing landscapes could also be thought of as different habitats an animal encounters throughout its life history. For example, different resource distributions could represent the differences between feeding and breeding habitats that an animal migrates between or the seasonally varying differences between summer and winter habitats. In this case, the total amount of resources may also differ, but the general conclusions in terms of the relative performance of different movement processes would still hold. Thus, if the model were tested in dynamic landscape conditions, it may be that memory is most useful in times of resource constraint, such as hard-to-find winter forage, after a habitat perturbation, or during times of more limited mobility, such as the breeding season. For example, ungulate species on the Isle of Rhum, Scotland all foraged on high-quality, high-biomass areas in the summer during high resource availability, but showed resource partitioning during the winter [66]. The decimation of foraging habitat caused by a cyclone and floods led to changes in dugong distributions and led to emigration to other areas [67].

In conclusion, we have developed a flexible framework for considering memory with a continuous-space and continuous-time movement model that could be applied to a variety of systems once calibrated and validated against field data. Our results suggest that the best environments to look for evidence of memory-driven search are those with sparse, contiguous patches of high-value resources that regenerate quickly located in an otherwise devoid landscape. Thus, there is a large payoff for finding a resource patch, whether in size, value, or locational difficulty. While separating memory-driven foraging from sensory-driven alternatives is difficult [11], our findings of disproportionate space use of higher value areas, higher consumption rates, and consumption variability all point to memory influencing the movement of animals in a variety of simulated contexts.

ACKNOWLEDGMENTS

The authors thank James J. Anderson, Renee Ha, Mark Kot, Aaron J. Wirsing, and two anonymous reviewers for helpful comments on this work.

REFERENCES

1. Morales JM, Moorcroft PR, Matthiopoulos J, Frair JL, Kie JG, Powell RA, et al. Building the bridge between animal movement and population dynamics. Phil Trans R Soc B. 2010;365(1550):2289–2301. pmid:20566505
2. McNamara JM, Fawcett TW, Houston AI. An adaptive response to uncertainty generates positive and negative contrast effects. Science. 2013;340(6136):1084–1086. pmid: 23723234
3. Goodwin RA, Politano M, Garvin JW, Nestler JM, Hay D, Anderson JJ, et al. Fish navigation of large dams emerges from their modulation of flow field experience. Proc Natl Acad Sci USA. 2014;111(14):5277–5282. pmid:24706826
4. Gil M, De Marco RJ. Honeybees learn the sign and magnitude of reward variations. J Exp Biol. 2009;212(17):2830–2834. pmid:19684218
5. Menzel R, Greggers U, Smith A, Berger S, Brandt R, Brunke S, et al. Honey bees navigate according to a map-like spatial memory. Proc Natl Acad Sci USA. 2005;102(8):3040–3045. pmid:15710880
6. Cruse H, Wehner R. No need for a cognitive map: decentralized memory for insect navigation. PLOS Comput Biol. 2011;7(3):e1002009. pmid:21445233
7. Reese E. Orientation behavior of butterflyfishes (family Chaetodontidae) on coral reefs: spatial learning of route specific landmarks and cognitive maps. In: Motta P, editor.

Developments in Environmental Biology of Fishes. vol. 9. Springer Netherlands; 1989. p. 79–86.
8. Regular PM, Hedd A, Montevecchi WA. Must marine predators always follow scaling laws? Memory guides the foraging decisions of a pursuit-diving seabird. Anim Behav. 2013;86(3):545–552.
9. Boyer D, Crofoot MC, Walsh PD. Non-random walks in monkeys and humans. J R Soc Interface. 2012;9(70):842–847. pmid:22031731
10. Kamil AC, Roitblat HL. The ecology of foraging behavior: implications for animal learning and memory. Annu Rev Psychol. 1985;36(1):141–169. pmid:3883888
11. Fagan WF, Lewis MA, Auger-Méthé M, Avgar T, Benhamou S, Breed G, et al. Spatial memory and animal movement. Ecol Lett. 2013;16(10):1316–1329. pmid:23953128
12. Van Moorter B, Visscher D, Herfindal I, Basille M, Mysterud A. Inferring behavioural mechanisms in habitat selection studies getting the null-hypothesis right for functional and familiarity responses. Ecography. 2013;36(3):323–330.
13. Shettleworth SJ. Animal cognition and animal behaviour. Anim Behav. 2001;61(2):277–286.
14. Mueller T, Fagan W, Grimm V. Integrating individual search and navigation behaviors in mechanistic movement models. Theor Ecol. 2011;4(3):341–355.
15. Avgar T, Deardon R, Fryxell JM. An empirically parameterized individual based model of animal movement, perception, and memory. Ecol Model. 2013;251:158–172.
16. Morales JM, Haydon DT, Frair J, Holsinger KE, Fryxell JM. Extracting more out of relocation data: building movement models as mixtures of random walks. Ecology. 2004;85(9):2436–2445.
17. Patterson TA, Thomas L, Wilcox C, Ovaskainen O, Matthiopoulos J. State-space models of individual animal movement. Trends Ecol Evol. 2008;23(2):87–94. pmid:18191283
18. Smouse PE, Focardi S, Moorcroft PR, Kie JG, Forester JD, Morales JM. Stochastic modelling of animal movement. Phil Trans R Soc B. 2010;365(1550):2201–2211. pmid:20566497
19. Watkins KS, Rose KA. Evaluating the performance of individual-based animal movement models in novel environments. Ecol Model. 2013;250(0):214–234.
20. Bennett DA, Tang W. Modelling adaptive, spatially aware, and mobile agents: elk migration in Yellowstone. Int J Geogr Inf Sci. 2006;20(9):1039–1066.
21. Boyer D, Walsh PD. Modelling the mobility of living organisms in heterogeneous landscapes: does memory improve foraging success? Phil Trans R Soc A. 2010;368(1933):5645–5659. pmid:21078640
22. Gautestad AO, Mysterud I. Spatial memory, habitat auto-facilitation and the emergence of fractal home range patterns. Ecol Model. 2010;221(23):2741–2750.
23. Gautestad AO. Memory matters: influence from a cognitive map on animal space use. J Theor Biol. 2011;287:26–36. pmid:21810430
24. Van Moorter B, Visscher D, Benhamou S, Börger L, Boyce MS, Gaillard JM. Memory keeps you at home: a mechanistic model for home range emergence. Oikos. 2009;118(5):641–652.
25. Spencer WD. Home ranges and the value of spatial information. J Mammal. 2012; 93(4):929–947.
26. Nabe-Nielsen J, Tougaard J, Teilmann J, Lucke K, Forchhammer MC. How a simple adaptive foraging strategy can lead to emergent home ranges and increased food intake. Oikos. 2013;122(9):1307–1316.
27. Grove M. The evolution of spatial memory. Math Biosci. 2013;242(1):25–32. pmid:23246804
28. Nathan R, Getz WM, Revilla E, Holyoak M, Kadmon R, Saltz D, et al. A movement ecology paradigm for unifying organismal movement research. Proc Natl Acad Sci USA. 2008;105(49):19052–19059. pmid:19060196
29. Bailey DW, Gross JE, Laca EA, Rittenhouse LR, Coughenour MB, Swift DM, et al. Mechanisms that result in large herbivore grazing distribution patterns. J Range Manage. 1996;49(5):386–400.

30. Kacelnik A, Krebs JR, Ens B. Foraging in a changing environment: an experiment with starlings (Stumus vulgaris). In: Commons ML, Kacelnik A, Shettleworth SJ, editors. Quantitative Analyses of Behavior VI: Foraging. Hillsdale, NJ: Erlbaum; 1987. p. 63–87.
31. Anderson JJ. An agent-based event driven foraging model. Nat Resour Model. 2002;15(1):55–82.
32. Anderson JJ, Bracis C, Goodwin RA. Pavlovian conditioning from a foraging perspective. In: Ohlsson, S, Catrambone, R, editors. Proceedings of the 32nd Annual Conference of the Cognitive Science Society. Austin: Cognitive Science Society; 2010.
33. Humston R, Olson DB, Ault JS. Behavioral assumptions in models of fish movement and their influence on population dynamics. Trans Am Fish Soc. 2004;133(6):1304–1328.
34. Charnov EL. Optimal foraging, the marginal value theorem. Theor Popul Biol. 1976;9(2):129–136. pmid:1273796
35. Gurarie E, Ovaskainen O. Towards a general formalization of encounter rates in ecology. Theor Ecol. 2013;6(2):189–202.
36. Butcher JC. Numerical methods for ordinary differential equations. Hoboken, NJ: J. Wiley; 2008.
37. Courant R, Friedrichs K, Lewy H. On the partial difference equations of mathematical physics. IBM J Res Dev. 1967;11(2):215–234.
38. Schlather M, Malinowski A, Oesting M, Boecker D, Strokorb K, Engelke S, et al. RandomFields: Simulation and Analysis of Random Fields; 2014. R package version 3.0.10. Available from: http://CRAN.R-project.org/package=RandomFields.
39. R Core Team. R: A Language and Environment for Statistical Computing. Vienna, Austria; 2014. Available from: http://www.R-project.org/.
40. Hollander M, Wolfe DA. Nonparametric Statistical Methods. 2nd ed. New York: John Wiley & Sons; 1999.
41. Hothorn T, Hornik K, van de Wiel MA, Zeileis A. A Lego System for Conditional Inference. Am Stat. 2006;60(3):257–263.
42. Hothorn T, Hornik K, van de Wiel MA, Zeileis A. Implementing a Class of Permutation Tests: The coin Package. J Stat Softw. 2008;28(8):1–23.
43. Benjamini Y, Hochberg Y. Controlling the False Discovery Rate: A Practical and Powerful Approach to Multiple Testing. JPhil Trans R Soc B. 1995 Jan;57(1):289–300.
44. Breiman L. Random forests. Mach Learn. 2001;45(1):5–32.
45. Strobl C, Boulesteix AL, Zeileis A, Hothorn T. Bias in random forest variable importance measures: illustrations, sources and a solution. BMC Bioinformatics. 2007;8(1):25. pmid:17254353
46. Strobl C, Malley J, Tutz G. An introduction to recursive partitioning: rationale, application, and characteristics of classification and regression trees, bagging, and random forests. Psychol Methods. 2009;14(4):323–348. pmid:19968396
47. Poldrack RA, Packard MG. Competition among multiple memory systems: converging evidence from animal and human brain studies. Neuropsychologia. 2003;41(3):245–251. pmid:12457750
48. Kolling N, Behrens TEJ, Mars RB, Rushworth MFS. Neural Mechanisms of Foraging. Science. 2012;336(6077):95–98. pmid:22491854
49. Johnson DS, Thomas DL, Ver Hoef JM, Christ A. A general framework for the analysis of animal resource selection from Telemetry Data. Biometrics. 2008;64(3):968–976. pmid:18047525
50. Gurarie E, Ovaskainen O. Characteristic spatial and temporal scales unify models of animal movement. Am Nat. 2011;178(1):113–123. pmid:21670582
51. McClintock B, Johnson D, Hooten M, Ver Hoef J, Morales J. When to be discrete: the importance of time formulation in understanding animal movement. Mov Ecol. 2014;2(1):21. pmid:25709830
52. Turchin P. Quantitative analysis of movement: measuring and modeling population redistribution in animals and plants. Sunderland, Mass.: Sinauer Associates; 1998.
53. Codling EA, Hill NA. Sampling rate effects on measurements of correlated and biased random walks. J Theor Biol. 2005;233(4):573–588. pmid:15748917

54. Avgar T, Baker JA, Brown GS, Hagens J, Kittle AM, Mallon EE, et al. Space-use behavior of woodland caribou based on a cognitive movement model. J Anim Ecol. 2015.
55. Lima SL. Putting predators back into behavioral predator-prey interactions. Trends Ecol Evol. 2002;17(2):70–75.
56. Mueller T, Olson KA, Fuller TK, Schaller GB, Murray MG, Leimgruber P. In search of forage: predicting dynamic habitats of Mongolian gazelles using satellite-based estimates of vegetation productivity. J Appl Ecol. 2008;45(2):649–658.
57. Eliassen S, Jørgensen C, Mangel M, Giske J. Quantifying the adaptive value of learning in foraging behavior. Am Nat. 2009;174(4):478–489. pmid:19694535
58. Esposito S, Incerti G, Giannino F, Russo D, Mazzoleni S. Integrated modelling of foraging behaviour, energy budget and memory properties. Ecol Model. 2010;221(9):1283–1291.
59. Vincenot CE, Mazzoleni S, Moriya K, Carteni F, Giannino F. How spatial resource distribution and memory impact foraging success: a hybrid model and mechanistic index. Ecological Complex. 2015;22:139–151.
60. Turner MG, Wu Y, Romme WH, Wallace LL. A landscape simulation model of winter foraging by large ungulates. Ecol Model. 1993;69(3?–4):163–184.
61. Oom SP, Beecham JA, Legg CJ, Hester AJ. Foraging in a complex environment: from foraging strategies to emergent spatial properties. Ecol Complex. 2004;1(4):299–327.
62. Sheppard JK, Lawler IR, Marsh H. Seagrass as pasture for seacows: landscape-level dugong habitat evaluation. Estuar Coast Shelf Sci. 2007;71(1–2):117–132.
63. Preen A. Impacts of dugong foraging on seagrass habitats: observational and experimental evidence for cultivation grazing. Mar Ecol Prog Ser. 1995;124(1):201–213.
64. Aragones LV, Lawler IR, Foley WJ, Marsh H. Dugong grazing and turtle cropping: grazing optimization in tropical seagrass systems? Oecologia. 2006;149(4):635–647. pmid: 16821013
65. Olson KA, Fuller TK, Mueller T, Murray MG, Nicolson C, Odonkhuu D, et al. Annual movements of Mongolian gazelles: nomads in the Eastern Steppe. J Arid Environ. 2010;74(11):1435–1442.
66. Gordon I, Illius A. Resource partitioning by ungulates on the Isle of Rhum. Oecologia. 1989;79(3):383–389. pmid:23921404
67. Preen A, Marsh H. Response of dugongs to large-scale loss of seagrass from Hervey Bay, Queensland Australia. Wildl Res. 1995;22(4):507–519.

CHAPTER 18

Lumpy skin disease of cattle: an emerging problem in the Sultanate of Oman

Mohamed Hassan Tageldin[1], David Brian Wallace[2,3], Gertruida Hermanna Gerdes[2,6], John Fraser Putterill[2], Roelf Rudolph Greyling[2,5], Maanda Noaxe Phosiwa[2], Rashied Mohammed Al Busaidy[1], Sultan Issa Al Ismaaily[4]

1 Department of Animal and Veterinary Sciences, College of Agricultural and Marine Sciences, Sultan Qaboos University, P.O Box 34, PC 123 Muscat, Sultanate of Oman

2 ARC-Onderstepoort Veterinary Institute, Private Bag X5, Onderstepoort 0110, South Africa

3 Department of Veterinary Tropical Diseases, Faculty of Veterinary Science, University of Pretoria, Private Bag X4, Onderstepoort 0110, South Africa

4 Ministry of Agriculture, P.O Box 467, PC 100 Muscat, Sultanate of Oman

5 Deltamune, 248 Jean Ave, Centurion 0157, South Africa

6 Churchill Mansions, Observatory, 7925 Cape Town, South Africa

ABSTRACT

Lumpy skin disease (LSD) is a highly infectious disease of cattle caused by a virus belonging to the *Capripoxvirus* genus of the family *Poxviridae*. The purpose of this study is to place on record the first confirmation of LSD in the Sultanate. The disease was diagnosed and confirmed using polymerase chain reaction, histopathology, transmission electron microscopy and serum neutralization testing. The epizootic occurred in 2009 involving a large number of animals and covering a wide area including Nezwa, Alqabel, Sohar, Saham and Burimi. Morbidity and mortality rates of 29.7 and 26.3 %, and 13.6 and 15.4 % were observed at Nezwa and Sohar, respectively. The clinical signs were much more severe in Holstein–Friesian cattle compared to indigenous breeds and were characterized by multiple skin nodules covering the neck, back, perineum, tail, limbs and genital organs. Affected animals also exhibited lameness, emaciation and cessation of milk production. Oedema of limbs and brisket, and superficial lymph node enlargement were highly prominent. It is not known from where the virus originated, or how it spread to the Sultanate. The disease has become endemic in the country and is liable to extend to other Gulf Cooperation Council Countries and cause a pandemic. It is of major concern to the Omani dairy industry. Due to the widespread presence of screw worm, serious economic losses can follow outbreaks.

Keywords: Histopathology, LSD, PCR, Serum neutralization, Sultanate of Oman, TEM

1. INTRODUCTION

Lumpy skin disease (LSD) is a serious disease of cattle caused by lumpy skin disease virus (type strain, Neethling), which, together with sheep and goat poxviruses, constitute the *Capripoxvirus* genus of the *Poxviridae* family (Davies and Otema 1981; Woods 1988). The disease is characterized by large skin nodules covering all parts of the body, fever, enlarged lymph nodes, nasal discharge and lachrymation, but the severity of clinical signs is highly variable (Davies 1991a). LSD causes significant economic losses due to permanent hide damage. Temporary or permanent infertility may occur in cows and bulls (Tuppurainen and Oura 2011). It leads to reduced milk yield and sometimes death due to secondary bacterial infections (Chihota et al. 2003). In addition, it disrupts the trade in cattle and their products from LSD endemic countries (Babiuk et al. 2008). LSD was initially restricted to countries in sub-Saharan Africa, but from 1984 to 1988, there were unconfirmed reports of the disease in cattle in Oman and Kuwait (House et al. 1990; Kumar 2011). In 1988, it was confirmed in Egypt where it subsequently became enzootic (Ali et al. 1990; House et al. 1990; Davies 1991a). The virus then apparently spread by insect transmission from Egypt to Israel in 1989 (Davies 1991b), causing this disease to occur in a number of dairy herds, with a second outbreak occurring in dairy herds in 2006 (Brenner et al. 2006). In both outbreaks, the implementation of vaccination, strict quarantine measures and slaughter policies was successful in eradicating the disease (Davies 1991a;

Yeruham et al. 1995; Brenner et al. 2006). LSD was also suspected in Saudi Arabia in 1989 in Arabian oryx (*Oryx leucoryx*) (Greth et al. 1992).

Clinical cases suggestive of LSD was observed in the Sultanate of Oman in 2009 (Kumar 2011). It was characterized by the appearance of multiple skin nodules covering the entire body associated with the persistent high fever and depression.

A comprehensive clinical report and follow-up treatment on the outbreak in one dairy herd in the Batinah (Sohar) region is available (Kumar 2011). Our report describes the first confirmation of LSD in Oman from Sohar and Nezwa outbreaks and highlights the associated pathological features.

2. MATERIALS AND METHODS

2.1 Clinical history and sample collection

In April 2009, a severe disease of cattle resembling LSD was reported from Nezwa (Interior), Alqabel (Eastern), Sohar, Saham (Batinah) and Burimi regions. The outbreaks involved seven herds (64 North Oman, Jersey and cross-bred cattle) and one herd (3,300 Holstein–Friesian dairy cows) at Nezwa and Sohar, respectively. Samples were collected from 22 and 38 cows from Nezwa and Sohar, respectively. Skin biopsies were collected for virus isolation, polymerase chain reaction (PCR), negative staining for transmission electron microscopy and histopathology. Sera were collected for serum neutralization testing (Beard et al. 2010) and necropsies were performed on two dead Holstein–Friesian animals. Biopsies and tissues collected at necropsy were fixed in 10 % buffered formalin, processed, sectioned and stained with either haematoxylin and eosin or phloxine–tartrazine stain (Bancroft and Gamble 2008).

2.2 Sample preparation for electron microscopic examination

Small tissue sections were excised from visible lesions on the affected tissue and homogenised using a mortar and pestle in sterile double-distilled water (ddH_2O). The suspension was centrifuged at low speed (1,000 × g) for 5 min to remove coarse debris. The supernatant was further centrifuged at 10,000 × g for 20 min and the supernatant fraction discarded. The pellet was gently washed twice with ddH_2O and suspended in phosphotungstic acid (pH 6.4). This suspension was then applied dropwise to a Formvar-coated copper grid, allowed to dry and viewed at 80 kV using a Jeol JEM-1200 transmission electron microscope (Japan).

2.3 Sample preparation for polymerase chain reaction analysis

A thin tissue section removed from each sample using sterile technique was chopped into 5 mm^3 cubes and transferred to a separate mortar. Sterile phosphate-buffered saline (PBS) (2 ml) was added and the pieces were ground with a pestle in carborundum powder. The mixtures were then transferred to Eppendorf tubes and allowed to stand for 3 min to precipitate large detritus. The supernatants were transferred to new Eppendorf tubes and sonicated using a Sonorex TK52 waterbath sonicator (Bandelin, Germany) at 35 kHz for 10 min. The mixtures were subsequently vortexed and centrifuged at 2,000 rpm (358 × *g*) for 2 min in an Avanti 30 Beckman benchtop centrifuge (Beckman, USA). The supernatants were transferred to new Eppendorf tubes and centrifuged at 16,000 rpm (22,897 × *g*) for 15 min to pellet the viral particles. The supernatants were discarded and the virus-containing pellets resuspended in 200 µl PBS for DNA extraction using a MagNA Pure LC Total Nucleic Acid Isolation Kit (Roche, Germany) on a MagNA Pure LC Instrument (Roche, Germany) according to the manufacturer's instructions.

2.4 PCR conditions

The primers were designed from sequence data derived from the South African Onderstepoort vaccine strain and Warm baths field isolate of LSDV (Genbank accession numbers AF409138 and AF409137, respectively) (Kara et al. 2003). Primer pair 1, consisting of primer DW-TK (5′- GCC GAT AAC ATA TAT AGA CCC –3′) and primer OP49 (5′- GTG CTA TCT AGT GCA GCT AT –3′), is used to amplify a 434-bp LSDV genomic fragment between positions 56698–57132, and primer pair 2, consisting of primer L132F (5′- CAC TTC CCT TTT AAG C –3′) and primer L132R (5′- CAT TCT ACA ATC TCC ATG CG –3′), amplifies a 492-bp fragment between genomic positions 119801–120292. The PCRs were performed using an Eppendorf Master Cycler® gradient thermo cycler (Merck, Germany) and 25 µl reaction volumes consisting of 2.5 µl 10× PCR buffer (containing 20 mM $MgCl_2$) (Takara Biomedical, Japan), 2 µl 2.5 mM dNTPs (Takara Biomedical, Japan), 0.25 U Taq DNA polymerase (TaKaRa Ex Taq™, Takara Biomedical, Japan), 20 nmoles of each primer (Gibco-Brl, Scotland), template DNA (~0.1 ng) and sterile ddH_2O. Template DNA was denatured initially for 90 s at 95 °C, followed by 35 cycles of denaturation (45 s at 95 °C), primer annealing (45 s at 56 °C) and strand extension (60 s at 72 °C), ending with a final strand extension step for 7 min at 72 °C. These conditions were used for both primer pairs.

3. RESULTS

3.1 Distribution and clinical signs

The morbidity and mortality rates of 29.7 and 26.3 %, and 13.6 and 15.4 % at Nezwa and Sohar, respectively. At Alqabel and Burimi, 17 and 19 cattle were infected with mortality rates of 5.9 and 5.3 %, respectively (data on morbidity rates were unavailable). At Sohar, 85 % of infected animals either died as the result of complications or were culled due to weakness and low milk productivity. Holstein–Friesian cows showed more severe clinical signs compared to local indigenous cattle. The clinical signs were characterized by nodules ranging in size from several millimetres to 2–5 cm in diameter on various parts of the body, particularly on the head, neck, back, perineum, udder, testicles, tail and limbs. Clinical signs also included salivation, lameness, severe emaciation, cessation of milk production and death. Oedema of the limbs, brisket and superficial lymph node enlargement were very prominent. Conjunctivitis, keratitis and corneal opacity were frequently seen. Necropsy showed nodules on the dorsum of the tongue. The skin nodules involved the epidermis, dermis, subcutaneous tissue and the musculature.

At the onset of the outbreak at Sohar, all non-infected cattle (2,851 animals) were vaccinated with live attenuated sheep and goat pox vaccine (Kenya strain, KS1). Five percent of these animals developed clinical signs of the disease.

3.2 Histopathology and TEM

The skin lesions were characterized by multifocal necrosis and inflammatory infiltration in the epidermis and/or dermis. In some lesions, the necrosis and inflammatory responses were limited to the dermis, while the overlying epidermis was largely intact. These lesions were observed near sweat glands, hair follicles or sebaceous glands. The inflammatory cells observed were mainly lymphocytes, with low numbers of macrophages and occasionally eosinophils. These cells were particularly prominent around blood vessels adjoining the necrotic lesions (Fig. 1). Intracytoplasmic inclusions were observed in mononucleated cells (Fig. 2). The vascular changes were very prominent and included vasculitis, perivasculitis and perivascular necrosis with concomitant thrombosis. Some arteries depicted thickening of tunica media associated with narrowing of the lumen (Fig. 3).

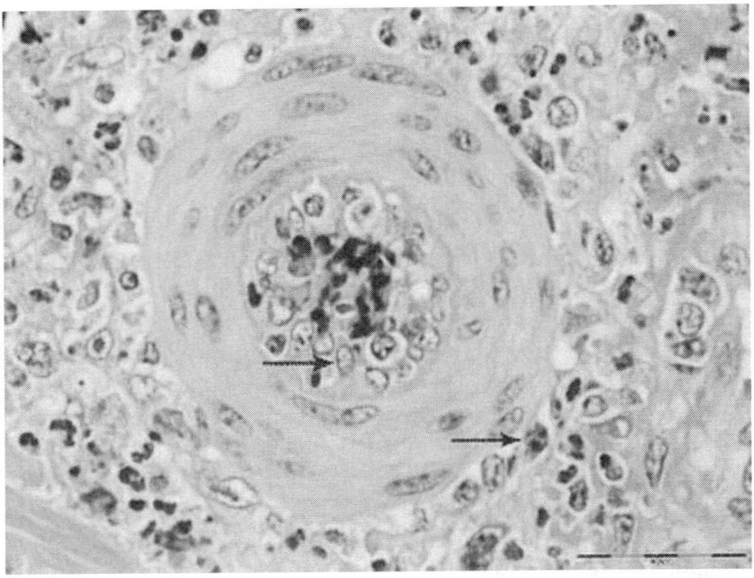

Figure. 1 A small artery in the vicinity of a skin nodule. Note the vasculitis indicated by the presence of inflammatory cells inside and around blood vessels (*arrows*)

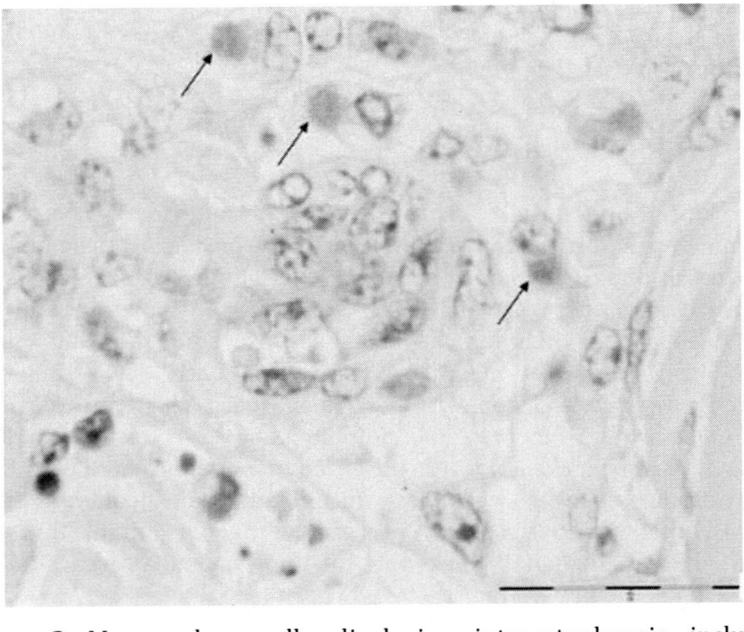

Figure. 2 Mononuclear cells displaying intracytoplasmic inclusions (*arrows*)

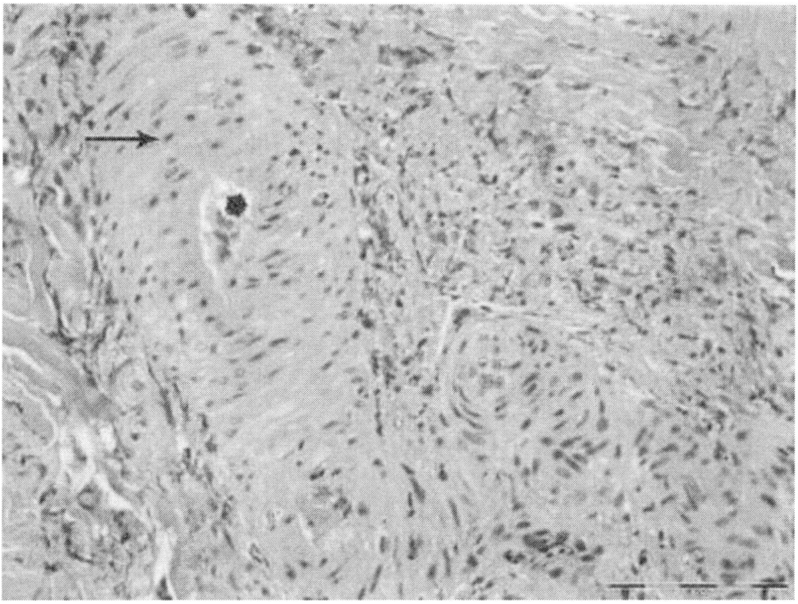

Figure. 3 A small artery showing marked thickening of the tunica media (*arrow*), as well as a narrow lumen (*star*)

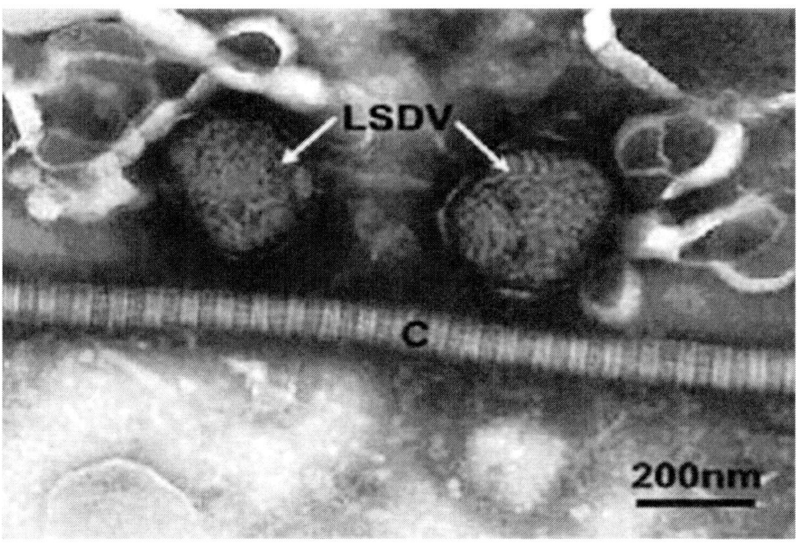

Figure. 4 Transmission electron micrograph of two negatively PTA-stained LSDV particles indicated (*arrows*) in close association with a collagen fibre (*C*). The particles show a typical thread-like structure on their surface and typical "brick-shaped" morphology (*arrows*)

Electron micrograph images of negatively stained preparations showed high densities of typical poxvirus particles (Fig. 4).

3.3 Serology

Sera were collected from Nizwa from infected Jersey, North Oman and North Oman crossed with Jersey cattle, whereas the healthy in contact with infected animals were North Oman cattle and North Oman cattle crossed with Jersey (Table 1). Sera collected from infected Holstein–Friesian cattle were from Sohar. All sera samples tested from infected cattle were positive for antibodies to LSD, whereas those from the healthy in contact with infected animals did not show any detectable antibodies (Table 1).

Table 1 LSD serum neutralization testing for local, cross and exotic breeds

No of serum sample	Area	Neutralization	Animal breed	Animal health
1	Nezwa	Positive	Jersey	Infected
3	Nezwa	Positive	North Oman cattle × Jersey	Infected
7	Nezwa	Positive	North Oman cattle	Infected
1	Nezwa	Negative	North Oman cattle	Healthy-in contact
3	Nezwa	Negative	North Oman × Jersey	Healthy-in contact
19	Sohar	Positive	Holstein–Friesian	Infected

Positive antibodies detected range from 1:4 to 1:24. Negative means no antibodies detected

3.4 PCR analysis

The two primer pairs used for virus identification are homologous to regions of the LSDV thymidine kinase (TK) and ORF132 genes, respectively. The TK gene is highly conserved among the capripoxviruses and thus primer pair 1 also binds to the TK genes of sheep pox and goat pox viruses. However, LSDV ORF132 is unique to LSDV and thus primer pair 2 only binds to LSDV DNA (data not shown).

The results of PCR amplification of DNA extracted from the Oman cattle samples using both primer pairs are shown in Fig. 5. Amplification products of the expected sizes for LSDV were obtained for all the samples, including the positive LSDV controls, when analysed using agarose gel electrophoresis and ethidium bromide staining under UV light, thus indicating that the samples contained a strain of LSDV.

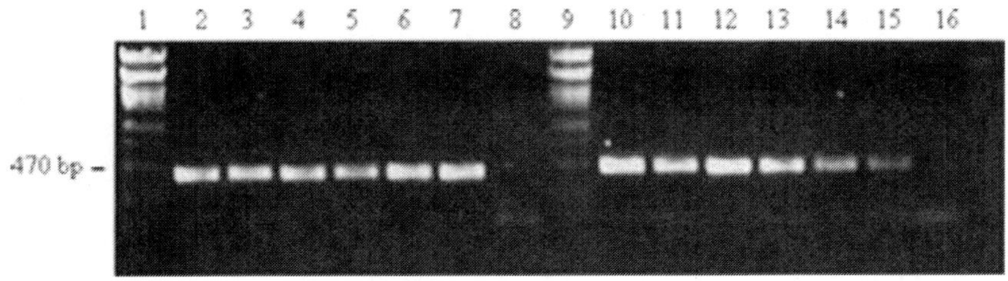

Figure. 5 Agarose gel (1 %) electrophoresis separation of PCR amplification products of Oman LSD samples. Note: DNA was stained using ethidium bromide and viewed and photographed under a UV light source. *Lanes 1 and 9*, Phage lambda PstI marker DNA; *lanes 2–8*, PCR products generated using primer pair 1; *lanes 10–16*, PCR products generated using primer pair 2. *Lane 2*, Oman 1; *lane 3*, Oman 2; *lane 4*, Oman 3; *lane 5*, Oman 4; *lane 6*, positive control (LD18 LSDV field isolate); *lane 7*, positive control (LSDV Onderstepoort vaccine); *lane 8*, sterile water negative control; *lane 10*, Oman 1; *lane 11*, Oman 2; *lane 12*, Oman 3; *lane 13*, Oman 4; *lane 14*, positive control (LD18 LSDV field isolate); *lane 15*, positive control (LSDV Onderstepoort vaccine); *lane 16*, negative control (sterile water)

4. DISCUSSION

Based on the clinical signs, histopathology, PCR, electron microscopy and serum neutralization testing, LSD was diagnosed in Omani cattle in the epizootic of 2009. This represents the first confirmation for the presence of LSD in the Sultanate of Oman. The disease is considered a transboundary animal disease due to its significant impact on trade and food security, and its ability to spread to other countries (Rossiter and Al Hammadi 2009). The real danger of the disease lies in the fact that it has continued to spread, extending its range to include all of Africa, with infrequent appearances in the Middle East (Woods 1988). However, LSD has been eradicated from Israel at a high cost through rapid diagnosis, slaughtering of all diseased and in-contact cattle and small ruminants, and vaccination (Davies 1991b). The LSD epizootic in the Oman involved a large number of animals and a wide area, including Eastern, Interior and Batinah regions. It is unclear how the disease is maintained during inter-epidemic periods (Hunter and Wallace 2001). The virus is thought to persist by unapparent infection cycling in cattle or in old lesions, and low-level persistence in wildlife cannot be excluded (Davies 1991b; Hunter and Wallace 2001). The epizootics observed in the Oman have shown more variation in morbidity and mortality and in the rate at which they spread, as have been previously reported (Ali et al.1990; House et al.1990; Davies, 1991a, b). Such differences are thought to reflect the transmission efficiency and population densities of potential vector populations (Davies 1991b). Holstein–Friesian cattle showed more severe skin lesions compared to the local breeds. It is not known what genetic factors influence the disease severity (Babiuk et al. 2008); however, it is claimed that very young calves, lactating cows and animals suffering from malnutrition generally develop the most severe infections, probably due to impaired cellular immunity (Hunter and Wallace 2001). High ambient temperatures, coupled with farming practices to produce high milk yields, could be deemed to stress the animals and contribute to the severity of the disease in Holstein–Friesian cattle.

The crossbred and North Oman cattle that were in contact with the infected animals remained healthy and failed to develop detectable neutralizing antibodies indicating that direct contact plays little role in the transmission of LSD (Davies 1991a; Carn and Kitching 1995; Babiuk et al. 2008).

The usual quarantine methods, including animal isolation and vaccination, were ineffective control measures in preventing the spread of LSD within the herd at the Sohar farm. This could have been due to the animals already incubating the virus, thus making it too late for the vaccine to afford protection (Hunter and Wallace 2001) or, as observed in Israeli dairy herds (Brenner et al. 2006), the possibility exists that the Kenya sheep and goat pox vaccine is underattenuated, causing clinical disease in a significant proportion of vaccinated animals.

The skin is the most susceptible organ for virus replication (Babiuk et al. 2008), explaining the high virus levels detected in the skin by electron microscopy. In contrast to previous findings (Prozesky and Barnard 1982; House et al. 1990), our skin histopathology results did not reveal the presence of microvesicles, hyperkeratosis or acanthosis. In addition, the presence of eosinophils and the prominent vascular changes were more conspicuous than those reported by Prozesky and Barnard (1982). This discrepancy could be attributed to differences in the stage of disease progression at which the samples were collected or virus strain differences.

All animals displaying clinical signs related to LSD, including the local crossbred and Holstein–Friesian cattle, revealed low levels of neutralizing antibody. This could be attributed to the collection of sera at the early stage of infection when antibody levels are still low or due to the presumed immunosuppressive properties of the virus (Kara et al. 2003).

Due to the widespread presence of screw worm (*Chrysomya bezziana*) in the Sultanate, the danger of co-existence of LSD with myiasis is another problem that could have a serious impact on the economy of the country.

The precise origin of the LSD virus responsible for the outbreak in the Sultanate is unknown. However, it is possible that the disease was introduced into the country by infected cattle imported from the African Horn countries including Somalia and Djibouti. The uncontrolled movement of infected animals may be a factor which increases the hazard of disease spreading to various regions of the Sultanate. Circumstantial evidence suggests that future epizootics are likely to occur. It is feared that the disease could cross boundaries into neighbouring countries or spread fairly rapidly throughout the region and become a pandemic. A new outbreak of LSD in Israel in July 2012 (ProMed report no. 20120728.1218484) attests to the fact that LSD is now firmly entrenched within the Middle East and widespread preventative measures should become routine practice.

ACKNOWLEDGMENTS

The authors wish to thank Professor John Soley (Faculty of Veterinary Science, University of Pretoria, South Africa) for critically reading the manuscript and for assisting with the histopathology data interpretation, and to the South African ARC for financial support.

REFERENCES

1. Ali, A.A., Esmat, M., Attia, H., Selim, A., Abdel-Humid, Y. M., 1990. Clinical and pathological studies on lumpy skin disease in Egypt. Veterinary Record, 127, 549–550.
2. Babiuk, S., Bowden, T. R., Boyle, D.B., Wallace, D. B., Kitching, R. P., 2008. Capripoxviruses: An Emerging Worldwide Threat to Sheep, goats and Cattle. Transboundary and Emerging Diseases, 55, 263 – 72.
3. Bancroft, J.D., Gamble, M., 2008. Theory and Practice of Histological Techniques. In: Bartlett, J.H. (ed), *Microorganisms*, 6th edn. Churchill Livingstone, Elsevier. pp 309–331.
4. Beard, P.M., Sugar, S., Bazarragchaa, E., Gerelmaa, O., Gerelmaa, U., Tserendorj, S. H., Tuppurainen, E., Sodnomdarjaa, R., 2010. A description of two outbreaks of capripox virus disease in Mongolia. Veterinary Microbiology, 142, 427–431.

5. Brenner, J., Haimovitz, M., Oron, E., Stram, Y., Fridgut, O., Bumbarov, V., Kuznetzova, L., Oved, Z., Waserman, A., Garazzi, S., Perl, S., Lahav, D., Edery, N., Yadin, H., 2006. Lumpy skin disease (LSD) in a large dairy herd in Israel, June 2006. Israel Journal of Veterinary Medicine, 61, 73–77.
6. Carn, V.M., Kitching, R. P., 1995. The clinical response of cattle following infection with lumpy skin disease (Neethling). Epidemiology and Infection, 114, 219–226.
7. Chihota, C.M., Rennie, L. F., Kitching, R. P., Mellor, P. S., 2003. Attempted mechanical transmission of lumpy skin disease virus by biting insects. Medical Veterinary Entomology, 17, 294–300.
8. Davies, F.G., 1991a.Lumpy skin disease of cattle: A growing problem in Africa and the Near East. World Animal Review, 68, 37 – 42.
9. Davies, F.G., 1991b. Lumpy skin disease, an African Capripoxviruses Disease of Cattle. British Veterinary Journal, 147, 489 – 502.
10. Davies, F.G., Otema, C., 1981. Relationships of capripox viruses found in Kenya with two Middle Eastern strains and some orthopox viruses. Research in Veterinary Science, 31, 253–255.
11. Greth, A., Gourreau, J. M., Vassart, M., Nguyen-Ba-Vy, M. W., Charies, P., 1992. Capripoxviruses disease in an Arabian Oryx (*Oryx leucoryx*) from Saudi Arabia. Journal of Wildlife Diseases, 28, 295–300.
12. House, J.A., Wilson, T. M.,. El Nakashly, S., Karim, I. A., Ismail, I., El Danaf, N., Moussa, A. M., Ayoub, N. N., 1990. The isolation of lumpy skin disease virus and bovine herpes virus-4 from cattle in Egypt. Journal of Veterinary Diagnostic Investigation, 2, 111–115.
13. Hunter, P., Wallace, D., 2001. Lumpy skin disease in Southern Africa: A review of the disease and aspects of control. Journal South Africa Veterinary Association, 72, 68–71.
14. Kara, P.D., Afonso, C. L., Wallace, D. B., Kutish, G. F., Abolnik, C., Lu, Z., Vreede, F. T., Taljaard, L. C., Zsak, A., Viljoen, G. J., Rock, D. L., 2003. Comparative sequence analysis of the South African Vaccine Strain and two virulent field Isolates of Lumpy Skin Disease Virus. Archives of Virology, 148, 1335–1356.
15. Kumar, S.M., 2011. An Outbreak of Lumpy Skin Disease in a Holstein Dairy Herd in Oman: A Clinical Report. Asian Journal of Animal and Veterinary Advances, 6, 851–859.
16. Prozesky, L., Barnard, B. J., 1982. A study of the pathology of lumpy skin disease in cattle. Onderstepoort Journal of Veterinary Research, 49, 167–175.
17. Rossiter, P.B., Al Hammadi, N., 2009. Living with transboundary animal diseases (TADs). Tropical Animal Health and Production, 41, 999–1004.
18. Tuppurainen, E.S.M., Oura, C.A.L., 2011. Review: Lumpy Skin Disease: An Emerging Threat to Europe, the Middle East and Asia. Transboundary and Emerging Diseases, 59, 40–48.
19. Woods, J.A., 1988. Lumpy skin disease—A Review. Tropical Animal Health and Production, 20, 11–17.
20. Yeruham, I., Nir, O., Braverman, Y., Davidson, M., Grinstein, H., Hymovitch, M., Zamir, O., 1995: Spread of lumpy skin disease in Israeli dairy herds. Veterinary Record, 137, 91–93.

CHAPTER 19

Disgust as an adaptive system for disease avoidance behaviour

Valerie Curtis*, Mícheál de Barra, Robert Aunger

The Hygiene Centre, London School of Hygiene and Tropical Medicine, London WC1E 7HT, UK

ABSTRACT

Disgust is an evolved psychological system for protecting organisms from infection through disease avoidant behaviour. This 'behavioural immune system', present in a diverse array of species, exhibits universal features that orchestrate hygienic behaviour in response to cues of risk of contact with pathogens. However, disgust is also a dynamic adaptive system. Individuals show variation in pathogen avoidance associated with psychological traits like having a neurotic personality, as well as a consequence of being in certain physiological states such as pregnancy or infancy. Three specialized learning mechanisms modify the disgust response: the Garcia effect, evaluative conditioning and the law of contagion. Hygiene behaviour is influenced at the group level through social learning heuristics such as 'copy the frequent'. Finally, group hygiene is extended symbolically to cultural rules about purity and pollution, which create social separations and are enforced as manners. Cooperative hygiene endeavours such as sanitation also reduce pathogen prevalence. Our model allows us to integrate perspectives from psychology, ecology and cultural evolution with those of epidemiology and anthropology. Understanding the nature of disease avoidance psychology at all levels of human organization can inform the design of programmes to improve public health.

Keywords: infectious disease; disgust; evolutionary psychology; adaptive variation;hygiene behaviour; manners

1. INTRODUCTION: THE PROBLEM OF PARASITES

Parasites are ubiquitous; in some ecosystems their biomass rivals that of predators [1]. Parasitic viruses, bacteria, protozoa, nematodes, helminthes and arthropods live in durable relationships with their hosts, from whom they draw energy, shelter, transport and reproductive opportunity. They damage their host's inclusive fitness by producing toxins, manipulating behaviour to their own ends, and spreading to kin and community [2]. The costs of infection constitute an important selection pressure, which all animals face. As a result, natural selection has designed elegant and interlocking solutions to protect animals from parasite damage, including a range of physiological barriers and a complex immune system [3]. Beyond these physiological defences, animals also defend themselves from infection through behaviour that functions as a 'behavioural immune system' [4].

Pathogen avoidance behaviour has a long evolutionary history, and can be found in a broad range of taxa. For example, eusocial insects manage faecal wastes [5], crustaceans avoid diseased conspecifics [6], herbivores forage selectively to avoid faeces [7,8] and grooming behaviour is found in a range of species [9]. The disgust system is a psychological mechanism for producing pathogen avoidant behaviour [10–12]. In previous work, we have stressed the universality of disgust, showing that there is much that is similar about disgust responses between animals and humans, between humans and over historical time [11,13]. However, there is also much that differs between individuals and between social groups. In this paper, we look at disgust and disease avoidance behaviour in human individuals and in human social groups as an adaptive system. Natural selection has produced a solution to the problem of hard-to-detect parasites by designing a system that is sensitive to local information about infection risk. This system responds to parasite pressure not just over evolutionary time, but over lifetimes, using what cues it can. This may be information about an individual's current state, its history of sickness and exposure to disgusting experiences, or what it has learnt from the local culture and from the hygiene practices of others.

In this paper, we begin by recapping evidence that disgust is a universal driver of pathogen avoidance behaviour in humans. We then turn to sources of variation in the disgust system at the individual and group level. We set out the links between the hygiene behaviour of individuals and of groups and between individual disgust and the content of cultures. This allows us to integrate perspectives from psychology, ecology and cultural evolution, as well as epidemiological perspectives on disease prevalence, anthropological perspectives on manners and morality, and the symbolism of purity and pollution. Our conclusions stress the importance of using an evolutionary perspective in combination with interdisciplinary sources to provide an integrated understanding of the set of human behaviours that have a foundation in pathogen avoidance. This, may, in turn, offer insights that are important to the practice of public health.

2. THE UNIVERSALITY OF DISGUST

Disgust is a fundamental part of human nature. Darwin was the first to propose that disgust is expressed universally [14] and many studies since then have supported this proposal [15,16]. Though there has been no systematic cross-cultural survey of the objects and events that elicit disgust in humans, the available data suggest that there is a universal set of disgust cues. These include bodily wastes, body contents, sick, deformed, dead or unhygienic people, some sexual behaviour, dirty environments, certain foods—especially if spoiled or unfamiliar—and certain animals [11,17,18]. Objects that have contacted any of the above can also become disgusting. Further, certain types of immoral acts are widely described as disgusting. Contact with disgust elicitors, real or imagined, is associated with (i) a characteristic facial expression that is recognizable across cultures [16,19], (ii) behaviour patterns that include withdrawal, distancing, stopping or dropping the object of disgust and shuddering [20,21], (iii) physiological changes including lowered blood pressure and galvanic skin response, recruitment of serotonin pathways, increased immune strength [22,23], and (iv) reports of negative affect including nausea.

Pathogen avoidance behaviour is universal across cultures, with all societies demonstrating individual and group-level hygiene behaviours. These include bodily, domestic and communal cleansing, avoidance of close contact or exchange of bodily fluids with others (with exceptions for mates and kin), and the avoidance of foods that are spoilt, contaminated or unfamiliar.

Explanations of disgust in the philosophical, anthropological, humanities and psychological literatures have been varied and inconsistent, reflecting this broad range of phenomena to be explained. The phenomenological philosopher Aurel Kolnai thought that disgust resulted from excess and surfeit: 'A surplus of life, ... an indifference to quality ... a desire towards death' [24, pp. 72–73]. Freud considered disgust a learned reaction formation that could be cultivated towards any activity through development [25]. The social anthropologist Mary Douglas argued that dirt and disgust are a product of culture, such that anomalous objects and events that do not fit the local cosmology have to be rejected, so as not to threaten social order [26]. In psychology, the dominant Rozin–Haidt school has disgust originating in the rejection of spoiled foods, but also serving to cope with the existential terror of being an animal and hence mortal [21,27]. A recent cultural study of disgust labels it 'the Hydra' since it seems too complex to explain [28].

An evolutionary perspective, however, provides a parsimonious explanation for the multiple elicitors and the behavioural tendencies of the disgust system. According to Darwinian thinking, disgust should be considered an adaptive system that drives the behavioural avoidance of infectious disease [11] Constant selection pressure from the ubiquitous presence of pathogenic parasites in animal and human ancestral environments [29] would have selected for those individuals with alleles disposing towards a 'behavioural immune system' preventing contact with, and incorporation of, pathogens [4]. Setting aside, for the present, the issue of moral disgust, it can be seen that all of disgust's basic elicitors (listed above) are implicated in the risk of transmission of infectious disease [11] and paired stimuli with, and without, disease risk show significant differences in disgust response [10,12]. This relationship between disgust elicitors and disease sources appears consistently across cultures and through the historical record [14,20,30]. Rats and other mammals display the characteristic gape expression after eating noxious food, and, as in humans, this reaction is dependent on the insular cortex [31] suggesting that disgust may be at least a pan-mammalian adaptation. Because behavioural immune systems are ubiquitous in animals and predate the evolution of modern humans, all humans should come equipped with a disgust system, rather than learning disgust as a product of culture (as Douglas, Freud and their followers have argued). It is also unlikely that disgust in humans originated in food distaste, because of the pan-vertebral need to avoid pathogens of all types, not just those using food as a vector of infection.

Brain imaging studies also show that there is a specific network associated with disgust. Viewing images of disgusting stimuli, or videos of people with disgusted expressions results in robust and recurring activation in specific brain areas—a neural network including the anterior insular cortex, basal ganglia, ventrolateral prefrontal cortex, anterior temporal cortex, medial prefrontal cortex and visual cortex [25,32,33]. Autobiographical recall of disgusting episodes activates the insular cortex and the basal ganglia [34], as does exposure to disgusting smells [33]. Two recent meta-analyses of functional neuroimaging studies of emotion found that activation of the basal ganglia was reliably associated with disgust [35].

3. THE VARIABILITY OF DISGUST

Although disgust has an ancient and universal function, the disgust system reacts with different levels of activation to the same stimulus between individuals and over the lifetime of the same individual [27,36].

Figure 1 shows global variation in disgust sensitivity from a web survey with 38 845 participants. They were asked to rate how disgusting they found a series of disease-relevant images such as a sick person, a plate of what looked like bodily fluids and a crowded underground train, on a Likert scale of 0–5 (see [10] for details of methods). The results showed a high degree of variability between individuals (average standard deviation = 0.83). Differences in variability between continents were not statistically significant, varying only between 3.77 and 3.94 (s.d. = 0.05). Though a multivariate ANOVA model controlling for age, gender and occupational differences found cultural region to be a significant predictor of disgust sensitivity, this result was largely explained by the low average disgust sensitivity in the Australia/Oceania sample.

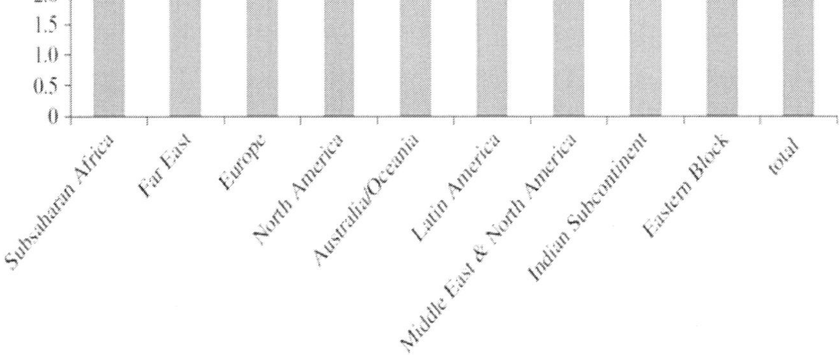

Figure 1. Mean and standard deviation of disgust scores for a sample of 38 845 individuals by region.

How do differences in disgust sensitivity arise? Evolutionary psychological approaches assume that psychological systems—including emotions—

represent solutions to adaptive problems repeatedly encountered during evolutionary history. Plasticity in these systems should reflect the degree of variability in these problems: low levels of variability over the lifespan and between generations will favour the evolution of highly constrained systems. Such systems reliably develop in a broad range of environments with little phenotypic variation and have the benefit of efficiency and rapid and reliable development [37]. Greater environmental variation, on the other hand, should favour more plastic systems that use environmental—including social—information to adaptively shape behaviour according to the challenges of the local environment [38].

Pathogen pressure has led to the selection of a disgust/hygiene behaviour system in individuals that is both universal and plastic to local environmental variation. Figure 2 schematizes the factors that influence the human disease avoidance system. While disgust motivates hygiene behaviour, disgust sensitivity varies between individuals as a trait and within individuals by their state and also through individual learning. Disgust and hygiene in ultrasocial humans facing shared pressure from pathogens are not, however, simply a matter for individuals. The lower half of the diagram depicts group effects in the adaptive system. *Culture*—the socially acquired information shared by a particular group—affects the individual system through social learning and group hygiene behaviour through norms about *manners*. Group hygiene can be symbolically extended into cultural ideas about purity and pollution and can also affect public health by influencing the prevalence and virulence of pathogens in the environment. Group hygiene also influences the hygiene behaviour of individuals via imitation. The content of culture is, itself, a product of the individual brains that support it and so reflects the predispositions of those brains. Finally, pathogens exert selective pressure on the whole system.

All of the components of the system, whether in brains (shown on the left of the diagram) or in behaviour (shown on the right) are affected by environmental factors extraneous to the disease avoidance system, including seasonality, climate, ecosystem, habitat and the particular structure of host populations (not shown). In the following sections of this paper, we describe the elements of this adaptive system and how it works. (For ease of reference, numbers in the figure refer to the relevant sections of the paper.)

4. DISGUST IN BRAINS: SOURCES OF DIVERSITY

Figure 2 shows pathogens exerting selection pressure on individual disgust and the hygiene behaviour it motivates. However, selection has not produced a constant level of disgust sensitivity. Variation can be accounted for by three kinds of phenomena: (i) innate trait differences; (ii) by plasticity, which allows for adaptation to current states; and (iii) the ability to learn from changes in the environment over the lifespan. These types of influence will be treated in turn.

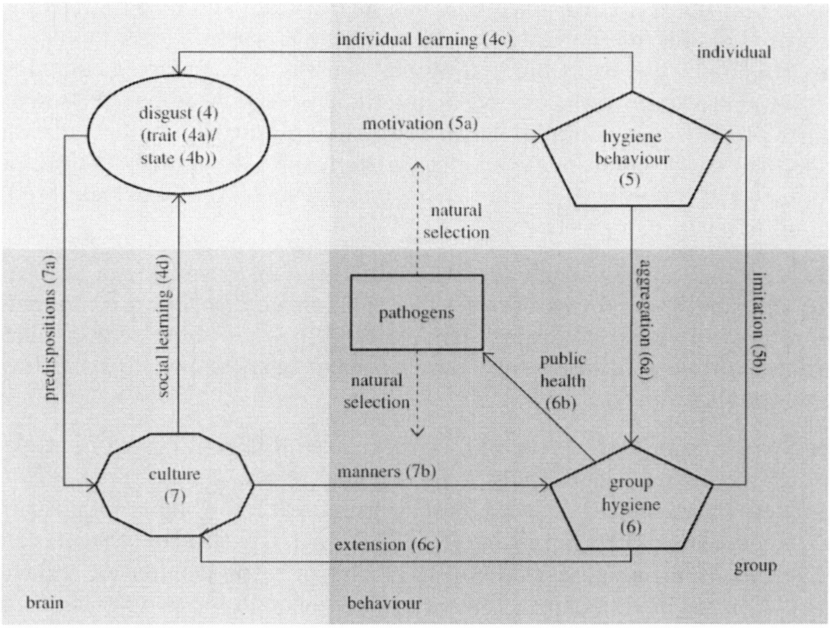

Figure 2. The disease avoidance adaptive system.

(a) Trait-based variation in disgust sensitivity

Individuals can consistently deviate from each other in behaviour because of stable trait-based differences. These arise from previous histories of adaptation in differing selective environments or trade-offs with other competing needs. A variety of traits are associated with variation in disgust sensitivity, including personality, gender and maternally inherited sensitivity.

The dominant theory of personality is the 'Big Five' approach, which suggests that stable, long-term patterns in behavioural proclivities can be summarized along five dimensions: extraversion, neuroticism, agreeableness, conscientiousness and openness [39,40]. Neuroticism is associated with a variety of deleterious traits such as an increased likelihood of experiencing negative emotions such as fear, sadness, anxiety and guilt [41]. High neuroticism is a strong predictor of psychiatric disorder in general, particularly depression [42], and is also associated with impaired physical health, presumably through chronic activation of stress mechanisms [43]. However, the reason for the persistence of such an apparently maladaptive personality trait may be that it helped to reduce the risk of predation and accidents in dangerous ancestral environments [44]. Given that there is covariance between neuroticism and disgust sensitivity scores [45] and that parasites were one of the biggest dangers in ancestral environments, it is probable that disgust sensitivity is, in fact, a component of the neuroticism trait. Indeed many animals display stable 'personality' traits [46,47]. For example, it has been shown that shy sunfish (*Lepomis gibbosus*) carry different types of parasite from bold sunfish, presumably because the differences in their behaviour mean that they are exposed to different levels of parasite risk [47].

Malfunctions of the disgust system can be seen as extreme trait variation. For example, some forms of obsessive compulsive disorder are thought to be related to excessive disgust sensitivity [48], while those with a genetic predisposition to Huntingdon's disease have been shown to have lower levels of disgust [49].

Men and women have recurrently experienced different costs of pathogen exposure. Disgust sensitivity varies consistently between males and females, with females consistently scoring substantially higher on measures of disgust sensitivity than males [10,27,50]. We hypothesized that this trait difference reflects women's differing history of responsibility for childcare [11]. Women, in effect, need to be disgusted enough for two people if they are to keep their dependent children free of disease.

(b) State variation in disgust sensitivity

While individuals are born with varying disgust sensitivity traits (which may or may not be adaptive in current environments), there are also adaptive advantages to being able to modify one's disgust sensitivity according to one's current physiological state. There are times when one may be more vulnerable to pathogens; upregulating disgust sensitivity and hence concern for hygiene may thus be adaptive. Equally, there may be states in which it is advantageous to lower one's disgust sensitivity—when hungry or short of suitable mates, for example.

When physiological immunity is compromised, the probable costs of infection are greater and disgust sensitivity should increase. This interaction between physiological immunity and the behavioural immune system has been termed the 'compensatory prophylaxis hypothesis' by Fessler & Navarrete [51]. In one recent study, participants who reported more frequent infections had both higher disgust sensitivity and more ruminations about contamination and disease [52]. Women undergo adaptive immunosuppression after ovulation and during the first trimester of pregnancy in order for the maternal immune system to be able to tolerate the paternal genetic material in the blastocyst. Fessler *et al.* found stronger disgust responses to disease-relevant stimuli during the first trimester of pregnancy [53]. Increased nausea during early pregnancy is argued to fulfil a similar function, limiting exposure to toxins and pathogens during this vulnerable stage [54–56]. Because progesterone is the endocrine cue for a downregulation of immune response, Fleischman & Fessler [57] found that salivary progesterone correlated positively with disgust sensitivity and the frequency of rumination about disease and contagion in normally cycling women. Similarly, Conway *et al.* [58] found an increased sensitivity to social cues of disease (disgust expression plus averted eyes in another individual's face) when progesterone levels peak in normally cycling women. One further study found that opiate users make fewer errors in identifying disgust faces than non-, or ex-users [59]. Opiates have long been recognized as a powerful immunosuppressant [60].

State changes can also result simply from metabolic activity: after some time, the need to acquire additional resources increases, producing a state of hunger. Though hunger may lead to increased vulnerability to disease, it also signals that the benefit of caloric intake has increased in relation to the benefit of disease avoidance. This trade-off in disgust sensitivity is illustrated by a recent experiment by Hoefling *et al.* [61]. They found that, compared with satiated controls, participants who had not eaten food for 15 h showed significantly reduced facial expressions of disgust when presented with disgusting food stimuli. Interestingly, food deprivation did not influence facial expression of disgust in response to other disease-salient stimuli such as body wastes. The effects of long-term food deprivation have not, to our knowledge, been studied. Anecdotally, however, extreme hunger has been associated with cannibalism and other behaviours that would induce disgust in satiated individuals. It could, therefore, be predicted that chronic malnutrition would downregulate disgust for food but upregulate disgust for other potential sources of pathogens. It could also be expected that individuals who have been unable to secure suitable matings might downregulate their level of disgust towards a mating opportunity, thus trading the possibility of reproductive success for a higher risk of infection.

The perceived vulnerability to disease (PVD) scale has been used to glean self-report data from participants about both their perceived susceptibility and their aversion to germs [62]. The perceived infectability subscale, which asks about prior history of disease (e.g. 'I have a history of susceptibility to infectious disease') correlates with two different disgust sensitivity measures, indicating that the disgust system can be calibrated by previous states of illness (i.e. disease exposure).

(c) Learning from hygiene behaviour

While disgust varies between individuals as a trait and within individuals according to their state, it is also important for the disgust system to be able to respond appropriately to specific features of the current environment. However, learning from pathogens is difficult; they are too small to be seen with the naked eye; they are costly to learn about through trial and error learning; they spread easily and imperceptibly; and their sources are highly diverse. Over evolutionary time, certain classes of stimuli have evolved warning 'flags'—e.g. the smell of rotting flesh, the taste of faeces and the sight of deformity—thanks to the unique association between previous disease threat and a particular cue. Three learning mechanisms help animals to tune this innate system to current threats.

The first mechanism used by humans, and other omnivores, for responding to environmental variation in disease threat is known as the Garcia effect. Here, the experience of illness following ingestion of a food results in the 'flagging' of that food as aversive [63]. The Garcia mechanism is known to 'misfire': nausea induced by chemotherapy can result in long-term aversion to foods consumed before treatment [64]. More importantly from an evolutionary perspective, Garcia learning can only occur following illness. It is thus an efficient, but risky, way of establishing what is, and is not, a disease threat in the local environment.

Evaluative conditioning is a second learning process that results in the creation of stable disgust flags. Here, a powerful disgust reaction results in a previously neutral stimuli acquiring a disgust label (e.g. the sight of lumpy milk may fail to cause a disgust response until the milk is tasted and the disgust response to the bad taste then creates an association with the sight of lumpy milk). Labels like this are quickly acquired and slow to fade [65,66]; the process is succinctly described as 'a sticky form of relational learning resistant to extinction' in the title of a paper by Olatunji *et al.* [66]. Indeed, Baumeister [67] argues that this kind of learning is the primary function of the 'hot' affective component of the emotion.

A third disgust-specific learning mechanism is the 'law of contagion'. When an object or stimulus flagged as disgusting touches another previously neutral object, this new object immediately acquires a disgust label, even if the contact is fleeting and no visible trace remains [20]. This learning system can thus track the spread of pathogenic contamination from surface to surface. Contagion-based labels are temporary and item specific: for example, a pen touched on a toilet seat will acquire a disgust label, but this will dissipate given time or washing, and other unrelated pens will not acquire this disgust flag [68].

(d) Social learning from culture

Humans are 'informavores', seeking information about the best way to behave both from what others say and from what others do. Conspecifics provide a rich source of information about what is disgusting in the local environment. Social learning can make use of that information, avoiding the need for risky learning from direct contact with pathogen cues. This body of knowledge can then be readily passed on vertically from parent to child, and can also spread horizontally (within generations), given the human fascination with the disgusting [28,69].

An important way to acquire knowledge of what to avoid is to pay attention to specific social cues such as expressions of disgust in conspecifics. The fear literature suggests that experience of the expression of fear in a conspecific may be necessary for genetically 'prepared' stimuli (such as snakes) to become fear inducing [70]. A similar social triggering is probably required for some cues to become disgusting as well. The expression of disgust in another human elicits activation of the same specialized neural circuits that are activated by disgusting stimuli [33]; presumably this facilitates the acquisition of new disgust 'flags'. A study by Bayliss *et al.* [71] examined this process by looking at how facial expressions influenced participants' evaluations of everyday objects. Objects paired with disgust expressions were rated more aversive than those paired with neutral or joyful faces. The ability of facial expressions of disgust to influence behaviour emerges early in life. Hertenstein *et al.* found that a 15 s verbal and visual display of disgust towards a novel object reduced infants' contact with a novel toy. In 14 month olds, the effects lasted at least 1 h [72].

Social learning may also be important in the reduction or removal of inappropriate disgust flags. In the domain of food choice, children are typically conservative in their preferences, with many foods classed as distasteful or disgusting. In an environment with adults and other individuals consuming food without disgust, negative evaluations fade and a more varied diet can develop [73]. This process can account for the wide cross-cultural differences in patterns of food preferences, as well as the ability to consume 'off'-smelling foods such as durian and blue cheese, or unfamiliar animal foods such as witchety grubs or balut eggs. In the absence of appropriate social learning, benign or even healthy foodstuffs such as oily fish or green vegetables may continue to elicit disgust throughout life [73]. Analogous processes are common in non-human animals; Norway rats become willing to try a new food if they smell traces of the food on the coat of a conspecific [74]. A similar process may play a role in the reduction of stigma towards—and disgust of—people with physical deformity, for example. Experiments with infants show that positive attitudes by parents result in the dissipation of avoidance behaviours [75].

5. HYGIENE BEHAVIOUR

In figure 2 we show individual hygiene behaviour as a product of the disgust system in individual brains and also a product of the copying of the hygiene behaviour of others.

(a) Disgust motivates hygiene behaviour

There is surprisingly little data on correlations between disgust sensitivity and hygiene behaviour. However, a recent study found that high disgust sensitivity predicted behaviour on only one of four behavioural avoidance tasks. The authors attributed this inconsistency to ceiling effects in their own experimental design. However, individual self reports of disgust during the task were a predictor of avoidance behaviour in each task [76]. Using more aversive behavioural tasks (touching urine, eating a cookie from the floor, etc.), Deacon & Olatunji [77] found that higher disgust sensitivity was associated with greater behavioural avoidance. Paul Rozin's many studies have demonstrated that participants who rated their subjective experience of disgust on paper as being particularly strong were more likely to refuse to touch a cockroach, less likely to eat unusual food items and less willing to touch body products like mucous [78].

(b) Imitation: copy the successful, copy the frequent

Section 4*c* described how learning in the form of 'do what I say' can influence the disgust system. It is also possible for individuals to learn from the behaviour of others directly—learning in the form of 'do what I do'. Humans use the behaviour of those around them as cues as to what are 'fit' strategies, using both the frequency of the behaviour [79] and the success of the model (e.g. wealth, high status and health) as guides [80].

We have some evidence that individuals do, in fact, imitate the hygiene behaviour of the group (though there are other effects here beyond blind copying). A study monitoring handwashing with soap in a motorway service station showed that it was more frequent if there were several people present in the toilet [81]. Similar effects have been found elsewhere [82].

Such blind copying can, of course be harmful. A review of handwashing studies in 11 countries found that handwashing with soap after defecation was rare (17% observed to do so on average over the studies). Respondents often claimed that handwashing with soap was not practiced because 'it is not what we do around here'. The copying of the unhealthy, but frequent, variant was thus perpetuating risky behaviour in the group as a whole [83].

6. GROUP HYGIENE BEHAVIOUR

So far we have looked at the effects of proximate factors on disgust sensitivity in brains and the hygienic behaviour of individual humans. Yet, humans are a social species; our behaviour reflects social as well as individual considerations. The lower half of figure 2 schematizes group-based effects on the disease avoidance system.

(a) Aggregation

Clearly group hygiene behaviour is the sum total of the hygiene behaviours of individuals; this is schematized by the 'aggregation' line in figure 2.

(b) Group-level hygiene behaviour and its effect on pathogens

Group hygiene behaviour is not, however, solely an amalgam of the behaviour of individuals. A social group acting together can engage in a wide range of cooperative strategies to reduce the transmission or the virulence of pathogens, or destroy their habitat. This is, indeed, the basic premise of public health. Individuals cooperate for the benefit of the group as a whole in, for example, sequestrating the sick, building and maintaining water and waste disposal infrastructure, adopting certain places and not others as defecation grounds, and in the provision of health services, such as vaccination, infectious disease treatment, the promotion of condoms and insecticide-treated bednets and the operation of food hygiene systems.

Public health is a common good: individuals who defect from paying the costs of contributing can still enjoy the benefits. Public health problems are, therefore, subject to free riding and the 'tragedy of the commons' [84]. The maintenance of cooperation with respect to public hygiene is notoriously difficult. In developing countries, community water supplies may work well for the first few years after they are constructed but then break down, for example. This is because it only takes a few individuals defecting from contributing to the on-going costs for cooperation to falter [85]. Various remedies have been tried. For example, in the 2009 H1N1 swine flu epidemic, healthcare personnel in the UK were implored to put the common good above their own fear of the individual effects of vaccination. Cooperative hygiene behaviour may need reinforcement in the form of sanctions or punishment—as we will discuss in §7*b*.

Group hygiene became especially important for humans living in close spatial proximity, as technological advances led to increasing urbanization [86]. While pathogens that spread directly from person to person may have tended to evolve towards lower virulence (it is to their own transmission advantage to keep their hosts alive and ambulatory), pathogens that spread through vectors such as mosquitoes or communal water supplies can evolve towards greater virulence. Ewald has presented evidence that one group hygiene behaviour, the chlorination of water supplies, has driven the cholera vibrion towards lower virulence because it now has to rely on interpersonal transmission rather than easy passage through faecally contaminated water [87].

(c) The cultural extension of group hygiene

We have seen how group hygiene behaviour emerges from the hygiene behaviour of individuals, and in the following section we will see how norms serve to reinforce hygiene as a public good. But does the hygiene behaviour of groups also affect the content of culture? The large anthropological literature on the symbolic use of hygiene in the separations that cultures make between the pure and the polluted, the in-caste and out-caste, the sacred and the profane, suggests that it does. [26,88].

In figure 2 we have labelled as *extension*, the process by which hygiene behaviours with a biological function come to serve other cultural functions. So, for example, the act of washing took on symbolic significance in the lustrations and purification rituals of the early Semitic and ancient Greek religions, and later in Hinduism, Buddhism, Islam and Christianity. In ancient Egypt, priests were a group set apart by their immaculate purity; washing from head to foot twice every day and twice every night [89]. The Semitic religions employ rituals such as 'Kippuru': the washing off of a specially applied paste, to remove both material and moral pollution. Greek texts show how purification rituals could 'cleanse' the pollutions not just of childbirth, death and sex, but also of murder [90]. Muslims must bathe after defecation and prior to leaving the house, as well as symbolically, before contact with the divine in prayer.

The Laws of Manu, sacred texts of Hindu scripture (circa 200 BCE), prescribe the avoidance of the 12 impurities of the body: 'Oily exudations, semen, blood, urine, faeces, the mucous of the nose, ear wax, phlegm, tears, the rheum of the eyes and sweat ... ' (Chapter V versus 135). While the biological basis for the avoidance of these substances is clear, the same text extends the argument to those whose work involves contact with polluting substances, making them a separate and polluted caste. These social distinctions may have their origins in biological avoidance strategies but then are made purely symbolic through invocation of the notion of 'purity'. Thus, groups in power can label other groups 'impure'. Untouchables can be argued to be infectious, hence disgusting and contagious. These associations are used to stigmatize the poor and the low caste. One's ability to stay physically clean—a measure of class—is extended to one's ability to remain morally pure.

The symbolic transformation from untouchable as 'dirty' to untouchable as 'impure' makes a group at one stroke not just biologically dangerous, but culturally dangerous. Rules of social distinction and separation become based on purely symbolic criteria, but are supported and sustained by the emotional resources of disgust and contagion [91]. With these rules of separation in place, one can then invoke moral indignation when they are violated, because such acts infect the social body with dangerous ideas of disease, sedition and corruption of the status quo, raising the possibility of the overturn of the social hierarchy.

Dietary laws can also be extended from biologically reasonable avoidances of particular animals as food to whole new classes of animals. Adherence to dietary rules then serves as symbols that demarcate members of one's own group from others [26]. The same logic of extension can be applied to relationships with outsiders: members of other groups can be both dirty (because they are likely to harbour unfamiliar diseases) and impure because they do not follow the same cultural prescriptions. Such xenophobia motivates solidarity within the group, which is helpful in inter-group conflicts, and so can be adaptive [92].

7. THE EFFECTS OF CULTURE ON DISGUST AND HYGIENE

Culture consists of the pool of skills, attitudes, beliefs and values which have been socially learned by the set of individuals in a population or group [93]. Figure 2 shows culture affecting disgust via social learning (which we discussed in §4*b*). Culture also affects the hygiene behaviours that are adopted by groups (and hence, the behaviour of individuals) through the power of norms and manners. We also show group culture as, at least partially, a product of the biases of individual brains through predispositions to take on board certain ideas and not others. In modern societies, culture has also come to contain scientific ideas about pathogens and disease, and microscopic pathogens have finally become 'visible'. We will discuss the consequences of these developments for the adaptive system at the end of §8.

(a) Psychological predispositions affect culture

The content of culture is determined, at least to some degree, by learning biases: predispositions to pay attention to certain facts and not others [94]. Information about what is disgusting is salient to individuals because it can inform about how to avoid disease. Hence, individuals will pay attention to such information and will also preferentially pass it on to other individuals, if this can benefit their own fitness directly or indirectly. For example, the germ theory of disease may have come to be widely accepted globally because the idea of the spread of invisible disease agents accords well with the law of contagion and the predisposition it creates to be aversive to contamination [13]. Urban myths and etiquette rules that involve body wastes, animals or other disgust elicitors have greater 'cultural fitness' than non-emotional alternatives [95,96]. Although disgust is a negatively valenced emotion, and people generally avoid environments or people that elicit the emotion, humans also exhibit a clear fascination for the grotesque, the lewd and the filthy.

(b) Manners

Groups have all sorts of cultural beliefs, attitudes and norms that can have major effects on behaviour. While beliefs and attitudes can be idiosyncratic, norms are commonly accepted rules for appropriate behaviour. Deviations from proper behaviour generally attract sanctions, such as shunning, fining or killing. Every culture has norms about hygiene behaviour, though cultures differ in which specific behaviours are regarded as hygienic, concerning, for example, table manners, willingness to engage in physical contact with others, bathing rituals, food taboos and caste divisions. Such rules determine the way cultures treat those who do 'dirty jobs' and the things they label as polluted. Culturally prescribed rules that limit contact with pathogenic substances, and thus support more instinctive avoidance behaviours, can be called 'manners'. Manners such as keeping oneself groomed and hence ecto-parasite free, sneezing into a handkerchief, practicing safe sex or not defecating in public spaces serve to keep one's own infectious material away from others. While the emotional resources of disgust underpin these norms—as Nichols' analysis of sixteenth-century manuals of etiquette showed [96]—they can also serve to demarcate groups by caste, class or as 'outsider'.

Hence individuals who do not maintain a level of individual hygiene behaviour sufficient to protect others have 'bad manners'. They are commonly labelled as disgusting, can become subject to social sanctions (e.g. shunned), and thus lose the benefits that accrue from social life. Societies employ 'hygiene police' such as environmental health officers and border control officials to detect and punish those who cheat on pro-social hygiene rules—e.g. by selling contaminated food or importing potentially rabid dogs. Those who do not pay their water rates are fined or lose access to the benefits of clean water, and failure to address the need to sanction non-cooperators is one of the key reasons why village level water supply systems so often fail in developing countries [85].

8. DISCUSSION: THE DISEASE AVOIDANCE SYSTEM

In this paper, we have set out the components of the adaptive system that produces disease avoidance behaviour. We have seen that natural selection has produced a system that has taught individuals how to behave hygienically. Pathogens are, however, not just an issue for individuals, but for groups, and as a highly social animal, humans have group responses to diseases. Hence we have postulated a parallel system whereby group brains (culture) affect group behaviour as regards hygiene, and vice versa, and these can have important effects on the prevalence of pathogens. These group effects affect the brain and behavioural responses of individuals.

While we have set out evidence from many types of source about the individual links, is there any evidence of it acting as a complete system? Some intriguing new studies provide evidence that this is the case. It has been shown that societies faced with high pathogen pressure have higher average scores on personality traits such as extraversion and openness to experience [97], higher average scores on collectivism versus individualism scales [98] and higher numbers of religions [99]. These effects could not be explained by latitude, climate or socio-economic status. Further, in the first two studies, historical disease prevalence was a better predictor of these psychological variables than contemporary prevalence, suggesting that disease risk is a plausible cause, rather than a consequence, of the cultural differences observed. Taken together, these results suggest that societies that have faced high pathogen threats tend to become more inwardly focused, leading to less mixing and hence less contact with potential pathogen threats in the form of individuals from other groups.

Figure 2 suggests three ways in which the system may be operating to generate these effects. First of all, natural selection may have been in operation in the past history of the individuals in these groups, leading those under high pathogen stress to evolve higher levels of hygiene, driven by higher disgust sensitivity. This could have translated into lower willingness to engage with other cultures, and greater conservatism (which has been shown to be associated with conservative values [100]). However, when we compared the same disease prevalence data with our own disgust sensitivity scores from the dataset shown in figure 1, we found no significant correlations, which argues against genetic differences between groups owing to varying histories of pathogen prevalence.

The second way of explaining the observed relationships between pathogens and group 'openness' is that individuals are responding to the perception that there is a lot of disease about, i.e. sickness in others causing a risk, which drives individual learning, which modifies disgust responses, which shows up in lower openness scores and enhanced xenophobia.

Navarrete & Fessler [101] observed that, not only does sensitivity to disgust predict more negative attitudes towards foreign peoples (xenophobia), but also predicts more positive attitudes towards one's own cultural in-group (ethnocentrism). They also showed that another disease-relevant individual difference variable—PVD—predicts both xenophobia and ethnocentrism. Further, xenophobic reactions to foreigners have been found to be stronger among people who feel personally vulnerable to germs and disease (as measured by the PVD scale) [102]. More particularly, results showed that higher levels of PVD predicted stronger anti-immigrant attitudes towards those from subjectively foreign locations. There was no such effect on attitudes towards culturally familiar immigrant populations. It thus seems probable that the intervening variable that can explain the observed relationship between pathogens and cultural characteristics is, in fact, disgust. A similar effect was observed in World of Warcraft (a web-based multi-player game). When a plague of 'Corrupt blood' began to kill up to half of the players, they began to avoid big cities [103].

A final possibility is that selection is operating on social groups. Those groups with the strongest collective anti-pathogen defences, such as strong norms about manners or avoiding strangers, outcompete those groups without. Using a model which carefully spells out the links in this adaptive system thus helps reasoning about the ways in which a complex set of results can be better understood.

The interrelated adaptive system we have described is not, of course, static. It exhibits positive feedback loops that can lead to interesting dynamics. For example, the proportion of hygiene cooperators in a population may fluctuate cyclically because as defectors from group-level hygiene activity become frequent, so will pathogens. When pathogen stress is high, it once again pays to cooperate, and the proportion of defectors should again decrease. The proportion of hygiene defectors may thus go up and down in response to fluctuations in the costs of cooperation. This dynamic has been formally modelled in the context of risky versus safe sexual behaviour: as the proportion of 'safe' strategists increases, the prevalence of sexually transmitted diseases (STDs) decreases. In this changed environment, 'risky' individuals may have some advantage. Oscillations of high disease/safe behaviour, low disease/risky behaviour thus occur [104].

There is also a positive feedback loop between culture and group hygiene by which cultural rules with real biological import (manners) can become subject to the symbolization principle we call 'extension'. When a set of cultural rules becomes symbolic, they can extend beyond behaviour with direct relevance to biological pathogens. An obvious extension is from rules governing contact with biological parasites to rules governing contact with 'social parasites' (i.e. individuals that claim an unfair proportion of social resources). In this way, systems of manners can be extended to become systems of moral rules. Violators of rules for the apportionment of socially produced resources can then be labelled as disgusting, and sanctioned accordingly.

It is probable that there have also been major shifts in the way the system has operated since the dawn of *Homo sapiens*. Changes in environment, social structures and technology have led to changes in pathogen prevalence, as well as in individual and group hygiene. While the system operated 'blindly' for most of human history, Leewenhoek's invention of the microscope 300 years ago finally allowed us to 'see' pathogens. Pasteur & Koch popularized the idea that invisible germs caused disease, and the idea has now caught on globally, perhaps because the notion of an invading parasitic life form is so exquisitely disgusting [13]. The scientific method has also allowed us to determine the most effective and cost-effective methods of disease prevention, so as to inform health policy internationally [105]. We have thus moved from the instinctive practice of safe hygiene, to more educated approaches. However, the ancient emotion of disgust retains its power; a recent hygiene promotion campaign in Ghana successfully elicited disgust to increase rates of handwashing nationally [106]. We now plan to test whether promoting hygiene as good manners is an effective disease control strategy.

9. CONCLUSIONS

We have seen that disgust, hygiene behaviour and culture form an interlinked adaptive system, which has long served to reduce the dangers of disease. We have argued that disgust in the brain and the disease avoidance behaviour that it motivates is universal in humans (and in other animals) and is a product of the selection pressure of pathogens in the environment. However, disgust is also plastic, being able to retune according to signals from within the body and from the social and biological environment. Heightened disgust sensitivity leads to heightened disease avoidance behaviour (hygiene). However, as a social species, group hygiene behaviour is not just an aggregate of individual behaviour but the consequence of individuals using each other as models to be imitated, and as resources for cooperation in healthy behaviours. Changes in group hygiene behaviour, including cooperative public health activities, can reduce pathogen prevalence. The tendency to defect on responsibilities to the group can be countered by norms imposed by culture. The content of culture is itself biased by learning mechanisms based in the disgust system in individual brains.

We have shown that humans are equipped with a series of highly specialized learning mechanisms and biases. Learning about disease (and disgust) is not the product of domain-general systems; features of these learning mechanisms clearly reflect the nature of the disease avoidance problem they interact with. But the informational inputs to these systems differ, both between people and across groups, as a consequence of local disease problems and the learning histories of the individuals in one's social group. Individual differences are therefore inevitable. This variation, however, is frequently presented as evidence that evolutionary explanations are inappropriate (e.g. [107,108]). On the contrary, variable responses in the disgust system are indicative of a rich evolved psychology capable of absorbing and integrating information from diverse sources, and generating adaptive behaviour in a wide range of environments.

The adaptive system depicted in figure 2 highlights the intricate interrelations between the content of individual brains and group brains (culture) and individual and group behaviour. We have used it to explore one human need—the avoidance of disease. However, humans have multiple evolved needs including a need for food, to avoid predators, to rear young, to maintain social affiliations, etc. [109]. Our framework should prove useful in integrating perspectives from across disciplines for other similar adaptive systems.

Infectious disease still remains a leading cause of mortality worldwide [110] and promoting safe hygiene may be one, if not the most, cost-effective means of preventing disease [105]. Evolutionarily informed work that can explain the causes of disease avoidant behaviour may offer vital clues as to how best to change environments and cultures so as to favour changes in group and individual behaviour, and hence to prevent this annual toll of infectious disease. We hope that those who seek to promote disease avoidant behaviours will recognize the power of using the disgust system's natural affinity for producing adaptive responses to disease threats when developing future public health programmes.

ACKNOWLEDGEMENTS

The authors would like to thank Diana Santos Fleischman, Tom Dickens and two anonymous reviewers who helped to shape the paper, as well as our funders: LSHTM, ESRC and Hindustan Unilever.

REFERENCES

1. Kuris, A. M. et al. 2008 Ecosystem energeticimplications of parasite and free-living biomass in three estuaries. Nature 454, 515–518. (doi:10.1038/nature06970)

2. Combes, C. 2001 Parasitism: the ecology and evolution ofintimate interactions. Chicago, IL: University of ChicagoPress.
3. Abbas, A. A., Lichtmann, A. & Pillai, S. 2007 Cellularand molecular immunology. Philadelphia, PA: Saunders.
4. Schaller, M. & Duncan, L. A. 2007 The behavioralimmune system. Its evolution and social psychological implications. In Evolution of the social mind: evolutionary psychology and social cognition (eds J. P. Forgas, M. G. Haselton & W. Von Hippel). New York, NY: Psychological Press.
5. Cremer, S., Armitage, S. A. O. & Schmid-Hempel, P. 2007 Social immunity. Curr. Biol. 17, R693–R702. (doi:10.1016/j.cub.2007.06.008)
6. Behringer, D. C., Butler, M. J. & Shields, J. D. 2006 Ecology: avoidance of disease by social lobsters.Nature 441, 421–421. (doi:10.1038/441421a)
7. Hutchings, M. R., Gordon, I. J., Kyriazakis, I. & Jackson, F. 2001 Sheep avoidance of faecescontaminated patches leads to a trade-off between intake rate of forage and parasitism in subsequent foraging decisions. Anim. Behav. 62, 955–964. (doi: 10.1006/anbe.2001.1837)
8. Hutchings, M. R., Kyriazakis, I., Anderson, D. H., Gordon, I. J. & Coop, R. L. 1998 Behavioural strategies used by parasitized and non-parasitized sheep to avoid ingestion of gastro-intestinal nematodes associated with faeces. Anim. Sci. 67, 97–106. (doi:10.1017/ S1357729800009838)
9. Moore, J. 2002 Parasites and the behaviour of animals. New York, NY: Oxford University Press.
10. Curtis, V., Aunger, R. & Rabie, T. 2004 Evidence that disgust evolved to protect from risk of disease. Proc. R. Soc. Lond. B 271, S131–S133. (doi:10.1098/ rsbl.2003.0144)
11. Curtis, V. & Biran, A. 2001 Dirt, disgust, and disease: is hygiene in our genes? Perspect. Biol. Med. 44, 17–31. (doi:10.1353/pbm.2001.0001)
12. Oaten, M., Stevenson, R. J. & Case, T. I. 2009 Disgust as a disease-avoidance mechanism. Psychol. Bull. 135, 303–321. (doi:10.1037/a0014823)
13. Curtis, V. 2007 Dirt, disgust and disease: a natural history of hygiene. J. Epidemiol. Community Health 61, 660–664. (doi:10.1136/jech.2007.062380)
14. Darwin, C. 1872 The expression of the emotions in man and animals. London, UK: Murray.
15. Brown, D. E. 1991 Human universals. New York, NY: McGraw-Hill.
16. Ekman, P. & Friesen, W. V. 1971 Constants across cultures in the face and emotion. J. Pers. Soc. Psychol. 17, 124–129. (doi:10.1037/h0030377)
17. Haidt, J., McCauley, C. & Rozin, P. 2007 Individualdifferences in sensitivity to disgust: a scale sampling seven domains of disgust elicitors. Pers. Individual Differ. 16, 701–713.
18. Rozin, P., Haidt, J. & McCauley, C. 2008 Disgust. In Hand book of emotion (ed. M. Lewis), 3rd edn. New York, NY: Guilford Press.
19. Mesquita, B. & Frijda, N. H. 1992 Cultural variations in emotions: a review. Psychol. Bull. 112, 179–204. (doi:10.1037/0033-2909.112.2.179)
20. Angyal, A. 1941 Disgust and related aversions. J. Abnorm. Soc. Psychol. 36, 393–412. (doi:10.1037/ h0058254)
21. Rozin, P. & Fallon, A. E. 1987 A perspective on disgust. Psychol. Rev. 94, 23–41. (doi:10.1037/0033-295X.94. 1.23)
22. Rubio-Godoy, M., Aunger, R. & Curtis, V. 2007 Serotonin—a link between disgust and immunity? Med. Hypotheses 68, 61–66. (doi:10.1016/j.mehy.2006. 06.036)
23. Schaller, M., Miller, G. E., Gervais, W. M., Yager, S. & Chen, E. 2010 Mere visual perception of others' disease symptoms facilitates a more agressive immune response. Psychol. Sci. (doi:10.1177/0956797610368064)
24. Kolnai, A., Korsmeyer, C. & Smith, B. 2004 On disgust. Chicago, IL: Open Court Publishing Company.
25. Phillips, M. L., Senior, C., Fahy, T. & David, A. S. 1998 Disgust—the forgotten emotion of psychiatry. Br. J. Psychiat. 72, 373–375. (doi:10.1192/bjp.172.5. 373)
26. Douglas, M. 2002 Purity and danger. New York, NY:Routledge.

27. Haidt, J., McCauley, C. & Rozin, P. 1994 Individualdifferences in sensitivity to disgust—a scale sampling 7 domains of disgust elicitors. Pers. Individual Differ. 16, 701–713. (doi:10.1016/0191-8869(94)90212-7)
28. Wilson, R. R. 2002 The Hydra's tale: imagining disgust. Alberta, Canada: University of Alberta Press.
29. Maynard-Smith, J. 1993 The theory of evolution. Cambridge, UK: Cambridge University Press.
30. Smith, V. S. 2007 Clean: a history of personal hygiene and purity. Oxford, UK: Oxford University Press.
31. Panksepp, J. 2007 Criteria for basic emotions: is DISGUST a primary emotion? Cogn. Emotion 21, 1819–1828. (doi:10.1080/02699930701334302)
32. Sprengelmeyer, R., Rausch, M., Eysel, U. T. & Przuntek, H. 1998 Neural structures associated with recognition of facial expressions of basic emotions. Proc. R. Soc. Lond. B 265, 1927–1931. (doi:10.1098/ rspb.1998.0522)
33. Wicker, B., Keysers, C., Plailly, J., Royet, J.-P., Gallese, V. & Rizzolatti, G. 2003 Both of us disgusted in my insula: the common neural basis of seeing and feeling disgust. Neuron 40, 655–664. (doi:10.1016/S0896- 6273(03)00679-2)
34. Fitzgerald, D. A., Posse, S., Moore, G. J., Tancer, M. E., Nathan, P. J. & Phan, K. L. 2004 Neural correlates of internally-generated disgust via autobiographical recall: a functional magnetic resonance imaging investigation. Neurosci. Lett. 370, 91–96. (doi: 10.1016/j.neulet.2004.08.007)
35. Murphy, T. K., Sajid, M. W. & Goodman, W. K. 2006 Immunology of obsessive-compulsive disorder. Psychiatr. Clin. North Am. 29, 445–469. (doi:10.1016/ j.psc.2006.02.003)
36. Olatunji, B. O., Haidt, J., McKay, D. & David, B. 2008 Core, animal reminder, and contamination disgust: three kinds of disgust with distinct personality, behavioral, physiological, and clinical correlates. J. Res. Pers. 42, 1243–1259. (doi:10.1016/j.jrp.2008. 03.009)
37. Tooby, J. & Cosmides, L. 1992 The psychological foundations of culture. In The adapted mind: evolutionary psychology and the generation of culture (eds J. H. Barkow & L. Cosmides & J. Tooby). Oxford, UK: Oxford University Press.
38. Geary, D. C. & Huffman, K. J. 2002 Brain and cognitive evolution: forms of modularity and functions of mind. Psychol. Bull. 128, 667–698. (doi:10.1037/ 0033-2909.128.5.667)
39. Costa Jr, P. T. & McCrae, R. R. 1992 Normal personality assessment in clinical practice: the NEO Personality Inventory. Psychol. Assess. 4, 5–13. (doi:10.1037/1040-3590.4.1.5)
40. Goldberg, L. R. 1990 An alternative 'description of personality': the Big-Five factor structure. J. Pers. Soc. Psychol. 59, 1216–1229. (doi:10.1037/0022-3514.59. 6.1216)
41. Gunthert, K. C., Cohen, L. H. & Armeli, S. 1999 The role of neuroticism in daily stress and coping. J. Pers. Soc. Psychol. 77, 1087–1100. (doi:10.1037/0022- 3514.77.5.1087)
42. Claridge, G. & Davis, C. 2001 What's the use of neuroticism? Pers. Indiv. Differ. 31, 383–400. (doi:10. 1016/S0191-8869(00)00144-6)
43. Neeleman, J., Sytema, S. & Wadsworth, M. 2002Propensity to psychiatric and somatic ill-health: evidence from a birth cohort. Psychol. Med. 32, 793–803. (doi:10.1017/ S0033291702005901)
44. Nettle, D. 2006 The evolution of personality variation in humans and other animals. Am. Psychol. 61,622–631. (doi:10.1037/0003-066X.61.6.622)
45. Hennig, J., Po¨ ssel, P. & Netter, P. 1996 Sensitivity to disgust as an indicator of neuroticism: a psychobiological approach. Pers. Indiv. Differ. 20, 589–596. (doi:10. 1016/0191-8869(95)00218-9)
46. Pervin, L. A. & John, O. P. 1999 Handbook of personality. Chicago, IL: Guilford Press.
47. Sih, A., Bell, A. & Johnson, J. C. 2004 Behavioral syndromes: an ecological and evolutionary overview. Trends Ecol. Evol. 19, 372–378. (doi:10.1016/j.tree. 2004.04.009)
48. Olatunji, B. O., Lohr, J. M., Sawchuk, C. N. & Tolin, D. F. 2007 Multimodal assessment of disgust in contamination-related obsessive-compulsive disorder. Behav. Res. Ther. 45, 263–276. (doi:10.1016/j.brat. 2006.03.004)

49. Sprengelmeyer, R., Schroeder, U., Young, A. W. &Epplen, J. T. 2006 Disgust in pre-clinical Huntington's disease: a longitudinal study. Neuropsychologia 44, 518– 533. (doi:10. 1016/j.neuropsychologia.2005.07.003)
50. Rozin, P., Haidt, J. & McCauley, C. 1999 Individual differences in disgust sensitivity: comparisons and evaluations of paper-and-pencil versus behavioural measures. J. Res. Pers. 33, 330–351. (doi:10.1006/ jrpe.1999.2251)
51. Fessler, D. M. T. & Navarrete, C. D. 2003 Domain specific variation in disgust sensitivity across the menstrual cycle. Evol. Hum. Behav. 24, 406–417. (doi:10.1016/S1090-5138(03)00054-0)
52. Stevenson, R. J., Case, T. I. & Oaten, M. J. 2009 Frequency and recency of infection and their relationship with disgust and contamination sensitivity. Evol. Hum. Behav. 30, 363–368. (doi:10.1016/j.evolhumbehav. 2009.02.005)
53. Fessler, D. M. T., Eng, S. J. & Navarrete, C. D. 2005 Elevated disgust sensitivity in the first trimester of pregnancy: evidence supporting the compensatory prophylaxis hypothesis. Evol. Hum. Behav. 26, 344–351.
54. Flaxman, S. M. & Sherman, P. W. 2008 Morningsickness: adaptive cause or non adaptive consequence of embryo viability? Am. Nat. 172, 54–62. (doi:10. 1086/588081)
55. Pepper, G. V. & Roberts, C. 2006 Rates of nausea and vomiting in pregnancy and dietary characteristics across populations. Proc. R. Soc. B 273, 2675–2679. (doi:10.1098/ rspb.2006.3633)
56. Sherman, P. W. & Flaxman, S. M. 2002 Nausea andvomiting of pregnancy in an evolutionary perspective.Am. J. Obstet. Gynecol. 186, 190–197. (doi:10.1067/ mob.2002.122593)
57. Fleischman, D. S. & Fessler, D. M. 2006 Luteal phase and progestin effects on OCD measures and disgust sensitivity. Paper presented at The Human Behavior and Evolution Society Conf., Philadelphia, PA, 7–11 June.
58. Conway, C. A., Jones, B. C., Debruine, L. M., Welling,L. L. M., LawSmith, M. J., Perrett,D. I., Sharp, M. A. &Al-Dujaili, E. A. S. 2007 Salience of emotional displays ofdanger and contagion in faces is enhanced when progesterone levels are raised. Horm. Behav. 51, 202–206. (doi:10.1016/j.yhbeh.2006.10.002)
59. Martin, L., Clair, J., Davis, P., O'Ryan, D., Hoshi, R. &Curran, H. V. 2006 Enhanced recognition of facial expressions of disgust in opiate users receiving maintenance treatment. Addiction 101, 1598–1605.
60. Roy, S. & Loh, H. H. 1996 Effects of opioids on the immune system. Neurochem. Res. 21, 1375–1386.(doi:10.1007/BF02532379)
61. Hoefling, A., Likowski, K. U., Deutsch, R., Ha¨fner, M.,Seibt, B., Mu¨ hlberger, A., Weyers, P. & Strack, F. 2009 When hunger finds no fault with moldy corn: food deprivation reduces food-related disgust. Emotion 9, 50. (doi:10.1037/a0014449)
62. Duncan, L. A., Schaller, M. & Park, J. H. 2009Perceived vulnerability to disease: development and validation of a 15-item self-report instrument. Pers. Indiv. Differ. 47, 541–546. (doi:10.1016/j.paid.2009. 05.001)
63. Garcia, J., Hankins, W. G. & Rusiniak, K. W. 1974Behavioral regulation of the milieu interne in man and rat. Food preferences set by delayed visceral effects facilitate memory research and predator control. Science 185, 824–831. (doi:10.1126/science. 185.4154.824)
64. Bernstein, I. L. 1978 Learned taste aversions in childrenreceiving chemotherapy. Science 200, 1302–1303. (doi:10.1126/science.663613)
65. De Jong, P. J. & Merckelbach, H. 1998 Blood-injection injury phobia and fear of spiders: domain specific individual differences in disgust sensitivity. Pers. Indiv. Differ. 24, 153–158. (doi:10.1016/S0191- 8869(97)00178-5)
66. Olatunji, B. O., Forsyth, J. P. & Cherian, A. 2006Evaluative differential conditioning of disgust: a sticky form of relational learning that is resistant to extinction. J. Anxiety Disord. 21, 820–834.
67. Baumeister, R. F., Vohs, K. D. & Nathan Dewall, C. 2007 How emotion shapes behavior: feedback, anticipation, and reflection, rather than direct causation. Pers. Soc. Psychol. Rev. 11, 167–203. (doi:10. 1177/1088868307301033)

68. Tolin, D. F., Worhunsky, P. & Maltby, N. 2004 Sympathetic magic in contamination-related OCD. J. Behav. Ther. Exp. Psychiat. 35, 193–205. (doi:10.1016/j.jbtep.2004.04.009)
69. Sartre, J. P. 1943 Being and Nothingness. New York, NY: Washington Square. (Trans. Hazel E. Barnes 1966).
70. Ohman, A. & Mineka, S. 2001 Fears, phobias, and preparedness: toward an evolved module of fear and fear learning. Psychol. Rev. 108, 483–522. (doi:10. 1037/0033-295X.108.3.483)
71. Bayliss, A. P., Frischen, A., Fenske, M. J. & Tipper, S. P. 2007 Affective evaluations of objects are influenced by observed gaze direction and emotional expression. Cognition 104, 644–653. (doi:10.1016/j.cognition. 2006.07.012)
72. Hertenstein, M. J. & Campos, J. J. 2004 The retention effects of an adult's emotional displays on infant behavior. Child Dev. 75, 595–613. (doi:10.1111/j. 1467-8624.2004.00695.x)
73. Benton, D. 2004 Role of parents in the determination of the food preferences of children and the development of obesity. Int. J. Obes. Relat. Metab. Disord. 28, 858–869. (doi:10.1038/sj.ijo.0802532)
74. Galef, B. G. 1996 Social enhancement of food preferences in Norway rats: a brief review. In Social learning in animals: the roots of culture (eds C. Hayes & B. G. Galef). San Diego, CA: Academic Press.
75. Egliston, K. A. & Rapee, R. M. 2007 Inhibition offear acquisition in toddlers following positive modeling by their mothers. Behav. Res. Ther. 45, 1871–1882. (doi:10. 1016/j.brat.2007.02.007)
76. Woody, S. R. & Tolin, D. F. 2002 The relationship between disgust sensitivity and avoidant behavior: studies of clinical and nonclinical samples. J. Anxiety Disord. 16, 543–559. (doi:10.1016/S0887-6185(02)00173-1)
77. Deacon, B. & Olatunji, B. O. 2007 Specificity of disgust sensitivity in the prediction of behavioral avoidance in contamination fear. Behav. Res. Ther. 45, 2110–2120. (doi:10.1016/j.brat.2007.03.008)
78. Rozin, P., Haidt, J., McCauley, C., Dunlop, L. & Ashmore, M. 1999 Individual differences in disgust sensitivity: comparisons and evaluations of paper-andpencil versus behavioral measures. J. Res. Pers. 33, 330–351. (doi:10.1006/jrpe.1999.2251)
79. Boyd, R. & Richerson, P. J. 1995 Why does culture increase human adaptability? Ethol. Sociobiol. 16, 125–143. (doi:10.1016/0162-3095(94)00073-G)
80. Henrich, J. & Gil-White, F. J. 2001 The evolution of prestige: freely conferred deference as a mechanism for enhancing the benefits of cultural transmission. Evol. Hum. Behav. 22, 165–196. (doi:10.1016/S1090- 5138(00)00071-4)
81. Judah, G., Aunger, R., Schmidt, W.-P., Michie, S., Granger, S. & Curtis, V. 2009 Experimental pretesting of hand-washing interventions in a natural setting. Am. J. Public Health 99, S405–S411. (doi:10.2105/ AJPH.2009.164160)
82. Suroviak, J. M. L. 2005 The effects of social norms on male hygiene. Indiv. Differ. Res. 3, 1541–1745. 83
83. Curtis, V. A., Danquah, L. O. & Aunger, R. V. 2009 Planned, motivated and habitual hygiene behaviour: an eleven country review. Health Educ. Res. 24, 655– 673. (doi:10. 1093/her/cyp002)
84. Hardin, G. 1968 The tragedy of the commons. Science 162, 1243–1248. (doi:10.1126/science.162.3859.1243)
85. Kleemeier, E. 2000 The impact of participation on sustainability: an analysis of the Malawi rural piped scheme program. World Dev. 28, 929–944. (doi:10. 1016/S0305-750X(99)00155-2)
86. Johnson, A. W. & Earle, T. K. 2000 The evolution of human societies: from foraging group to agrarian state. Stanford, CA: Stanford University Press.
87. Ewald, P. W. 1995 The evolution of virulence: a unifying link between parasitology and ecology. J. Parasitol. 81, 659–669. (doi:10.2307/3283951)
88. MacQueen, S. 1995 Anthropology and germ theory. J. Hosp. Infect. 30, 116–126. (doi:10.1016/0195- 6701(95)90012-8)
89. Burkert, W. 1985 Greek religion. Cambridge, UK: Harvard University Press.

90. Parker, R. 1983 Miasma: pollution and purity in early Greek religion. Oxford, UK: Clarendon Press.
91. Dumont, L. 1966 Homo Hierarchicus: the caste system and its implications. Chicago, IL: University of Chicago Press.
92. Falomir-Pichastor, J. M., Mun˜oz-Rojas, D., Invernizzi, F. & Mugny, G. 2004 Perceived in-group threat as a factor moderating the influence of in-group norms on discrimination against foreigners. Eur. J. Soc. Psychol. 34, 135–153. (doi:10.1002/ejsp.189)
93. Boyd, R. & Richerson, P. J. 1988 Culture and the evolutionary process. Chicago, IL: University of Chicago Press.
94. Sperber, D. 1994 The modularity of thought and the epidemiology of representations. In Mapping the mind: domain specificity in cognition and culture (eds L. A. Hirschfeld & S. A. Gelman), pp. 39–67. New York, NY: Cambridge University Press. (doi:10.1017/CBO 9780511752902.003)
95. Heath, C., Bell, C. & Steinberg, E. 2001 Emotional selection in memes: the case of urban legends. J. Pers. 81, 1028–1041. (doi:10.1037/0022- 3514.81.6.1028)
96. Nichols, S. 2002 On the genealogy of norms: a case for the role of emotion in cultural evolution. Phil. Sci. 69, 234–255. (doi:10.1086/341051)
97. Schaller, M. & Murray, D. R. 2008 Pathogens, personality, and culture: disease prevalence predicts worldwide variability in socio sexuality, extraversion, and openness to experience. J. Pers. Soc. Psychol. 95, 212–221. (doi:10.1037/0022-3514.95.1.212)
98. Fincher, C. L., Thornhill, R., Murray, D. R. & Schaller, M. 2008 Pathogen prevalence predicts human cross cultural variability in individualism/collectivism. Proc. R. Soc. B 275, 1279–1285. (doi:10.1098/rspb. 2008.0094)
99. Fincher, C. L. & Thornhill, R. 2008 Assortative sociality, limited dispersal, infectious disease and the genesis of the global pattern of religion diversity. Proc. R. Soc. B 275, 2587–2594. (doi:10.1098/rspb.2008.0688)
100. Inbar, Y., Pizarro, D. A. & Bloom, P. 2009 Conservatives are more easily disgusted than liberals.400 V. Curtis et al. Disgust as an adaptive system Phil. Trans. R. Soc. B (2011) Cogn. Emotion 23, 714–725. (doi:10.1080/026999308 02110007)
101. Navarrete, C. D. & Fessler, D. M. T. 2006 Disease avoidance and ethnocentrism: the effects of disease vulnerability and disgust sensitivity on intergroup attitudes. Evol. Hum. Behav. 27, 270–282. (doi:10.1016/ j.evolhumbehav.2005.12.001)
102. Faulkner, J., Schaller, M., Park, J. H. & Duncan, L. A. 2004 Evolved disease-avoidance mechanisms and contemporary xenophobic attitudes. Group Processes Intergroup Behav. 7, 333–353. (doi:10.1177/ 1368430204046142)
103. Lofgren, E. T. & Fefferman, N. H. 2007 The untapped potential of virtual game worlds to shed light on real world epidemics. Lancet Infect. Dis. 7, 625–629. (doi:10.1016/S1473-3099(07)70212-8)
104. Boots, M. & Knell, R. 2002 The evolution of risky behaviour in the presence of a sexually transmitted disease. Proc. R. Soc. Lond. B 269, 585–589. (doi:10. 1098/rspb.2001.1932)
105. Jamieson, D., Bremen, J., Measham, A., Alleyne, G. & Claeson, M. 2006 Disease control priorities in developing countries. Oxford, UK: Oxford University Press.
106. Scott, B., Schmidt, W., Aunger, R., Garbrah-Aidoo, N. & Animashaun, R. 2007 Marketing hygiene behaviours: the impact of different communications channels on reported handwashing behaviour of women in Ghana. Health Educ. Res. 22, 225–233.
107. Miller, W. I. 1998 The anatomy of disgust. Cambridge, MA: Harvard University Press.Zuk, M. 2007 Riddled with life. New York, NY: Harcourt.
108. Aunger, R. & Curtis, V. 2008 Kinds of behaviour. Biol.Philos. 23, 317–345. (doi:10.1007/s10539-007-9108-4)
109. WHO 2004 The 10 leading causes of death by broadincome group. Factsheet no. 310. Geneva, Switzerland:WHO.

CHAPTER 20

Modelling influenza A H5N1 vaccination strategy scenarios in the household poultry sector in Egypt

Ihab El Masry[1], Jolianne Rijks[2], Marisa Peyre[3], Nick Taylor[4], Juan Lubroth[5], Yilma Jobre[1]

1 Emergency Centre for Transboundary Animal Disease, Food and Agriculture Organization of the United Nations (FAO), P.O. Box 2223, Cairo, Egypt

2 Dutch Wildlife Health Centre, Yalelaan 1, 3584CL Utrecht, The Netherlands

3 Animal et Gestion Intégrée des Risques, Centre International de Recherche en Agriculture pour le Développement, 34398 Montpellier, France

4 Veterinary Epidemiology and Economics Research Unit, Reading University, School of Agriculture, Policy and Development, Berkshire, UK

5 Food and Agriculture Organization of the United Nations (FAO), Rome, Italy

ABSTRACT

Highly pathogenic avian influenza (AI) due to H5N1 virus was first reported in Egypt in February 2006; since then, the government has allowed avian influenza vaccination in poultry. The present study evaluated the impact of AI vaccination in terms of cumulative annual flock immunity (CAFI): the percentage of bird × weeks protected by immunity. This evaluation took account of the combined effects of vaccination coverage, vaccine efficacy (VE), and different characteristics of household poultry production on the effectiveness of the adopted vaccination strategy (VS), and provided alternative options for improvement. The evaluation used a population and vaccination model that calculates the CAFI. Participatory approaches were employed in 21 villages to develop the vaccination and flock parameters required for the model. The adopted VS were compared in the model with three alternative VS scenarios in terms of the CAFI. Vaccination coverage varied among villages but was generally low (between 1 and 48 %; median 14 %). Under the adopted VS, the CAFI predicted for the villages ranged from 2 to 31 %. It was concluded that despite the enormous effort put into rural household poultry AI vaccination by the Egyptian government, village CAFI is unlikely to be maintained at the levels required to significantly reduce the virus load and restrict transmission. In HPAI-endemic countries that consider AI vaccination as one of the disease control options, the high cost of mass AI vaccination campaigns and their achievable benefits must be compared with other available control measures, which may include targeted vaccination. Achievable vaccination coverage, VE and the different characteristics of commercial and household (village) poultry production are key parameters determining the feasibility and cost-effectiveness of different AI vaccination strategies.

Keywords: Avian influenza, HPAI H5N1, Egypt, Vaccination coverage, Flock immunity, Vaccination strategy

1. INTRODUCTION

Highly pathogenic avian influenza (HPAI) H5N1 virus was detected in the Middle East (Egypt, Iraq, Iran, Israel, Jordan, Kuwait, Palestinian Territories) between November 2005 and March 2006. The virus was detected in Sudan and Djibouti in April 2006 and in Saudi Arabia in March 2007 (Williams and Peterson 2009). Of these countries, Egypt has been most severely affected. Poultry production in Egypt is present in many diverse forms, from large- and small-scale commercial sectors, to 'traditional' household poultry (HP) keeping.

HPAI due to the H5N1 subtype was first reported in poultry in Egypt in February 2006 (World Organization for Animal Health (OIE), 2006), and the first human case was reported in March 2006. The government allowed avian influenza (AI) vaccination in poultry shortly thereafter. During the period covered by this study (July 2007 to April 2009), AI vaccination was the main tool used to control HPAI in HP in Egypt. Some countries such as India and Bangladesh refrained from using vaccination as part of their control and prevention strategy (Food and Agriculture Organization of the United Nations (FAO), 2011). The General Organization for Veterinary Services (GOVS) is responsible for AI vaccination campaigns in Egypt. During the study period, four bi-annual house-to-house mass vaccination campaigns were implemented. In total, 212.5 million doses of inactivated AI vaccines, mainly of H5N1 but also H5N2 viruses, were used. The objectives of these campaigns were to reduce the virus load and transmission.

The combined impact of the vaccination coverage (VC), vaccine efficacy (VE), and different types of HP production on overall vaccination effectiveness, in terms of protection of the population, were neither sufficiently studied nor well documented in Egypt. Proper understanding of these factors, however, provides an insight into probable flock immunity levels and thereby, a clue to the effectiveness of the vaccination strategy (VS) in terms of attaining the desired objectives.

Participatory approaches were employed to collect data to be fed into a "population and vaccination model" version 8.1 (FAO 2008—unpublished consultancy report), to calculate the cumulative annual flock immunity (CAFI): the percentage of bird × weeks protected by immunity.

The objective of the present study was to compare the current AI VS adopted for the HP sector in Egypt with three alternative strategies in terms of the CAFI, and recommend a suitable AI VS.

2. MATERIALS AND METHODS

2.1 The model

The 'population and vaccination model' was developed to estimate the impact of different vaccination strategies on CAFI. It simulates the dynamic poultry population of a village, updating the status of the population at weekly intervals. The model also simulates development of immunity in the population in response to vaccination. A key output of this model is CAFI, which is calculated for any time period and for all or a subset of the population (species, breed, age group) as follows:

$$CAFI \frac{\sum \text{nSib}}{\sum \text{nTnb}}$$

Where ΣnSib = the sum of successfully immunised bird × weeks (i.e., considered protected) within a given poultry population across *n* weeks;

$\Sigma nTnb$ = the sum of all present bird × weeks within the same poultry population across *n* weeks (irrespective of their immunity status). Because the model recalculates both the total number of birds present and birds immune on a weekly basis, the units in the CAFI equation are bird × weeks. In this paper, the CAFI result is presented as a percentage, and the period of calculation is 52 weeks.

In mixed-village flocks (with multiple species/breeds presenting different production and vaccination response characteristics), the flock was modelled using separate models for each sub-population. The overall CAFI is an aggregate of the CAFIs of each sub-population weighted by the relative proportion of each sub-population in the total village flock.

2.2 Study population and data collection

The characteristics of HP production in different localities was studied based on data collected in 34 randomly selected (out of 413) villages in Dakahlia (DK) governorate. Additionally, 12 villages were purposively selected from Minya, Asyut and Qena governorates. DK governorate is located in Lower Egypt where several village typologies are present. The remaining governorates are from Upper Egypt (UE) where there is more evident heterogeneity in village typology.

A group of public veterinarians, having received training on participatory epidemiology, collected raw data that were analyzed and used as model inputs, providing village flock production (Table 1) and vaccination parameters. Vaccination parameters included frequency and timing of vaccination campaigns per year, VC, number of recommended vaccinations per species per annum, number of vaccinations implemented per species per annum, VE and the duration of immunity (DOI) following the initial and booster shots.

The process of data collection involved interviews with village key informants, a transect walk involving 15–20 house visits with individual interviews, observation of household birds and finally, group meetings with key informants and producers, where preliminary findings were verified and data gaps were filled in. Of the 46 studied villages, due to data quality concerns, information collected from only 21 villages (14 in DK and 7 in UE) were used in the model simulations.

Table 1 Values for flock parameters, based on average values from villages in DK governorate (values in brackets are specific for UE governorates)

Flock parameters	BC	Sasso chicken	SD	Pekin duck	Geese	Turkey
Starting and brought-in flock						
Approximate percentage of adults per breed in village	0.32 (0.67)	–	0.18 (0.49)	0.02 (0)	0.33	–
Ratio adult male to total adults	0.09	–	0.30	0.40 (0.22)	0.40 (0.22)	0.33
Seasonality						
Egg laying	No	No	No	No	Winter	No
Egg setting (to hatch)	No	No	No	No	Winter	No
Adult female population	No	No	No	No	No	No
Brought-in stock	No	No	No	No	No	No
Egg laying and hatching						
Average eggs laid/year per bird	180	–	41 (26)	150	20	31
Proportion of eggs consumed or sold	0.98	–	0.05	0.80	0.05	0.05
Egg incubation (weeks)	3	3	4	4	4	4
Hatchability (proportion)	0.70	0.70	0.73	0.73	0.70	0.81
Sales proportion of hatchlings sold out as day-old birds	–	–	0.05	–	–	–
Age stages						
Chicks						
Age limit (final week in stage)	6	3	8	3	12	8
Typical stage mortality	0.15	0.2	0.25	0.18	0.10	0.40
Sales proportion of surplus chicks sold as reared chicks	–	–	–	0.05	–	0.05
Growers						
Average age at removal (final week in stage)	21	10	26	12	30	24
Typical stage mortality	0.04	0.2	0.04	0.18	0.10	0.04
Young breeders						
Average age at sexual maturity	26	26	30	30	52	38
Typical stage mortality	0.04	0.04	0.04	0.04	0.10	0.04
Sales proportion of survivors sold out as breeders	0.05	0.05	0.05	0.05	0.05	–
Adults						
Average age at removal	156	104	104	104	260	104
Typical annual adult mortality	0.01	0.01	0.01	0.01	0.01	0.01

BC Balady Chicken, *SD* Sudani Duck

2.3 Estimation of VC

VC was estimated using the following formula:

$$VC = \frac{number\ of\ birds\ vaccinated\ during\ the\ latest\ campaign}{model\ estimate\ of\ poultry\ number\ in\ the\ village\ \text{during}\ the\ same\ period}$$

The relationship between village size and VC was statistically analyzed using Pearson's correlation coefficient (*r*).

2.4 Modelling different vaccination strategies

The DOI was a variable vaccination parameter considered in the model. As prescribed by vaccine manufacturers, the value used as DOI for chicken and turkey was 24 weeks. For duck and geese, however, the DOI is either 4 or 16 weeks depending up on the absence or presence of a booster dose. The fixed vaccination parameters were as follows: (1) two vaccination campaigns per year with starting times set at week 1 and week 27, respectively; (2) the duration of the campaign lasting 1 week and (3) 80 % VE. The VE level used in the model is the minimal required protection level (i.e., 80 %) as determined by the Central Laboratory for Evaluation of Veterinary Biologics for the registration of AI vaccines in Egypt. The vaccination parameters and the village flock production parameters were fed into the model and four VS scenarios (Table 2) were simulated.

Table 2 Parameters of the modelled four VS scenarios

	Scenario 1	Scenario 2	Scenario 3	Scenario 4
Number of campaigns per year	2	2	2	2
VC per campaign	Real VC	Real VC	70 %	70 %
Number of dose per campaign	1	1	1	1
booster vaccination for waterfowl	No	No	Yes	Yes
VC of brought-in stocks	0 %	80 %	0 %	80 %

3. RESULTS

3.1 Poultry flock composition and dynamics

Six major categories of species and breeds with distinct production parameters (Table 1) were found in village flocks. These were indigenous (Balady) chicken (BC); Sasso and other hybrid chickens; Sudani ducks (SD) and Pekin and other meat ducks, geese and turkey. In general, BC was the largest category present in villages, and sometimes homebred but generally bought from outside sources including hatcheries, peddlers, markets or farms. The age at which the BCs were bought by households ranged from 1–30 days. These birds were raised either for meat and/or egg production. Conversely, home breeding was the main source of SD and geese. Sasso and other hybrid chickens, Pekin and other meat ducks were often brought to households at age of 1–21 days from the fore-mentioned sources of BC, these types of chicken and ducks are kept for short duration as growers, and slaughtered at 2.5–3 months old.

3.2 Estimation of VC

The estimated VC showed great variation between villages. It ranged between 1–48 % ($n = 21$, median 14 %). Village size was negatively correlated with VC (Fig. 1; $n = 21$, $r = -0.49$, $P = 0.026$).

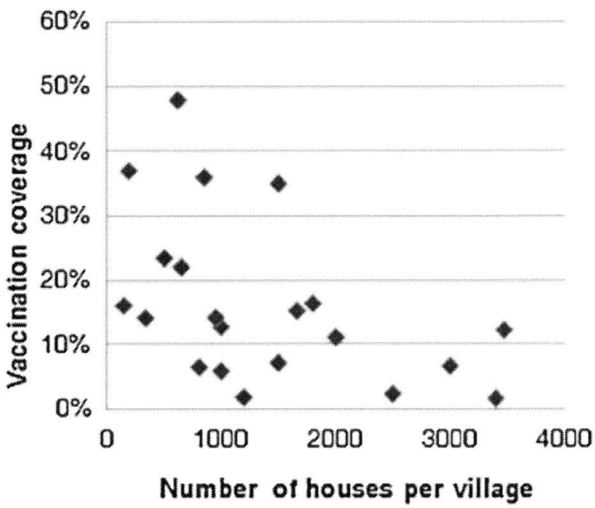

Figure. 1 Negative correlation between estimated village size and VC

3.3 CAFI

The CAFI predictions for the 21 villages under the four VS scenarios are presented in Table 3.

Table 3 Predicted CAFI in 21 villages under the four VS scenarios

Village number	Average number of poultry per household	Percentage of waterfowl (%)	Predicted CAFI			
			Scenario 1 (%)	Scenario 2 (%)	Scenario 3 (%)	Scenario 4 (%)
1	23	41	18	38	49	66
2	47	45	22	51	37	63
3	46	50	16	38	37	62
4	27	33	11	44	39	67
5	38	31	31	64	42	72
6	48	29	15	50	36	67
7	60	35	5	42	37	68
8	56	43	21	57	31	70
9	14	35	17	40	49	70
10	67	29	7	23	41	61
11	18	27	14	48	40	71
12	29	26	2	37	39	66
13	37	58	21	45	36	62
14	30	62	5	33	35	66
15	17	12	17	54	47	75
16	105	52	10	44	33	67
17	48	66	11	41	44	84
18	14	18	4	30	52	70
19	25	9	8	51	46	73
20	34	9	2	23	46	62
21	43	39	11	43	41	69
Median			11	41	40	67

3.4 Flock characteristics and CAFI

Statistical analysis revealed that CAFI is negatively correlated with standing flock size, and the proportion of waterfowl in the village. When VC is fixed at 70 %, a booster dose for waterfowl is given and when none of the brought-in stock are immune, the predicted CAFI decreases significantly as the average number of birds kept per poultry household increased ($n = 21$, $r = -0.640$, $P = 0.002$) and as the proportion of waterfowl increased ($n = 21$, $r = -0.524$, $P = 0.015$) (Table 3).

3.5 VC and CAFI

The CAFI for each village under scenario 1 has been plotted against the estimated VC in each village (Fig. 2). The model calculates the number of doses used under the different scenarios. The ratio between the numbers of doses in scenario 3 and the numbers of doses in scenario 1 was plotted against the estimated VC (Fig. 3), the resulting curve indicates for different levels of real VC the multiplying factor for the number of vaccine doses that would be required to evolve from scenarios 1 to 3.

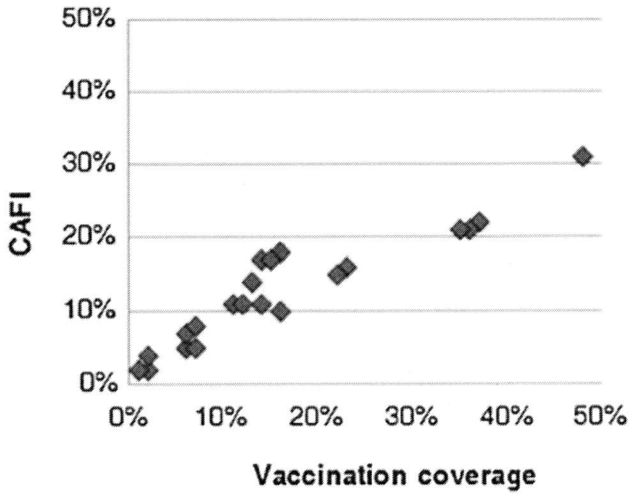

Figure. 2 Predicted CAFI in 21 villages under real VC (scenario 1)

4. DISCUSSION

The countries in which HPAI H5N1 virus is currently considered endemic include Bangladesh, China, Egypt, Indonesia, Vietnam as well as the Indian State of West Bengal (Hinrichs and Otte 2012). Waves of AI outbreaks may occur in poultry populations in endemic areas whenever large parts of the population are left with a low level of immunity as a result of low VC (Magalhaes et al. 2006).

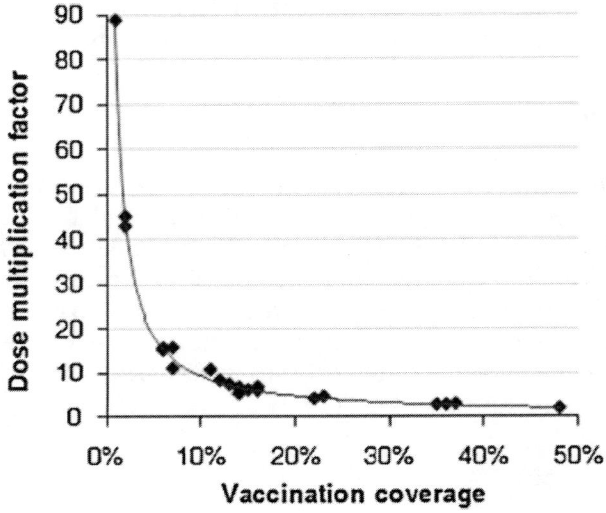

Figure. 3 The multiplying factor for the number of vaccine doses required to evolve from scenario 1 to scenario 3

The present study represents an assessment of the interaction of VS with poultry production parameters that affect population dynamics on the CAFI. The level of village CAFI depended on the VE and VC achieved. The latter varied between villages but was generally low. The 80 % VE level is in agreement with results obtained by Terregino et al. (2010), who confirmed that conventional vaccine is sufficient to prevent infection in the majority of birds challenged with a contemporary H5N1 Egyptian strain, and results obtained by Lebdah and Shahin (2010), who found an 80–92 % protection level in chickens vaccinated with either H5N1 or H5N2 vaccines.

CAFI is also affected by the flock composition and production parameters. The six categories of poultry species and breeds distinguished at village level differed in terms of sources, numbers, turnover and production goals. These differences influence the results achieved by the adopted VS. SD and geese are homebred so they can only be immunised by vaccination at household level, and in that sense, the adopted VS is appropriate for them. However, the modelled CAFI is questionable because the required booster shot for waterfowl is rarely given. Also, unlike the seasonal hatching of geese, hatching of Sudani ducklings is continuous which creates a steady increase in the proportion of unvaccinated ducks in the time between campaigns, with subsequent quick decline in the CAFI in the SD population.

For Pekin duck, the CAFI modelled under the adopted strategy was limited because the required booster shot was not given and the short cycle (4–6 weeks) results in rapid removal of vaccinated Pekin ducks (to be replaced by unvaccinated ones). For BC, the level of CAFI achieved was largely dependent on the reason for keeping them. When the focus was solely on egg production, hens were kept for years with few new introductions. In such flocks, high levels of CAFI could be maintained with the adopted VS if VC was good. In contrast, when BCs were mainly kept as growers, turnover was greater and introductions occurred throughout the year, and levels of CAFI declined under the adopted VS, unless brought-in stock was also vaccinated.

Possible reasons for the negative correlation between standing flock size and CAFI in scenario 3 include that keeping high numbers of poultry requires more stock to be brought in (not immune in this scenario), while the negative correlation between proportion of waterfowl and CAFI can be attributed to the fact that DOI in waterfowl is shorter than the campaign interval (especially if booster shots are not delivered).

AI vaccination in free-range poultry flocks is usually implemented as periodic campaigns and new hatched chicks are generally not vaccinated between vaccination campaigns (Lesnoff et al. 2009). The high and recurrent costs, technical difficulties, and epidemiological drawbacks of large-scale, open-ended and blanket vaccination programmes call for careful targeting of vaccination (Hinrichs et al. 2010). This study demonstrated the key beneficial influence on CAFI that may be achieved by vaccination of brought-in birds at source. In Egypt, it is common for village households to purchase birds at 3 or 4 weeks old that are reared at nursery farms. Targeting vaccination efforts at these nursery farms, all year round, would improve CAFI of HP populations.

5. CONCLUSION

Few countries are in a position to support government-funded mass vaccination campaigns (Food and Agriculture Organization of the United Nations 2011). The strategy of intensive countrywide vaccination as used in Vietnam was not implemented in Egypt and other biosecurity measures are not yet fully implemented (Kim et al. 2010). Nevertheless, and particularly, in HPAI-endemic countries, AI vaccination remains one of the main disease control options. This study highlights the importance of understanding key parameters such as VE, VC and campaign planning based on the different characteristics of the village poultry production in order to design and implement suitable AI vaccination strategies. The modelling outputs highlighted the difficulties in maintaining good flock immunity in HP with two vaccination rounds per year (Desvaux et al. 2012). The modelling approach may have a potential for extrapolation and could serve as a basis for similar studies on other major poultry diseases. Despite the enormous effort put into rural household poultry AI vaccination by the Egyptian government, village CAFI is unlikely to be maintained at the levels required to significantly reduce the virus load and restrict transmission. Reducing HPAI H5N1 viral load and transmission requires maintenance of high levels of flock immunity. This will require massive additional financial means, and it is questionable if it can be logistically feasible. The adopted AI VS of HP in Egypt has possibly become a net negative risk factor in the spread of the infection (Peyre et al. 2009). This is due to the fact that the positive impact in terms of raised CAFI is minimal while at the same time, the risk of transmitting HPAI among villages by the vaccination teams is a distinct possibility. AI vaccination absorbed 80–90 % of the budget for HPAI control put at the disposal of GOVS (Peyre et al. 2009). Targeted surveillance is needed in order to ensure that the protection achieved by mass vaccination is converted into a reduction in the likelihood of HPAI outbreaks occurring (Walker et al. 2010). The poor CAFI predicted by the model under the adopted VS suggests that more holistic packages of prevention and control measures should be explored. A package of surveillance and outbreak management measures, plus possibly targeted vaccination strategies (e.g., vaccination of birds at nursery farms) could be more effective. Before deciding on the continued use of vaccination as a tool for controlling HPAI, the outcome and the high cost of the vaccination campaigns must be compared with other available control measures in terms of cost-effectiveness and feasibility.

ACKNOWLEDGEMENTS

The authors would like to express their gratitude to the staff of the veterinary services; Dr. Rob de Rooij (former ECTAD Team Leader), Mr. Toni Ettel (ECTAD/EGY Programs Operation Officer) and Dr. Tony Wilsmore and Dr. Paolo Pagani (FAO Consultants) for their support during the field study. This work was part of the FAO project (OSRO/EGY/701/USA) funded by the United States Agency for International Development (USAID), for which the authors are grateful.

REFERENCES

1. Desvaux, S., Grosbois, V., Pham, T.T., Dao, D.T., Nguyen, T., Fenwick, S., Roger, F., Ellis, T. and Peyre,M., 2012. Evaluation of the vaccination efficacy against H5N1 in domestic poultry in the Red River Delta in Vietnam. Epidemiology and Infection.
2. FAO (Food and Agriculture Organization of the United Nations), 2011. Approaches to controlling, preventing and eliminating H5N1 Highly Pathogenic Avian Influenza in endemic countries. Animal Production and Health Paper No. 171. Food and Agriculture Organization of the United Nations, Rome, Italy.
3. Hinrichs, J. and Otte, J., 2012. Large-scale vaccination for the control of avian influenza: epidemiological and financial implications. In Zilberman D, Otte J, Roland-Holdst, Pfeiffer E (ed) Health and Animal Agriculture in Developing Countries, New York: Springer. 36:207-231.
4. Hinrichs, J., Otte, J. and Rushton, J., 2010. Technical, epidemiological and financial implications of large-scale national vaccination campaigns to control HPAI H5N1. CAB Reviews: Perspectives in Agriculture, Veterinary Science, Nutrition and Natural Resources Vol. 5 No. 021 pp. 1–20.
5. Kim, J. K., Kayali, G., Walker, D., Forrest, H. L., Ellebedy, A. H., Griffin, Y. S., Rubrum, A., Bahgat, M. M., Kutkat, M. A., Ali, A., Aldridge, J. R., Negovetich, N. J., Krauss, S., Webby, R. J. and Webster, R. G., 2010. Puzzling inefficiency of H5N1 influenza vaccines in Egyptian poultry. Proceedings of the National Academy of Sciences of the United States of America, 15 June 2010, vol. 107, no. 24, pp. 11044–11049.
6. Lebdah, M. A. and Shahin, A. M., 2010. Evaluation of avian influenza vaccines used in broiler flocks in Egypt. Journal of American Science, 6(12).
7. Lesnoff, M., Peyre, M., Duarte, P. C., Renard, J. F. and Mariner, J. C., 2009. A simple model for simulating immunity rate dynamics in a tropical free-range poultry population after avian influenza vaccination. Epidemiology and Infection, 137, pp 1405–1413.
8. Magalhaes, R., Pfeiffer, D., Wieland, B., Dung, D. and Otte, J., 2006 . Commune-level simulation model of HPAI H5N1 poultry infection and control in Vietnam. Pro-Poor Livestock Policy Initiative, Research Report, FAO, Rome, Italy, 8 pp.
9. OIE (World Organization for Animal Health), 2006. World animal health information database, summary of immediate notifications and follow-ups.
10. Peyre, M., Samaha, H., Makonnen, Y.J., Saad, A., Abd-Elnabi, A., Galal, S., Ettel, T., Dauphin, G., Lubroth, J., Roger, F. and Domenech, J., 2009. Avian influenza vaccination in Egypt: limitations of the current strategy. Journal of Molecular and Genetic Medicine, 3(2):198–204.
11. Terregino, C., Toffan, A., Cilloni, F., Monne, I., Bertoli, E., Castellanos, L., Amarin, N., Mancin, M. and Capua, I., 2010. Evaluation of the protection induced by avian influenza vaccines containing a 1994 Mexican H5N2 LPAI seed strain against a 2008 Egyptian H5N1 HPAI virus belonging to clade 2.2.1 by means of serological and in vivo tests. Avian Pathology, 39(3):215–22.
12. Walker, P. G., Cauchemez, S., Métras, R., Dung, H., Pfeiffer D. and Ghani, A. C., 2010. A Bayesian approach to quantifying the effects of mass poultry vaccination upon the spatial and temporal dynamics of H5N1 in Northern Vietnam. PLOS Computational Biology, 19;6(2).
13. Williams, R.A.J., and Peterson, T.A., 2009. Ecology and geography of avian influenza (HPAI H5N1) transmission in the Middle East and northeastern Africa. International Journal of Health Geographics, 8:47.

CHAPTER 21

Behavioral Defects in Chaperone-Deficient Alzheimer's Disease Model Mice

Juhi Ojha[1], Rajalakshmi V. Karmegam[1], J. Gunasingh Masilamoni[1], Alvin V. Terry Jr.[2], Anil G. Cashikar[1*]

[1]Center for Molecular Chaperones and Radiobiology, Medical College of Georgia, Augusta, Georgia, United States of America,

[2]Department of Pharmacology and Toxicology, Medical College of Georgia, Augusta, Georgia, United States of America

ABSTRACT

Molecular chaperones protect cells from the deleterious effects of protein misfolding and aggregation. Neurotoxicity of amyloid-beta (Aβ) aggregates and their deposition in senile plaques are hallmarks of Alzheimer's disease (AD). We observed that the overall content of αB-crystallin, a small heat shock protein molecular chaperone, decreased in AD model mice in an age-dependent manner. We hypothesized that αB-crystallin protects cells against Aβ toxicity. To test this, we crossed αB-crystallin/HspB2 deficient ($CRYAB^{-/-}HSPB2^{-/-}$) mice with AD model transgenic mice expressing mutant human amyloid precursor protein. Transgenic and non-transgenic mice in chaperone-sufficient or deficient backgrounds were examined for representative behavioral paradigms for locomotion and memory network functions: (i) spatial orientation and locomotion was monitored by open field test; (ii) sequential organization and associative learning was monitored by fear conditioning; and (iii) evoked behavioral response was tested by hot plate method. Interestingly, αB-crystallin/HspB2 deficient transgenic mice were severely impaired in locomotion compared to each genetic model separately. Our results highlight a synergistic effect of combining chaperone deficiency in a transgenic mouse model for AD underscoring an important role for chaperones in protein misfolding diseases.

1. INTRODUCTION

Accumulation of misfolded and aggregated proteins is a pathological hallmark of several neurodegenerative diseases. In Alzheimer's disease (AD), aggregated amyloid-beta peptide (Aβ) is a primary component of the senile plaques and is thought to be central to the associated neurotoxicity. Because molecular chaperones have evolved to protect cells against protein misfolding and aggregation, their importance in protein aggregation diseases must be understood. Members of the small heat shock proteins (sHsps) family, whose expression is regulated by the heat shock transcription factor 1 (HSF1), protect cells from a variety of environmental conditions such as heat and oxidative stress by antagonizing protein aggregation. *In vivo*, sHsps play an important role in enhancing stress resistance, regulating actin and intermediate filament dynamics, and inhibiting apoptosis [1], [2]. sHsps share a conserved α-crystallin domain of 80–100 amino acids at their C-terminus whereas their N-terminal regions are highly variable in sequence and length [3], [4]. The sHsps form dynamic oligomeric structures ranging from 9–50 subunits associating as either homo- or hetero-multimeric complexes [5], [6]. It has been proposed that the ATP-independent sHsps aid in refolding of denatured proteins by holding them in a reactivation-competent state and target them to subsequent refolding or degradation with the help of ATP-dependent chaperones like Hsp70 [2], [7].

The human and mouse genomes code for 10 genes for sHsps differing between 45 and 85% in sequence [8], [9]. Of these Hsp27 (Hsp25 in mouse), αB-crystallin, HspB6 and HspB8 are ubiquitously expressed [8]. Point mutations in human sHsps lead to several aggregation diseases [10] – for example, mutations in αA-crystallin leads to cataract [11], mutations in αB-crystallin leads to desmin-related myopathy [12], [13], [14], missense mutations in HSP27 is associated with Charcot-Marie-Tooth disease [15]. Over-expression of Hsp27 is highly protective against toxicity induced by αSyn [16] or polyglutamine [17] in cell culture models. Hsp27 and αB-crystallin have been found in proteinaceous inclusions of Alzheimer's and Parkinson's disease [18], [19], [20], [21], [22], [23], [24]. The sHsps are associated with senescence and longevity in worms [25], [26] and flies [27] suggesting their importance in aging-related diseases. Induction of αB-crystallin has been noted in Alexander's disease [28], Creutzfeldt-Jacob disease [22] and Alzheimer's disease [29]. Interestingly, the amyloid precursor protein (APP) central to AD was found to interact with αB-crystallin in worm [30] and mammalian cell models [31]. These observations underscore the importance of αB-crystallin (sHsps in general) in diseases of aging and protein aggregation, but their importance in mouse models has not been examined. To this end, we examined behavioral deficits when AD model mice were crossed mice lacking αB-crystallin/HspB2.

Mice knocked-out for αB-crystallin and HspB2 (genes are *CRYAB* and *HSPB2*, respectively) [32] are viable suggesting that it is either non-essential or redundant during development. Due to the close proximity (<1 kb) of the *HSPB2* gene to *CRYAB* and the sharing of promoter elements [33], the αB-crystallin knockout mice are also deficient for a second sHsp, HspB2. The knockout mice showed no significant difference in development and growth in comparison to wild-type mice until 40 weeks of age. Older knockout mice consistently lost weight and body fat and subsequently developed severe spine curvature (kyphosis) and skeletal muscle degeneration [32].

The importance of sHsps in mice experiencing an increased proteotoxic stress was previously not studied. In this report, we inter-crossed between mice lacking αB-crystallin/HspB2 [32] and Tg2576 transgenic mice [34], which over-produces the highly aggregation-prone amyloid-beta. We observed that chaperone-deficient transgenic mice developed severe locomotion defects compared to mice with either chaperone deficiency alone or transgene expression alone. Our results show that exacerbation of protein aggregation and loss of chaperones produced a new synthetic phenotype of motor defects in mice.

2. MATERIALS AND METHODS

2.1 Mice and Animal Care

One male B6;SJL-Tg(APPSWE2576Kha) transgenic mouse (Tg2576) [34] was purchased from Taconic Farms. The Tg2576 mice over-expresses the 695 amino acid isoform of amyloid precursor protein (APP) harboring Swedish familial mutations (K670N, M67IL) under the PrP promoter. These mice show impairment in learning and behavior by 9 months and develop extracellular amyloid plaques by 12 months. This mouse was out-crossed to

129SvEv wild-type mice for 4 generations. The resulting litters contained non-transgenic or transgenic mice. One breeding pair of *CRYAB*$^{-/-}$/*HSPB2*$^{-/-}$ mice in 129SvEv background [32] were gifted kindly by Dr. Eric Wawrousek, National Eye Institute, NIH, Bethesda, MD. Mice were maintained in a facility and program accredited by the Association for Assessment and Accreditation of Laboratory Animal Care International (AAALAC) with the approval number A3307-01. The animal use protocols were authorized by the Institutional Animal Care and Use Committee (IACUC) of the Medical College of Georgia.

All behavioral experiments were carried out at the Small Animal Behavior Core facility of the Medical College of Georgia in rooms equipped with white noise generators (San Diego Instruments, San Diego, CA) set to provide a constant background level of 70 dB and ambient lighting of approximately 25–30 Lux (lumen/m^2). Test subjects were handled daily for several minutes (each) for at least one week prior to experimentation. Animals were transferred (in their home cages) to the behavioral testing rooms each morning approximately 30 min before the beginning of experiments. Measures were taken to minimize pain or discomfort in accordance with the National Institute of Health Guide for the Care and Use of Laboratory Animals (NIH Publications No. 80-23) revised 1996. Significant efforts were also made to minimize the total number of animals used while maintaining statistically valid group numbers. All procedures employed during this study were reviewed and approved by the Medical College of Georgia IACUC and are consistent with AAALAC guidelines.

2.2 Generation of the required genotypes

To generate αB-crystallin/HspB2 deficient mice expressing the transgene, transgenic mice in 129SvEv background were crossed with the *CRYAB*$^{-/-}$ mice. The resulting litters contained *CRYAB*$^{+/-}$/*HSPB2*$^{+/-}$, mice with equal numbers of non-transgenics and transgenics. The *CRYAB*$^{+/-}$/*HSPB2*$^{+/-}$ transgenic mice were crossed with *CRYAB*$^{+/-}$/*HSPB2*$^{+/-}$ littermates to generate (i) *CRYAB*$^{+/+}$/*HSPB2*$^{+/+}$, *Tg*$^{0/0}$ (hereafter referred to as WT); (ii) *CRYAB*$^{+/+}$/*HSPB2*$^{+/+}$, *Tg*$^{+/0}$ (hereafter referred to as WTTg); (iii) *CRYAB*$^{-/-}$/*HSPB2*$^{-/-}$, *Tg*$^{0/0}$ (hereafter referred to as KO) and (iv) *CRYAB*$^{-/-}$/*HSPB2*$^{-/-}$, *Tg*$^{+/0}$ (hereafter referred to as KOTg) in addition to parental genotypes *CRYAB*$^{+/-}$/*HSPB2*$^{+/-}$, *Tg*$^{0/0}$ and *CRYAB*$^{+/-}$/*HSPB2*$^{+/-}$, *Tg*$^{+/0}$, which were not used further. Genotypes of the mice were confirmed by PCR of tail genomic DNA using appropriate primers as described in Brady et al [32] or in Hsiao et al [34]. The use of littermates for all genotypes minimizes the effect of out-crossing the mice. The genetic variations between the two parental mouse strains are expected to affect all four experimental genotypes identically.

2.3 Immunoblotting

To prepare brain lysates mice were euthanized using carbon dioxide followed by perfusion with PBS. Brains were dissected out from appropriate mice, frozen on liquid nitrogen and stored at -80°C until use. Lysates were prepared by homogenization in lysis buffer containing 50 mM Tris.HCl pH 7.5, 5 mM EDTA, 1% Triton-X100 supplemented with HALT protease inhibitor cocktail (Pierce). Lysate were cleared by centrifugation and the protein concentration was estimated using the BCA kit (Pierce). Samples were boiled in loading dye, separated by SDS-PAGE and transferred to PVDF membrane. The membrane was blocked using 10% milk in PBS with 0.1% tween-20. Primary antibodies were anti-Hsp27 (Sigma), Anti-αBcrystallin (Stressgen), anti-Hsp70 (Stressgen) and anti-Aβ (Sigma). Appropriate secondary antibodies (rabbit anti-mouse or goat anti-rabbit) conjugated with horseradish peroxidase were used followed by development of the immunoblot by chemiluminescence (Danville).

2.4 Open Field Activity

Mouse open field activity monitors (27.9 cm ×27.9 cm, Med Associates St Albans, VT) were used for these experiments. The following parameters were recorded for the 25 min test session: horizontal activity (horizontal photobeam breaks or counts), number of stereotypical movements, and vertical activity (vertical photobeam breaks). Thus, spontaneous locomotor activity, olfactory activity (rearing and sniffing movements) and stereotypical movements were assessed.

2.5 Hot Plate test (Supraspinal Nociception)

Individual mice were placed in the glass enclosed section of a Hot-Plate Analgesia Meter, Accuscan Instruments, Inc., Columbus OH. The temperature of the heating surface was elevated by 3°C per minute from a beginning temperature of 42°C to a maximum temperature of 49°C. The time elapsed (latency) before the subject lifted and/or licked a hind paw or jumped was recorded as a measure of nociception. Each mouse was given 3 trials separated by a 30-minute (minimum) inter-trial interval.

2.6 Fear Conditioning

Shock Threshold. A sequence of single foot shocks were delivered to wild type mouse subjects placed on the same electrified grid used for fear conditioning (see below) in order to assess the shock (i.e., sensory perception) threshold. Initially, a 0.1-mV shock was delivered for 1 sec, and the animals' behavior was evaluated for flinching, jumping, and vocalization. At 30-s intervals the shock intensity was increased by 0.1 mA up to 0.7 mA and then returned to 0 mA in 0.1-mA increments at 30-sec intervals. Threshold to vocalization, flinching, and then jumping were quantified for each animal by averaging the shock intensity at which each animal manifested a behavioral response to the foot shock. This average was determined to be 0.5 mA and used in subsequent experiments.

Training procedure. Separate groups of mice were subsequently evaluated for contextual and cued fear conditioning using the mA shock levels determined in the experiments described above. A two-pairing method of auditory cue and mild foot shock were conducted during a 7 min test session. Animals were initially allowed to freely explore the apparatus (MED-VFC-NIR-M, Med Associates, St Albans, VT) for 3 min, after which a 30-sec acoustic-conditioned stimulus (CS; white noise, 80 dB) was delivered. At the end of the CS, a 2-sec shock unconditioned stimulus (US) was applied to the grid floor. The CS-US pairing was delivered again at the 5-min mark.

Context-dependent Freezing. To evaluate contextual fear learning, test subjects were returned to the freeze monitor (i.e., same context) 24 hr after the training procedure described above, and freezing behavior was scored for 5 min.

Cue-dependent Freezing. To evaluate cued fear learning, the animals were placed in a different context (novel odor, lighting, cage floor, and visual cues) 24 hours following contextual testing. Baseline behavior was scored for 2 minutes, and then the CS was presented for a period of 2 minutes, followed by another 2 minutes period in which the auditory stimulus was absent.

Scoring Procedure. Subject movements in the freeze monitor apparatus were determined via the Med Associates Video-Tracking and scoring software. Data are expressed as percent freezing in 60-s epochs, each epoch divided into 6 or 12 5-s bins.

2.7 Statistical Analysis

All data were analyzed and graphed using SigmaSTAT and SigmaPLOT software programs, respectively. Open field cumulative counts were compared using 2-way repeated measure ANOVA (genotype vs. time) with pair-wise multiple comparison procedures using Holm-Sidak method. Open filed zone-time and distance traveled data were analyzed by one way ANOVA. Data from fear conditioning experiments were analyzed using two-way repeated measure ANOVA with comparison between wild-type versus knockout and non-transgenic versus transgenic genotypes. Data from hot plate tests (temperature and time) were analyzed using one-way ANOVA. Comparisons producing a p value <0.05 were considered significant.

3. RESULTS

3.1 Age-dependent decrease in the expression of αB-crystallin in transgenic mice

As guardians against protein misfolding and aggregation, chaperones protect the cells from the deleterious effects of aggregation. Stressful conditions such as heat shock cause an acute increase in misfolded proteins resulting in an increase in demand for chaperones, which is normally met by the stress-inducible expression of chaperones. On the other hand, a steady increase in damaged proteins (oxidized and nitrosylated) takes place due to aging, leading to a chronic demand for chaperones and accumulation of damaged proteins. Such a reduction in chaperone availability is thought to play a central role in triggering the protein aggregation cascade of natively unstructured and amyloidogenic polypeptides associated with several neurodegenerative diseases. Because misfolding and aggregation of the amyloid-beta (Aβ) peptide is a major hallmark of AD, we premised that an age-dependent decrease in available chaperones might exacerbate the increased demand for proteostasis mechanisms. We investigated this possibility by examining the expression of Hsp70, Hsp27 and αB-crystallin in the brains of non-transgenic and transgenic mice at young (3 months) and older (7 months) ages by immunoblotting (Figure 1A) and densitometric analysis of the band intensities (Figure 1B). Due to the muscle-specific expression of HspB2 and the unavailability of appropriate antibodies recognizing mouse HspB2, we did not examine its expression in the mouse brain. The expression of the human APP transgene was examined using an antibody specific to human Aβ. As expected, only transgenic mice showed expression of APP and the band intensities were significantly above the intensities in the same molecular weight range in non-trangenic mice ($p<0.05$, two-sample t-test). The expression of Hsp70 and Hsp27 were unchanged at the ages tested irrespective of the transgene expression or age of mice. A marginal elevation in the expression of Hsp27 and Hsp70 observed in Figure 1B was not significantly different between the two ages or between the two genotypes ($p>0.05$, two-sample t-test). The equal expression levels of these proteins also serve as control for equal protein loading. Interestingly, the expression of αB-crystallin was significantly decreased in older transgenic mice but not in non-transgenic mice. Densitometric measurement of band intensities showed that αB-crystallin levels of 3-month old transgenic mice were comparable to that in non-transgenic mice. The levels of αB-crystallin in 7 month-old transgenic mice decreased significantly compared to 3-month old transgenic mice ($p<0.05$, two-sample t-test). This suggests that αB-crystallin may be an early indicator of protein misfolding defects in the brains of transgenic mice.

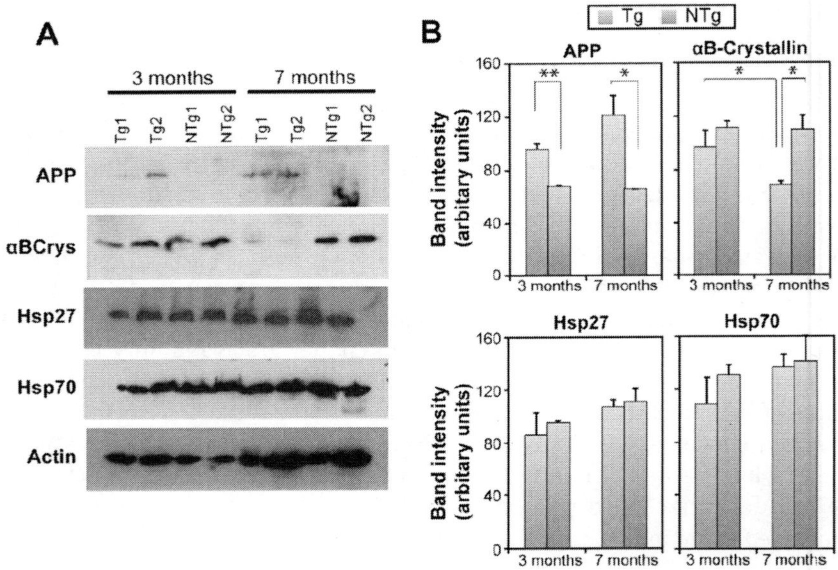

Figure 1. Chaperone levels in AD model mice. (A) Immunoblots showing age-dependence of αB-crystallin expression in mice expressing mutant human APP transgene. Brain lysates were prepared from two individual non-transgenic (NTg1 & NTg2) or transgenic (Tg1 & Tg2) mice each at 3 months or 7 months of age. Samples were immunoblotted for αB-crystallin, Hsp70, Hsp27, Aβ and actin. 40 μg of total protein was analyzed. Similar levels of Hsp70, Hsp27 and actin in all samples show equal protein loading. (B) Densitometric analysis of band intensities in A. Intensities were compared by two-sample Student's t-test for statistical signifcance. APP expression in Tg mice were significantly greater than NTg mice as expected ($p=0.006$ at 3 months, $p=0.017$ at 7 months). αB-crystallin expression was significantly lower at 7-months compared to 3-months old transgenic mice brain ($p=0.044$). αB-crystallin levels in 7-month old transgenic mice was also significantly lower than non-transgenic mice at similar age ($p=0.018$). All other comparisons showed no significant differences.

3.2 Generation of transgenic AD mice lacking αB-crystallin/HspB2

Based on our observations of reduced αB-crystallin in older transgenic but not in non-transgenic mice, we hypothesized that if this chaperone is protective, then its loss may show an exaggerated phenotype. To examine if decreased αB-crystallin levels were important for the manifestation of the behavioral phenotypes in the AD model mice we crossed Tg2576 mice (B6/SJL background) with wild-type ($CRYAB^{+/+}HSPB2^{+/+}$) or αB-crystallin/HspB2 knockout ($CRYAB^{-/-}HSPB2^{-/-}$) mice (129Sv background). The breeding scheme is shown in Figure 2A. This intercross resulted in litters with equal numbers of non-transgenic ($Tg^{0/0}$) or hemizygous transgenic ($Tg^{+/0}$) mice, which were heterozygous for αB-crystallin/HspB2 ($CRYAB^{+/-}HSPB2^{+/-}$). The litters were inbred to obtain mice of the following genotypes – (i) $CRYAB^{+/+}HSPB2^{+/+}$, $Tg^{0/0}$; (ii) $CRYAB^{+/+}HSPB2^{+/+}$, $Tg^{+/0}$; (iii) $CRYAB^{-/-}HSPB2^{-/-}$, $Tg^{0/0}$ and (iv) $CRYAB^{-/-}HSPB2^{-/-}$, $Tg^{+/0}$. From this point on, these mice are denoted as WT, WTTg, KO and KOTg, respectively. (Mice heterozygous for CRYAB (namely, $CRYAB^{+/-}HSPB2^{+/-}$, $Tg^{0/0}$; and $CRYAB^{+/-}HSPB2^{+/-}$, $Tg^{+/0}$) that were also generated in this cross were not used further.) By comparing only littermates for all experimental genotypes, we minimized the influence of genetic variation resulting from the parental strains. Further, any residual effects of genetic variation are expected to be evenly distributed between the four genotypes. The chaperone expression in each genotype was confirmed by western-blotting two sets of brain lysates from 7-month old mice for αB-crystallin, and Hsp27 (Figure 2B). The expression of Hsp27 was equal in all mice suggesting that the expression of Hsp27 was not increased to compensate for the loss of αB-crystallin/HspB2. The KO and KOTg samples were devoid of αB-crystallin as expected. WT samples showed higher amounts of αB-crystallin than WTTg samples as observed in Figure 1. Expression of APP was observed only in the transgenic samples (KOTg and WTTg) as expected (not shown). Although the WTTg mice express low levels of αB-crystallin, they are distinct from KOTg mice, which are devoid of αB-crystallin since embryogenesis.

3.3 Behavioral analysis

The decrease in αB-crystallin expression in older transgenic mice brains suggested that the over-produced APP may affect proteostasis not only by increased demand for chaperone function but also reduced expression. We premised that if αB-crystallin protected the mice from Aβ-mediated neuronal dysfunction, then in its absence these deficits should manifest at an earlier age. It was demonstrated that the Tg2576 mice showed behavioral changes as early as 6–7 months of age [34]. Since the mice deficient for αB-crystallin/HspB2 showed debilitating phenotypes about 9-10 months of age [32], we decided to investigate the behavioral phenotypes of WT (n=9), WTTg (n=7), KO (n=15) and KOTg (n=7) mice when they were 7–8 months (or 28–32 weeks).

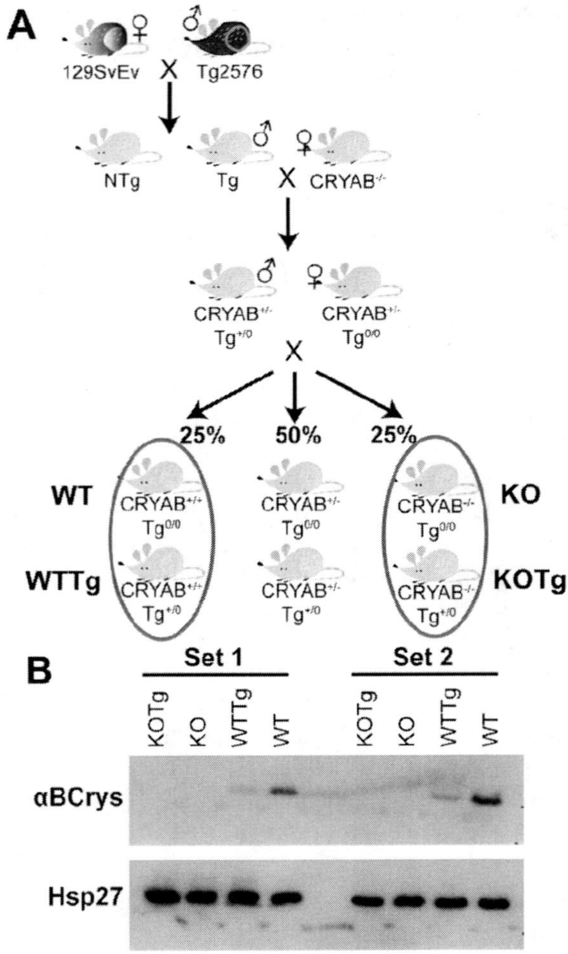

Figure 2. Generation of the required genotypes of mice. (A) Schematic diagram showing the mouse crosses that lead to the required genotypes. (B) Immunoblots showing αB-crystallin and Hsp27 expression in two sets of mice at 7 months of age. WT is $CRYAB^{+/+}HspB2^{+/+}$, $Tg^{0/0}$; WTTg is $CRYAB^{+/+}HspB2^{+/+}$, $Tg^{+/0}$; KO is $CRYAB^{-/-}HspB2^{-/-}$, $Tg^{0/0}$ and KOTg is $CRYAB^{-/-}HspB2^{-/-}$, $Tg^{+/0}$.

The following tests were designed to examine both the effects of Aβ on locomotion defects in αB-crystallin/HspB2 deficient mice and the effects of αB-crystallin/HspB2 loss on memory network defects in Tg2576 mice.

Open field Test. Open field test is one of the most common tests to monitor general motor activity, exploratory behavior and measures of anxiety **[35]**. Normal behavior in mice is to seek the protection of the periphery rather than the vulnerability of the center. Mice that are less anxious are expected to spend more time in the center. Mice that show signs of motor deficits are expected to show lower horizontal distances and vertical movements. Mice were individually placed in an open arena equipped with infrared photobeams to monitor mice behavior for 30 minutes without the experimenter being present in the room. The photobeams recorded the following activities – (i) counts of ambulatory (walking), stereotypic (repetitive activity, eg, grooming) and vertical (rearing up) movements; (ii) distance travelled in the center versus periphery and (iii) zone times in center versus periphery were monitored.

First of all, WT and KO groups behaved similarly in terms of vertical counts, stereotypic counts and ambulatory counts indicating that loss of αB-crystallin/HspB2 alone had negligible effect on locomotion at this age group (Figure 3A, B & C). Age associated muscle degeneration reported with the knockout mice was not manifested in terms of locomotion defects at this age group as all 3 parameters counts of KO group of animals was comparable to WT. Similarly, the WTTg group was comparable to WT suggesting that the expression of mutant APP in mice did not significantly affect locomotion by itself. Interestingly, when the transgene expression was combined with αB-crystallin/HspB2 deficiency in the KOTg group, the effect was synergistically enhanced compared to KO group ($P<0.004$, 2-way repeated measure ANOVA) in ambulatory activity and vertical activity. Differences were not statistically significant in stereotypic activity.

The total distance traveled by WT and KO mice were also very similar underscoring the lack of locomotion deficits in αB-crystallin/HspB2 knockout mice at this age. Distance traveled (Figure 3D) and zone times in the center and peripheral regions (Figure 3E) of the open field were also comparable (differences were not statistically significant). Although the total distance traveled by WTTg mice was significantly lower than the WT group, the zone times and distance traveled in the center and periphery were statistically similar. Similarly, KOTg mice traveled only about a third the distance traveled by KO mice but differences between center and periphery were not significant. Total distance traveled by KOTg mice was also significantly lower than that of WTTg mice highlighting the synergistic effects observed in ambulatory activity counts (Figure 3A). These results suggest that the differences between transgenic (WTTg and KOTg) and non-transgenic (WT and KO) mice were due to locomotion defects rather than defects in anxiety or exploratory behavior. The KOTg mice appeared to be unable to distinguish between center and periphery compared the WT group (Figure 3E) suggesting that they were significantly less anxious than WT mice. Although, direct comparisons between KOTg and WT groups are complicated, our data suggests synthetic effects of αB-crystallin/HspB2 deficiency and APP transgene expression.

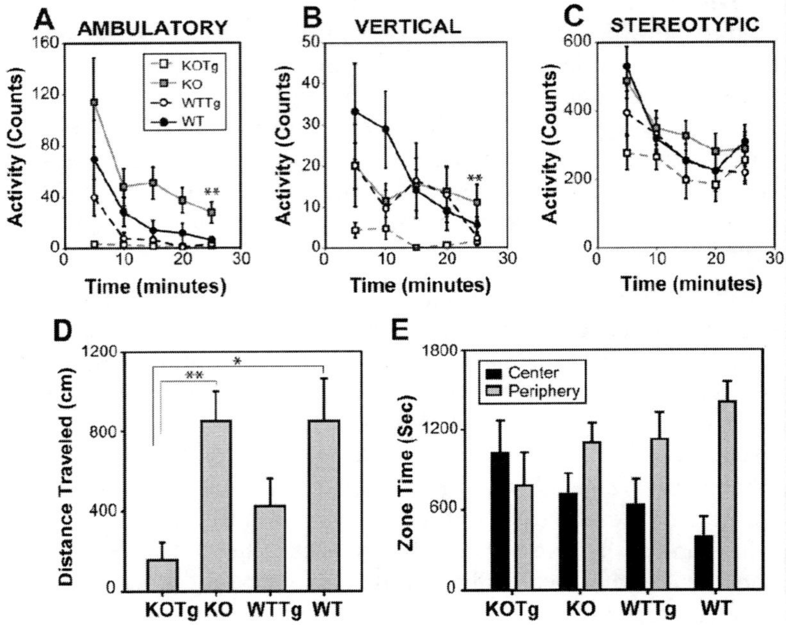

Figure 3. Open field test. Counts for Ambulatory (A), Vertical (B) and Stereotypic (C) activities of mice over a period of 25 minutes in the open field. Symbol representations are - KOTg (open grey square; n=7), KO (closed grey square; n=15), WTTg (open black circle; n=7) and WT (closed black circle; n=9). Mean values ± SEM are plotted. Data were analyzed by two-way repeated measure ANOVA, which indicated that KOTg group showed significantly less (p=0.004) than other groups. Differences were not statistically significant between other groups. (D) Total distance (in cm) traveled by the mice over a period of 30 minutes. Mean values ± SEM are plotted. Data were compared by t-tests. KOTg was significantly different from KO (p=0.007) and WT (p=0.017). (E) Zone time in center (black bars) and periphery (grey bars) for the mice during the 30 minutes. Data were compared by t-tests. Center zonetimes of both KO and KOTg were significantly different from WT (p=0.023 for KO and p=0.0006 for KOTg). Peripheral zonetimes were similar for all groups. (n=7 for KOTg; n=15 for KO; n=7 for WTTg and n=9 for WT).

Hot plate test. To understand whether the observed locomotion defects were coupled with sensory defects, we examined the mice for pain sensitivity to thermal stimulus (nociception). Each mouse was placed on a hot plate and the temperature of the heating surface was raised by 3°C per minute. The minimum time and temperature required to elicit response (paw licking or jumping) was noted (Figure 4). Somatosensation and its response as motor activity were scored to test for spinal cord related reflexes. The KO, WTTg and WT mice showed comparable responses and the differences were not significant. Maximum latency and higher temperatures were required for the KOTg group to elicit a response and were significantly greater than all other groups of mice with respect to both time and temperature. This result suggests that the combined effect of αB-crystallin/HspB2 loss and transgene expression caused a decreased sensory function not observed with each genotype individually, thus underscoring the synergism.

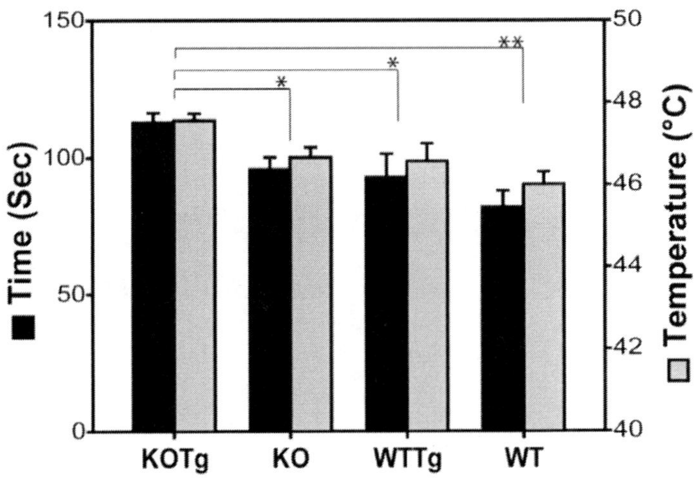

Figure 4. Hot plate test. The duration of time to elicit a thermo-sensitive reflex response in mice is shown on the left axis (black bars). The temperature at which the response was recorded is shown on the right axis (grey bars). Mean values ± SEM are plotted. Data for different groups were compared by t-tests. Both the time and temperature required elicit a response in KOTg were significantly higher than those for KO (p=0.022 and 0.031, respectively), WT (p=0.001 and 0.001, respectively) and WTTg (p=0.05 and 0.05, respectively). All other groups were statistically similar. WT is *CRYAB*$^{+/+}$, *Tg*$^{0/0}$; WTTg is *CRYAB*$^{+/+}$, *Tg*$^{+/0}$; KO is *CRYAB*$^{-/-}$, *Tg*$^{0/0}$ and KOTg is *CRYAB*$^{-/-}$, *Tg*$^{+/0}$. (n=7 for KOTg; n=15 for KO; n=7 for WTTg and n=9 for WT).

Fear conditioning test. In order to examine whether learning behavior was impaired in transgenic mice lacking αB-crystallin/HspB2, we performed fear-conditioning test. In this test paradigm the associative learning of a neutral cue (eg, sound tone) or a neutral context (eg, environment) with a brief aversive stimulus (eg, mild electric shock) is measured by monitoring the freezing behavior in mice (Figure 5). Fear conditioning test was performed by placing a mouse in a box equipped with a mechanism for monitoring the freezing behavior of the animal by recording photobeam breaks. The mice were first trained to associate the surroundings or a sound pulse (cue) with the mild aversive stimulus. Cue-dependent freezing was tested in a novel environment (i.e., one with different lighting, and olfactory and visual cues) and the freezing behavior associated with the tone was measured. Context-dependent freezing was monitored to evaluate the learned aversion of an animal for the environment associated with the mild aversive stimulus.

No significant differences in freezing behavior were obvious during the training period (Figure 5A) indicating that the locomotion defects observed with WTTg and KOTg mice (Figure 3) did not significantly affect the freezing response to the mild aversive stimulus. During training, mice did not exhibit freezing during the first cue-foot shock pair at 3 minutes, but freezing increased in all mice during the second cue-foot shock pair at 5 minutes. The subsequent increase in freezing at the end of 6-minute period suggested an anticipatory behavior for a third cue-foot shock pair or could also suggest decreased exploratory behavior due to familiarity with the surrounding.

In response to a tone cue (Figure 5B), the mice behaved similarly with no significant differences in the freezing behavior. Mice did not exhibit freezing and explored their novel environment during the first 2 minutes. All mice responded readily to the tone cue presented to them for 2 minutes by freezing. The mice mostly stayed frozen beyond the 4-minute time point. These results indicated good associative learning in all mice.

Interestingly, in response to contextual associative learning there were notable differences (Figure 5C). First, WT and WTTg mice froze similarly at the start of the experiment but differed significantly at later time points ($p=0.003$). While the WT mice rapidly froze subsequently and reached maximum freezing at the 2-minute time point, the WTTg mice only showed a gradual increase in freezing till the 5-minute time point. This suggested that context-dependent associative learning may be one of the first manifestations of APP transgene expression. Second, KO and KOTg mice showed a high degree of freezing behavior from the start of the experiment in comparison to their wild-type counterparts. This is an intriguing result – mice lacking αB-crystallin/HspB2 mice were seemingly better at contextual learning than their wild-type counterparts.

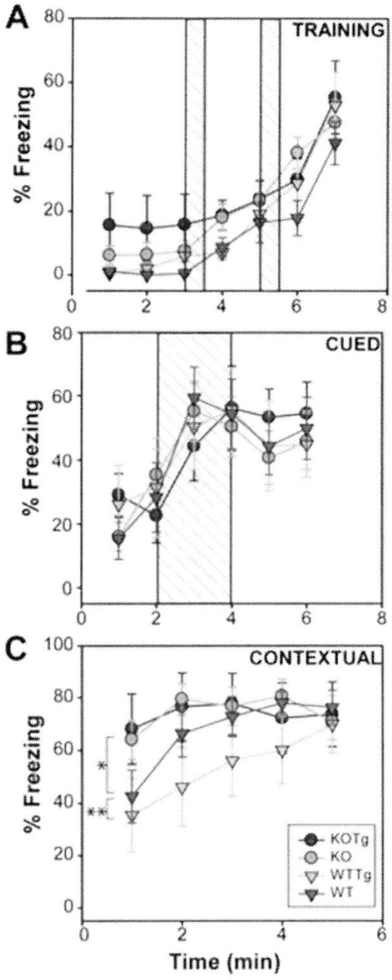

Figure 5. Fear conditioning test. Percent freezing of mice over time during Training (Top), Cued test (middle) and Contextual test (bottom). Mean values ± SEM are plotted. Data were analyzed by two-way repeated measure ANOVA. Differences between groups were not significant for training and cued tests. In contextual tests, WT performed significantly better than WTTg ($p=0.003$); difference between WTTg and KOTg was significant ($p=0.014$); other groups were not significantly different. Blue hatched boxes represent two separate auditory signals (80dB) of 30 seconds each during training and a single auditory signal for 2 minutes during cued test followed by 2 sec foot shock. Symbol representations are - KOTg (red circle), KO (orange circle), WTTg (yellow triangle) and WT (green triangle). (n=7 for KOTg; n=15 for KO; n=7 for WTTg and n=9 for WT).

This result cannot be satisfactorily explained by the locomotion defects in the knockout mice because the same mice are normal in the cued experiment. This leads us to suggest that the mice lacking αB-crystallin/HspB2 were perhaps hyper-sensitive to a particular contextual signal (such as auditory, visual, olfactory or tactile), which may have allowed them to associate this signal as a strong unitary cue with the aversive stimulus. The nature of this cue is unclear.

4. DISCUSSION

In this study, we aimed to investigate the effect of chaperone-deficiency in a mouse model for AD. When expression of APP was combined with loss of αB-crystallin/HspB2, we observed that new phenotypes involving locomotion and sensory function deficits were revealed. Neither loss of αB-crystallin/HspB2 nor transgenic expression of APP by themselves produced these phenotypes. The synthetic sick phenotypes underscore a negative synergy between expression of APP, which is thought to increase the production of the aggregation-prone Aβ peptide and reduced chaperone (αB-crystallin and HspB2) function.

When misfolding-prone proteins are overproduced, cellular health may depend on its ability to also overproduce chaperones. In AD brains, where the load of misfolded proteins is high, others have observed that αB-crystallin expression was high in glial cells in areas surrounding plaques and tangles [20], [36]. However, we observed that αB-crystallin levels in brain lysates decreased in an age-dependent manner in transgenic AD model mice compared to non-transgenic littermates. The contrasting observations may be attributed to the differences in methods - while others have monitored relatively small regions by immunohistochemistry in human AD brains [20], [36], we monitored the expression levels by immunoblotting lysates from the entire brains of a mouse model for AD. In this regard, we would like to point out that other studies have observed no significant differences in the expression of αB-crystallin in various regions of AD brains by immunoblotting methods [37], by mass spectrometry based protein identification methods [38] and by immunohistochemical methods [39]. Further, we speculate that the difference between the human and mouse results could be due to differential cellular responses towards the entire gamut of pathologies in AD brains versus the predominantly plaque pathology in the AD model mice used in this study. However, the exact reasons for these observations are presently unclear and require further experimentation.

We were unable to examine the effect of αB-crystallin/HspB2 loss on the deposition of Aβ plaques in the mice because the KOTg and the KO mice were considerably debilitated by the age of 9–10 months as previously reported for αB-crystallin/HspB2-deficient mice [32]. At this age, WTTg mice do not have significant levels of plaque pathology and plaques develop only in mice older than 10-12 months of age [34]. Immunohistochemical examination of mice brains at 7 months revealed no plaques under our experimental conditions (data not shown). However, it is thought that plaques may not be causative and others have shown that decreased dendritic spine density, impaired long-term potentiation (LTP), and behavioral deficits occurred many months before detectable plaques [40].

We observed a modest locomotor deficit in WTTg mice by open field tests (Figure 4D) suggesting that the expression of hAPP affected muscle function. In Tg2576 mice, expression of the APP transgene has been observed in muscle tissue [41] where the function of αB-crystallin and HspB2 is expected to be crucial [32]. The presence of amyloid oligomers in muscle cells has been shown to have potent toxicity [42]. Thus, the high degree of synergistic toxic effects of αB-crystallin/HspB2 loss and transgene expression in the skeletal muscle precluded the examination of plaque pathology at later ages. Because sHsps block the fibrillization and toxicity of Aβ [43], [44] it is suggested that in the absence of αB-crystallin/HspB2 enhanced Aβ toxicity may contribute to the manifestation of the synergistic phenotypes. An alternative explanation for the locomotion defect is that since APP is expressed throughout the brain in this mouse, it might contribute to deficits in the brain motor centers. The locomotion defects observed in the WTTg mice are important because these transgenic mice are routinely used in learning and memory tests that involve locomotion (eg, swimming in water maze). In agreement with our observations, motor deficits in Tg2576 have been observed at 6–7 months in a recent publication [45]. No significant perturbation in the motor reflexes was observed in WTTg mice as the mice could perceive and respond to thermal stimulus in the supraspinal nociception (hot-plate) test. However, reduced chaperone levels in the transgenic mice produced a sensory defect suggesting that this synergism also affected nociception.

In the WTTg mice, we also observed a significant limitation in their ability to associate mild aversive stimulus with the environmental context but not with an auditory cue. This difference in associative learning can be interpreted as follows – Contextual learning may be based on weak associations between the footshock and diverse signals (including visual, olfactory, auditory and sensory stimuli) and thereby differences between WT and WTTg are manifested significantly [46]. It has been thought that animals may associate only a subset of contextual elements, which leads to weak associative learning. The inability of WTTg to associate and integrate the complex contextual signals with the aversive stimulus indicates a reduced hippocampal function. Contextual associative learning deficit in Tg2576 has been previously observed [40], [45]. On the other hand, associative learning may be reliably paired with footshock when a strong unitary cue (such as an auditory tone) is provided and a Pavlovian response is established [47], which may overcome the minor differences in learning between different genotypes. The ability of WTTg mice to normally associate the fear of footshock with an auditory cue indicates that there is no obvious impairment in amygdala under these conditions. An alternative explanation for the superior contextual learning in KO and KOTg mice is a possibility that αB-crystallin and/or HspB2 may be involved in the folding or degradation of proteins that modulate this learning.

Because molecular chaperones prevent excessive misfolding and aggregation, cellular demand for chaperones are greater in aged and stressed cells, which accumulate damaged proteins. Our results demonstrating synergistic effects of reduced chaperones in the context of a mouse model for AD highlights the in vivo importance of sHsps in diseases characterized by protein misfolding and aggregation.

ACKNOWLEDGMENTS

The authors wish to thank Dr. Eric Wawrousek at the National Eye institute for the kind gift of the knockout mice.

REFERENCES

1. Narberhaus F (2002) Alpha-crystallin-type heat shock proteins: socializing minichaperones in the context of a multichaperone network. Microbiol Mol Biol Rev 66: 64–93.
2. Haslbeck M (2002) sHsps and their role in the chaperone network. Cell Mol Life Sci 59: 1649–1657.
3. MacRae TH (2000) Structure and function of small heat shock/alpha-crystallin proteins: established concepts and emerging ideas. Cell Mol Life Sci 57: 899–913.
4. Arrigo AP, Landry J (1994) Expression and function of the low-molecular-weight heat shock proteins. In: Morimoto RI, ed. The Biology of Heat Shock Proteins and Molecular Chaperones: Cold Spring Harbor Laboratory Press, Plainview, NY. 335 p.
5. Kim KK, Kim R, Kim SH (1998) Crystal structure of a small heat-shock protein. Nature 394: 595–599.
6. Haley DA, Horwitz J, Stewart PL (1998) The small heat-shock protein, alphaBcrystallin, has a variable quaternary structure. J Mol Biol 277: 27–35.
7. Horwitz J (1992) Alpha-crystallin can function as a molecular chaperone. Proc Natl Acad Sci U S A 89: 10449–10453.
8. Taylor RP, Benjamin IJ (2005) Small heat shock proteins: a new classification scheme in mammals. J Mol Cell Cardiol 38: 433–444.
9. Franck E, Madsen O, van Rheede T, Ricard G, Huynen MA, et al. (2004) Evolutionary diversity of vertebrate small heat shock proteins. J Mol Evol 59: 792–805.
10. Clark JI, Muchowski PJ (2000) Small heat-shock proteins and their potential role in human disease. Curr Opin Struct Biol 10: 52–59.

11. Litt M, Kramer P, LaMorticella DM, Murphey W, Lovrien EW, et al. (1998) Autosomal dominant congenital cataract associated with a missense mutation in the human alpha crystallin gene CRYAA. Hum Mol Genet 7: 471–474.
12. Vicart P, Caron A, Guicheney P, Li Z, Prevost MC, et al. (1998) A missense mutation in the alphaB-crystallin chaperone gene causes a desmin-related myopathy. Nat Genet 20: 92–95.
13. Treweek TM, Rekas A, Lindner RA, Walker MJ, Aquilina JA, et al. (2005) 120G alphaB-crystallin promotes the unfolding of reduced alpha-lactalbumin and is inherently unstable. Febs J 272: 711–724.
14. Bova MP, Yaron O, Huang Q, Ding L, Haley DA, et al. (1999) Mutation R120G in alphaB-crystallin, which is linked to a desmin-related myopathy, results in an irregular structure and defective chaperone-like function. Proc Natl Acad Sci U S A 96: 6137–6142.
15. Evgrafov OV, Mersiyanova I, Irobi J, Van Den Bosch L, Dierick I, et al. (2004) Mutant small heat-shock protein 27 causes axonal Charcot-Marie-Tooth disease and distal hereditary motor neuropathy. Nat Genet 36: 602–606.
16. Zourlidou A, Payne Smith MD, Latchman DS (2004) HSP27 but not HSP70 has a potent protective effect against alpha-synuclein-induced cell death in mammalian neuronal cells. J Neurochem 88: 1439–1448.
17. Wyttenbach A, Sauvageot O, Carmichael J, Diaz-Latoud C, Arrigo AP, et al. (2002) Heat shock protein 27 prevents cellular polyglutamine toxicity and suppresses the increase of reactive oxygen species caused by huntingtin. Hum Mol Genet 11: 1137–1151.
18. Shinohara H, Inaguma Y, Goto S, Inagaki T, Kato K (1993) Alpha B crystalline and HSP28 are enhanced in the cerebral cortex of patients with Alzheimer's disease. J Neurol Sci 119: 203–208.
19. van Rijk AF, Bloemendal H (2000) Alpha-B-crystallin in neuropathology. Ophthalmologica 214: 7–12.
20. Renkawek K, Voorter CE, Bosman GJ, van Workum FP, de Jong WW (1994) Expression of alpha B-crystallin in Alzheimer's disease. Acta Neuropathol 87: 155–160.
21. Renkawek K, Stege GJ, Bosman GJ (1999) Dementia, gliosis and expression of the small heat shock proteins hsp27 and alpha B-crystallin in Parkinson's disease. Neuroreport 10: 2273–2276.
22. Renkawek K, de Jong WW, Merck KB, Frenken CW, van Workum FP, et al. (1992) alpha B-crystallin is present in reactive glia in Creutzfeldt-Jakob disease. Acta Neuropathol 83: 324–327.
23. Dabir DV, Trojanowski JQ, Richter-Landsberg C, Lee VM, Forman MS (2004) Expression of the small heat-shock protein alphaB-crystallin in tauopathies with glial pathology. Am J Pathol 164: 155–166.
24. Head MW, Goldman JE (2000) Small heat shock proteins, the cytoskeleton, and inclusion body formation. Neuropathol Appl Neurobiol 26: 304–312.
25. Hsu AL, Murphy CT, Kenyon C (2003) Regulation of aging and age-related disease by DAF-16 and heat-shock factor. Science 300: 1142–1145.
26. McElwee JJ, Schuster E, Blanc E, Thomas JH, Gems D (2004) Shared transcriptional signature in Caenorhabditis elegans Dauer larvae and long-lived daf-2 mutants implicates detoxification system in longevity assurance. J Biol Chem 279: 44533–44543.
27. Liao PC, Lin HY, Yuh CH, Yu LK, Wang HD (2008) The effect of neuronal expression of heat shock proteins 26 and 27 on lifespan, neurodegeneration, and apoptosis in Drosophila. Biochem Biophys Res Commun 376: 637–641.
28. Iwaki T, Kume-Iwaki A, Liem RK, Goldman JE (1989) Alpha B-crystallin is expressed in non-lenticular tissues and accumulates in Alexander's disease brain. Cell 57: 71–78.
29. Lowe J, McDermott H, Pike I, Spendlove I, Landon M, et al. (1992) alpha B crystallin expression in non-lenticular tissues and selective presence in ubiquitinated inclusion bodies in human disease. J Pathol 166: 61–68.
30. Fonte V, Kapulkin V, Taft A, Fluet A, Friedman D, et al. (2002) Interaction of intracellular beta amyloid peptide with chaperone proteins. Proc Natl Acad Sci U S A 99: 9439–9444.

31. Cottrell BA, Galvan V, Banwait S, Gorostiza O, Lombardo CR, et al. (2005) A pilot proteomic study of amyloid precursor interactors in Alzheimer's disease. Ann Neurol 58: 277–289.
32. Brady JP, Garland DL, Green DE, Tamm ER, Giblin FJ, et al. (2001) AlphaBcrystallin in lens development and muscle integrity: a gene knockout approach. Invest Ophthalmol Vis Sci 42: 2924–2934.
33. Iwaki A, Nagano T, Nakagawa M, Iwaki T, Fukumaki Y (1997) Identification and characterization of the gene encoding a new member of the alphacrystallin/ small hsp family, closely linked to the alphaB-crystallin gene in a headto- head manner. Genomics 45: 386–394.
34. Hsiao K, Chapman P, Nilsen S, Eckman C, Harigaya Y, et al. (1996) Correlative memory deficits, Abeta elevation, and amyloid plaques in transgenic mice. Science 274: 99–102.
35. Wilson RC, Vacek T, Lanier DL, Dewsbury DA (1976) Open-field behavior in muroid rodents. Behav Biol 17: 495–506.
36. Wilhelmus MM, Otte-Holler I, Wesseling P, de Waal RM, Boelens WC, et al. (2006) Specific association of small heat shock proteins with the pathological hallmarks of Alzheimer's disease brains. Neuropathol Appl Neurobiol 32: 119–130.
37. Kato K, Inaguma Y, Ito H, Iida K, Iwamoto I, et al. (2001) Ser-59 is the major phosphorylation site in alphaB-crystallin accumulated in the brains of patients with Alexander's disease. J Neurochem 76: 730–736.
38. Yoo BC, Kim SH, Cairns N, Fountoulakis M, Lubec G (2001) Deranged expression of molecular chaperones in brains of patients with Alzheimer's disease. Biochem Biophys Res Commun 280: 249–258.
39. Anthony SG, Schipper HM, Tavares R, Hovanesian V, Cortez SC, et al. (2003) Stress protein expression in the Alzheimer-diseased choroid plexus. J Alzheimers Dis 5: 171–177.
40. Jacobsen JS, Wu CC, Redwine JM, Comery TA, Arias R, et al. (2006) Earlyonset behavioral and synaptic deficits in a mouse model of Alzheimer's disease. Proc Natl Acad Sci U S A 103: 5161–5166.
41. Kawarabayashi T, Younkin LH, Saido TC, Shoji M, Ashe KH, et al. (2001) Age-dependent changes in brain, CSF, and plasma amyloid (beta) protein in the Tg2576 transgenic mouse model of Alzheimer's disease. J Neurosci 21: 372–381.
42. Chen Q, Liu JB, Horak KM, Zheng H, Kumarapeli AR, et al. (2005) Intrasarcoplasmic amyloidosis impairs proteolytic function of proteasomes in cardiomyocytes by compromising substrate uptake. Circ Res 97: 1018–1026.
43. Raman B, Ban T, Sakai M, Pasta SY, Ramakrishna T, et al. (2005) AlphaBcrystallin, a small heat-shock protein, prevents the amyloid fibril growth of an amyloid beta-peptide and beta2-microglobulin. Biochem J 392: 573–581.
44. Wilhelmus MM, Boelens WC, Otte-Holler I, Kamps B, de Waal RM, et al. (2006) Small heat shock proteins inhibit amyloid-beta protein aggregation and cerebrovascular amyloid-beta protein toxicity. Brain Res 1089: 67–78.
45. Rustay NR, Cronin EA, Curzon P, Markosyan S, Bitner RS, et al. (2010) Mice expressing the Swedish APP mutation on a 129 genetic background demonstrate consistent behavioral deficits and pathological markers of Alzheimer's disease. Brain Res 1311: 136–147.
46. Rudy JW, O'Reilly RC (1999) Contextual fear conditioning, conjunctive representations, pattern completion, and the hippocampus. Behav Neurosci 113: 867–880.
47. Sanders MJ, Wiltgen BJ, Fanselow MS (2003) The place of the hippocampus in fear conditioning. Eur J Pharmacol 463: 217–223.

INDEX

G

H

I

L

M

N

P

R

S

T

V

Y

Z